RF Energy Harvesting and Wireless Power Transfer for IoT

RF Energy Harvesting and Wireless Power Transfer for IoT

Guest Editors

Onel Luis Alcaraz López
Katsuya Suto

Basel • Beijing • Wuhan • Barcelona • Belgrade • Novi Sad • Cluj • Manchester

Guest Editors

Onel Luis Alcaraz López
University of Oulu
Oulu
Finland

Katsuya Suto
The University of Electro-Communications
Tokyo
Japan

Editorial Office
MDPI AG
Grosspeteranlage 5
4052 Basel, Switzerland

This is a reprint of the Special Issue, published open access by the journal *Sensors* (ISSN 1424-8220), freely accessible at: https://www.mdpi.com/journal/sensors/special_issues/M1TOBAT3Q5.

For citation purposes, cite each article independently as indicated on the article page online and as indicated below:

Lastname, A.A.; Lastname, B.B. Article Title. *Journal Name* **Year**, *Volume Number*, Page Range.

ISBN 978-3-7258-3027-5 (Hbk)
ISBN 978-3-7258-3028-2 (PDF)
https://doi.org/10.3390/books978-3-7258-3028-2

© 2025 by the authors. Articles in this book are Open Access and distributed under the Creative Commons Attribution (CC BY) license. The book as a whole is distributed by MDPI under the terms and conditions of the Creative Commons Attribution-NonCommercial-NoDerivs (CC BY-NC-ND) license (https://creativecommons.org/licenses/by-nc-nd/4.0/).

Contents

About the Editors vii

Onel Luis Alcaraz López and Katsuya Suto
RF Energy Harvesting and Wireless Power Transfer for IoT
Reprinted from: *Sensors* **2024**, 24, 7567, https://doi.org/10.3390/s24237567 1

Mohamed Aboualalaa, Hesham A. Mohamed, Thamer A. H. Alghamdi and Moath Alathbah
A Pattern Reconfigurable Antenna Using Eight-Dipole Configuration for Energy Harvesting
Applications
Reprinted from: *Sensors* **2023**, 23, 8451, https://doi.org/10.3390/s23208451 5

**Zaed S. A. Abdulwali, Ali H. Alqahtani, Yosef T. Aladadi, Majeed A. S. Alkanhal,
Yahya M. Al-Moliki, Khaled Aljaloud and Mohammed Thamer Alresheedi**
A High-Performance Circularly Polarized and Harmonic Rejection Rectenna for
Electromagnetic Energy Harvesting
Reprinted from: *Sensors* **2023**, 23, 7725, https://doi.org/10.3390/s23187725 21

Sebastià Galmés
Statistical Characterization of Wireless Power Transfer via Unmodulated Emission
Reprinted from: *Sensors* **2022**, 22, 7828, https://doi.org/10.3390/s22207828 39

Toi Le-Thanh and Khuong Ho-Van
Performance Analysis of Wireless Communications with Nonlinear Energy Harvesting under
Hardware Impairment and κ-μ Shadowed Fading
Reprinted from: *Sensors* **2023**, 23, 3619, https://doi.org/10.3390/s23073619 63

Nadica Kozić, Vesna Blagojević, Aleksandra Cvetković and Predrag Ivaniš
Performance Analysis of Wirelessly Powered Cognitive Radio Network with Statistical CSI and
Random Mobility
Reprinted from: *Sensors* **2023**, 23, 4518, https://doi.org/10.3390/s23094518 78

Vieeralingaam Ganapathy, Ramanathan Ramachandran and Tomoaki Ohtsuki
Resource Allocation for Secure MIMO-SWIPT Systems in the Presence of Multi-Antenna
Eavesdropper in Vehicular Networks
Reprinted from: *Sensors* **2023**, 23, 8069, https://doi.org/10.3390/s23198069 105

Varada Potnis Kulkarni and Radhika D. Joshi
Modeling and Performance Analysis of LBT-Based RF-Powered NR-U Network for IoT
Reprinted from: *Sensors* **2024**, 24, 5369, https://doi.org/10.3390/s24165369 128

**Janis Eidaks, Romans Kusnins, Ruslans Babajans, Darja Cirjulina, Janis Semenjako and
Anna Litvinenko**
Efficient Multi-Hop Wireless Power Transfer for the Indoor Environment
Reprinted from: *Sensors* **2023**, 23, 7367, https://doi.org/10.3390/s23177367 149

Pedro E. Gória Silva, Nicola Marchetti, Pedro H. J. Nardelli and Rausley A. A. de Souza
Enabling Semantic-Functional Communications for Multiuser Event Transmissions via Wireless
Power Transfer
Reprinted from: *Sensors* **2023**, 23, 2707, https://doi.org/10.3390/s23052707 176

Aleksandra Cvetković, Vesna Blagojević, Jelena Anastasov, Nenad T. Pavlović and Miloš Milošević
Outage Analysis of Unmanned-Aerial-Vehicle-Assisted Simultaneous Wireless Information and Power Transfer System for Industrial Emergency Applications
Reprinted from: *Sensors* **2023**, *23*, 7779, https://doi.org/10.3390/s23187779 . . . . **196**

Jarne Van Mulders, Jona Cappelle, Sarah Goossens, Lieven De Strycker and Liesbet Van der Perre
UAV-Based Servicing of IoT Nodes: Assessment of Ecological Impact
Reprinted from: *Sensors* **2023**, *23*, 2291, https://doi.org/10.3390/s23042291 . . . . **214**

Ana Beatriz Rodrigues Costa de Mattos, Glauber Brante, Guilherme Luiz Moritz and Richard Demo Souza
Human and Small Animal Detection Using Multiple Millimeter-Wave Radars and Data Fusion: Enabling Safe Applications
Reprinted from: *Sensors* **2024**, *24*, 1901, https://doi.org/10.3390/s24061901 . . . . **245**

About the Editors

Onel Luis Alcaraz López

Onel Luis Alcaraz López is an Associate Professor (tenure track) in Sustainable Wireless Communications Engineering at the Centre for Wireless Communications (CWC), Oulu, Finland, and an Associate Editor of the *IEEE Transactions on Communications* and *IEEE Wireless Communications Letters*. His research interests include sustainable IoT, energy harvesting, wireless RF energy transfer, wireless connectivity, machine-type communications, and cellular-enabled positioning systems. He received his B.Sc. (1st class honors, 2013), M.Sc. (2017), and D.Sc. (with distinction, 2020) degrees in Electrical Engineering from the Central University of Las Villas (Cuba), the Federal University of Paraná (Brazil), and the University of Oulu (Finland), respectively. From 2013 to 2015, he served as a Specialist in Telematics for a Cuban telecommunications company (ETECSA). In 2020, he was a Postdoctoral Researcher in a joint project between the University of Oulu and Nokia Oulu, Finland. He completed on a six-month research visit to Rice University and the University of Houston, Texas, USA, in 2024. He was a collaborator to the 2016 Research Award given by the Cuban Academy of Sciences, a co-recipient of the 2019 and 2023 IEEE European Conference on Networks and Communications (EuCNC) Best Student Paper Award, and the recipient of both the 2020 Best Doctoral Thesis Award, granted by Academic Engineers and Architects in Finland TEK and Tekniska Föreningen in Finland TFiF in 2021, and the 2022 Young Researcher Award in the field of technology in Finland. He co-authored two books entitled, "Wireless RF Energy Transfer in the massive IoT era: towards sustainable zero-energy networks", Wiley, 2021, and "Ultra-Reliable Low-Latency Communications: Foundations, Enablers, System Design, and Evolution Towards 6G", Now Publishers, 2023.

Katsuya Suto

Katsuya Suto is an Associate Professor at the Graduate School of Informatics and Engineering, University of Electro-Communications, Tokyo, Japan. His research interests include semantic communications and protocols, vision-aided communications, radio propagation, and spectrum sharing. He received his B.Sc. degree in Computer Engineering from Iwate University, Morioka, Japan, in 2011, and his M.Sc. and Ph.D. degrees in Information Science from Tohoku University, Sendai, Japan, in 2013 and 2016, respectively. From 2016 to 2018, he worked as a Postdoctoral Fellow for Research Abroad, supported by the Japan Society for the Promotion of Science, in the Broadband Communications Research Lab at the University of Waterloo, Ontario, Canada. He currently serves as the Vice-Chair of the Technical Affairs Committee (TAC) for the IEEE Asia/Pacific Region Board. He has also served as a Guest Editor for several prestigious journals, including *IEEE Wireless Communications, IEEE Journal on Miniaturization for Air and Space Systems, IEEE Transactions on Vehicular Technology, Springer Peer-to-Peer Networking and Applications, IEEE Transactions on Cognitive Communications and Networking*, and the *International Journal of Distributed Sensor Networks*. He is currently an Associate Editor of the *International Journal of Computers and Applications*. He has been recognized with multiple accolades, including the Best Paper Award at IEEE VTC2013-Spring, IEEE/CIC ICCC2015, IEEE ICC2016, and IEEE Transactions on Computers in 2018.

Editorial

RF Energy Harvesting and Wireless Power Transfer for IoT

Onel Luis Alcaraz López [1,*] and Katsuya Suto [2]

[1] Faculty of Information Technology and Electrical Engineering, University of Oulu, 90570 Oulu, Finland
[2] Graduate School of Informatics and Engineering, The University of Electro-Communications, Tokyo 183-8585, Japan; k.suto@uec.ac.jp
* Correspondence: onel.alcarazlopez@oulu.fi

The rapid proliferation of the Internet of Things (IoT) has transformed modern living by interconnecting billions of devices across industrial, commercial, and domestic sectors. These pervasive IoT systems offer unprecedented capabilities, such as real-time monitoring, automation, and intelligent decision-making. However, the widespread deployment of IoT devices faces a critical challenge: sustaining device operations through sustainable energy provision and management. Traditional battery-powered solutions, with their finite lifespans and the logistical burden of periodic replacement, often prove impractical, especially in dense or remote networks. In response, radio frequency (RF) energy harvesting (EH) and wireless power transfer (WPT) technologies have emerged as pivotal innovations, enabling wirelessly powered systems that extend devices' lifetimes and reduce their maintenance costs. These technologies may provide scalable solutions for powering sensors, actuators, and edge nodes in numerous scenarios, fostering a more autonomous, resilient, and sustainable IoT infrastructure.

Research in RF-EH and RF-WPT, hereinafter referred to as EH and WPT for simplicity, has evolved significantly, driven by advances in antenna/circuit design and the development of protocols to optimize energy transfer across dynamic conditions. Still, challenges persist in the efficient, scalable, and ecologically sensitive deployment of these systems. This Special Issue, comprising 12 research papers from authors across the globe, addresses these challenges through diverse perspectives, exploring innovations and practical applications while critically examining their trade-offs and environmental impacts. The papers are grouped under five key topics that encapsulate the far-ranging approaches and novel contributions presented in this Special Issue. Together, these works aim to advance our understanding and unlock the transformative potential of RF-based energy solutions within the evolving IoT landscape, contributing to the development of sustainable and resilient wireless ecosystems.

1. Advanced Antenna and Rectenna Designs

Innovative antenna and RF circuit designs are crucial for maximizing EH efficiency and is the scope of the work carried out in [1,2]. In [1], Aboualalaa et al. contribute by developing a pattern-reconfigurable antenna that is optimized for EH applications. The antenna operates in the 4.17 to 4.5 GHz range and is capable of electronically steering its radiation pattern across 360 degrees using an RF switch matrix, even switching between directional and omnidirectional modes. The proposed solution achieves high performance, flexibility, and integration capability for IoT environments. Meanwhile, in [2], Abdulwali et al. introduce a high-performance circularly polarized rectenna for EH at 2.45 GHz. The design offers compact design, harmonic rejection, high radiation efficiency (80–91%), stable output for resistive loads, 7.2 dBi directivity, and a conversion efficiency ranging from 36% to 70% for low-input power levels (-10 to 0 dBm).

2. Statistical Analysis of EH

Assessing the performance statistics of EH in a variety of scenarios and channel conditions, as reported in [3–5], is crucial for predicting and later optimizing a system's

performance in real-world conditions. For instance, the energy harvested from an unmodulated carrier is statistically characterized by Galmés in [3] under generalized-K wireless propagation conditions, capturing the path loss, shadowing, and fading effects. The study provides exact closed-form expressions for the mean and variance of energy harvested over time, offering insights for optimizing WPT in large-scale IoT networks. The result emphasize the need for energy devices that can dynamically handle wide input ranges. Meanwhile, in [4], Le-Thanh et al. derive analytical expressions for the outage probability and throughput of a point-to-point wireless communication system in κ-μ shadowed fading channels, wherein the source harvests RF energy from a dedicated energy transmitter. They reveal that EH nonlinearity has a more severe impact on a system's performance than hardware imperfections and that communication reliability can be substantially improved through careful tuning of parameters such as the time-splitting factor. Finally, in [5] Kozić et al. analyze cognitive radio networks wherein secondary transmitters are powered by a dedicated energy transmitter and coexist with a primary network under strict interference constraints, factoring in statistical channel state information and random mobility. The study derives analytical expressions for the system's outage probability, outage capacity and ergodic capacity, considering the mobility modeled using the random waypoint model in a Nakagami-m fading environment. Their key findings highlight the trade-offs between harvested energy, interference limits, and throughput, emphasizing system optimization through parameters such as the interference threshold, fading characteristics, and mobility patterns, all validated using simulations.

3. Optimization and Resource Management in RF-Powered Communication Networks

Ensuring effective utilization of limited system resources is essential to enable more reliable and scalable WPT and EH applications in communication networks, and this is the focus in [6–9]. Specifically, the integration of WPT into vehicular and IoT communication networks in the presence of multi-antenna eavesdroppers is explored by Ganapathy et al. in [6]. They optimize the power allocation and splitting ratios to enhance both security and EH and highlight key trade-offs under different transmission conditions. Meanwhile, Kulkarni and Joshi provide a modeling framework for RF-powered New Radio Unlicensed networks in [7]. They leverage a three-dimensional Markov chain to characterize and analyze key performance parameters, such as throughput, collision transmission, and outage probabilities for IoT nodes that are wirelessly powered by a base station. Interestingly, it is shown that the node density and network configuration significantly impact the performance metrics, but the energy storage levels have a minimal effect. Notably, Eidaks et al. demonstrate multi-hop indoor WPT in [8], showcasing practical scenarios and efficiency gains. As their experiments indicate, multi-hop networks based on signal amplification can improve the power reception in line-of-sight and non-line-of-sight scenarios, especially by employing sub-GHz frequencies for a greater range. The authors emphasize the potential of such configurations to optimize power distribution without additional infrastructure, reducing the network complexity and improving coverage. Finally, combining semantic–functional communication with EH is proposed by Silva et al. in [9] to enable efficient multiuser event-driven sensor networks. In a nutshell, sensors transmit meaningful events using energy-based signaling without conventional demodulation, allowing for simultaneous data transmission and battery recharging in an efficient and integrated manner. Their simulation results show the approach can balance energy consumption, event detection, and communication efficiency, evincing promise for future WPT applications requiring real-time responsiveness.

4. UAV-Based Solutions and Ecological Assessments

Unmanned aerial vehicles (UAVs) offer potential avenues for power delivery to distributed IoT networks, particularly in challenging environments. However, such potential still requires further assessment to ensure sustainable and practical deployment, especially in terms operational constraints, performance trade-offs, and ecological impacts, several of which are addressed in [10,11]. Specifically, the use of UAVs is considered by Cvetković et al. [10] to restore communication and power in an industrial system during emergencies. They showcase the utility of UAV-assisted simultaneous wireless information and power transfer systems for industrial emergency applications, where maintaining power and data flow under challenging conditions is critical. They also assess how the system's performance depends on the UAV's positioning, power allocation, and EH efficiency. For this, and also related to the "EH Statistical Analysis" topic, analytical expressions for the system outage probability and throughput are derived considering Nakagami-m and Fisher–Snedecor fading channels. Meanwhile, Van Mulders et al. present a comprehensive study in [11] on using UAVs to service energy to IoT nodes via recharging or battery replacement. They compare energy efficiencies and ecological impacts between UAV-based recharging, battery replacement, and alternative methods such as remote WPT. For this, the energy consumption of a small license-free UAV is formally characterized, expressions for system efficiencies are derived, and efficient designs/deployments are investigated. The authors show that UAV-based servicing is more energy-efficient over greater distances compared with other methods, with battery swapping generally outperforming recharging due to reduced hovering time. The ecological implications, such as toxic materials and e-waste, are thoroughly assessed, offering insights into sustainable device maintenance practices.

5. RF Exposure Mitigation

Mitigating RF exposure is critical for the safe deployment of WPT systems, as high levels of electromagnetic fields may pose health risks and interfere with biological processes. Detecting potential humans/animals in the environment can help in this regard, as this allows for real-time monitoring and dynamic deactivating, beam steering, or adjusting the power transfer when needed, thus ensuring compliance with safety regulations while maintaining efficient energy delivery. A step in this direction is reported by Mattos et al. in [12], wherein multiple millimeter-wave (mmWave) radars are explored for detecting humans and small animals in indoor environments. The authors propose and evaluate different data fusion and radar positioning strategies, demonstrating that combining data from multiple radars significantly enhances the detection sensitivity and precision compared with a single radar.

Acknowledgments: We extend our deepest appreciation to the contributing authors, peer reviewers, and the editorial team, whose efforts made this Special Issue possible. We hope that it inspires and informs ongoing research and innovation in this vital field.

Conflicts of Interest: The authors declare no conflict of interest.

References

Aboualalaa, M.; Mohamed, H.; Alghamdi, T.; Alathbah, M. A Pattern Reconfigurable Antenna Using Eight-Dipole Configuration for Energy Harvesting Applications. *Sensors* **2023**, *23*, 8451. [CrossRef] [PubMed]

Abdulwali, Z.; Alqahtani, A.; Aladadi, Y.; Alkanhal, M.; Al-Moliki, Y.; Aljaloud, K.; Alresheedi, M. A High-Performance Circularly Polarized and Harmonic Rejection Rectenna for Electromagnetic Energy Harvesting. *Sensors* **2023**, *23*, 7725. [CrossRef] [PubMed]

Galmés, S. Statistical Characterization of Wireless Power Transfer via Unmodulated Emission. *Sensors* **2022**, *22*, 7828. [CrossRef] [PubMed]

Le-Thanh, T.; Ho-Van, K. Performance Analysis of Wireless Communications with Nonlinear Energy Harvesting under Hardware Impairment and κ-μ Shadowed Fading. *Sensors* **2023**, *23*, 3619. [CrossRef] [PubMed]

Kozić, N.; Blagojević, V.; Cvetković, A.; Ivaniš, P. Performance Analysis of Wirelessly Powered Cognitive Radio Network with Statistical CSI and Random Mobility. *Sensors* **2023**, *23*, 4518. [CrossRef] [PubMed]

Ganapathy, V.; Ramachandran, R.; Ohtsuki, T. Resource Allocation for Secure MIMO-SWIPT Systems in the Presence of Multi-Antenna Eavesdropper in Vehicular Networks. *Sensors* **2023**, *23*, 8069. [CrossRef] [PubMed]

Potnis Kulkarni, V.; Joshi, R. Modeling and Performance Analysis of LBT-Based RF-Powered NR-U Network for IoT. *Sensors* **2024**, *24*, 5369. [CrossRef] [PubMed]

Eidaks, J.; Kusnins, R.; Babajans, R.; Cirjulina, D.; Semenjako, J.; Litvinenko, A. Efficient Multi-Hop Wireless Power Transfer for the Indoor Environment. *Sensors* **2023**, *23*, 7367. [CrossRef] [PubMed]

Silva, P.; Marchetti, N.; Nardelli, P.; de Souza, R. Enabling Semantic-Functional Communications for Multiuser Event Transmissions via Wireless Power Transfer. *Sensors* **2023**, *23*, 2707. [CrossRef] [PubMed]

10. Cvetković, A.; Blagojević, V.; Anastasov, J.; Pavlović, N.; Milošević, M. Outage Analysis of Unmanned-Aerial-Vehicle-Assisted Simultaneous Wireless Information and Power Transfer System for Industrial Emergency Applications. *Sensors* **2023**, *23*, 777. [CrossRef] [PubMed]
11. Van Mulders, J.; Cappelle, J.; Goossens, S.; De Strycker, L.; Van der Perre, L. UAV-Based Servicing of IoT Nodes: Assessment of Ecological Impact. *Sensors* **2023**, *23*, 2291. [CrossRef] [PubMed]
12. Mattos, A.; Brante, G.; Moritz, G.; Souza, R. Human and Small Animal Detection Using Multiple Millimeter-Wave Radars and Data Fusion: Enabling Safe Applications. *Sensors* **2024**, *24*, 1901. [CrossRef] [PubMed]

Disclaimer/Publisher's Note: The statements, opinions and data contained in all publications are solely those of the individual author(s) and contributor(s) and not of MDPI and/or the editor(s). MDPI and/or the editor(s) disclaim responsibility for any injury to people or property resulting from any ideas, methods, instructions or products referred to in the content.

Communication

A Pattern Reconfigurable Antenna Using Eight-Dipole Configuration for Energy Harvesting Applications

Mohamed Aboualalaa [1,*], Hesham A. Mohamed [1], Thamer A. H. Alghamdi [2,3,*] and Moath Alathbah [4]

1. Microstrip Department, Electronics Research Institute, Cairo 11843, Egypt; hesham_280@eri.sci.eg
2. Wolfson Centre for Magnetics, School of Engineering, Cardiff University, Cardiff CF24 3AA, UK
3. Electrical Engineering Department, School of Engineering, Albaha University, Albaha 65779, Saudi Arabia
4. Department of Electrical Engineering, College of Engineering, King Saud University, Riyadh 11421, Saudi Arabia; malathbah@ksu.edu.sa
* Correspondence: mohamed.ali@ejust.edu.eg (M.A.); alghamdit1@cardiff.ac.uk (T.A.H.A.)

Abstract: A pattern reconfigurable antenna, composed of eight elements, is proposed for energy harvesting applications. Pattern reconfigurable antennas are a promising technique for harvesting from different wireless sources. The radiation pattern of the proposed antenna can be steered electronically using an RF switch matrix, covering an angle range from 0 to 360 degrees with a step size of 45 degrees. The proposed antenna primarily consists of an eight-dipole configuration that shares the same excitation. Each dipole is excited using a balun comprising a quarter-wavelength grounded stub and a quarter-wavelength open-circuit stub. The proposed antenna operates in the frequency range of 4.17 to 4.5 GHz, with an impedance bandwidth of 7.6%. By switching between the different switches, the antenna can be steered with a narrower rotational angle. In addition, the antenna can work in an omnidirectional mode when all switches are in the ON state simultaneously. The results demonstrate a good agreement between the numerical and experimental findings for the reflection coefficient and radiation characteristics of the proposed reconfigurable antenna.

Keywords: dipole antenna; electronically steering; energy harvesting; reconfigurable antenna; RF switch matrix

1. Introduction

Reconfigurable antennas are a type of antenna that can change their operating parameters, such as frequency, polarization, radiation pattern, and impedance, in real-time or under specific conditions [1–8]. This reconfigurability allows the antenna to adapt to different communication standards, frequency bands, or user requirements. Reconfigurable antennas are gaining popularity due to their flexibility and potential to enhance the performance of wireless communication systems. They can improve network coverage, increase the data rate, reduce interference, and provide seamless connectivity. This technology has significant implications for emerging applications, such as the Internet of Things (IoT), 5G networks, and satellite communication.

Pattern reconfigurable antennas are capable of dynamically changing their radiation patterns to accommodate varying operational requirements [9–12]. This type of antenna holds the advantage of electronically adjusting its radiation beams in different directions, eliminating the need for physical movement. With their ability to adapt to different directions, pattern reconfigurable antennas are gaining popularity in wireless communication systems, offering a flexible solution for improving network coverage, enhancing signal quality, and mitigating interference.

In this study, we propose a design for a pattern reconfigurable antenna specifically tailored for energy harvesting applications. Energy harvesting involves converting ambient energy from the surrounding environment, such as electromagnetic waves, into electrical energy that can power small electronic devices [13–20]. Consequently, reconfigurable antennas offer several advantages over conventional antennas when it comes to energy harvesting. They can be optimized to efficiently harvest energy from various sources and frequencies. By adjusting the antenna's radiation pattern, polarization, or frequency response, the harvested energy can be maximized even in dynamic and varying environments. Moreover, reconfigurable antennas can be integrated with other energy harvesting components, including rectifiers, impedance matching networks, and energy storage devices, to form a comprehensive energy harvesting system. Considering these aspects, the utilization of reconfigurable antennas in energy harvesting applications has the potential to revolutionize wireless power transfer and facilitate the development of self-powered and autonomous devices.

Multiple antennas are typically required to harvest from different RF sources [21–23], which results in a larger system size. However, a pattern reconfigurable antenna offers a suitable solution for harvesting energy from various sources located at different positions. Depending on the signal intensity, the antenna can steer its direction using a dedicated DSP control unit. The operational concept of using a pattern reconfigurable antenna for energy harvesting is demonstrated in Figure 1. The reconfigurable antenna is connected to a control unit, which in turn can determine the direction in which the antenna operates. Additionally, the antenna is connected in parallel with a rectifying circuit through a matching circuit, forming the rectenna structure. This rectenna can be connected to a rechargeable battery or directly to a load. Reconfigurable antennas hold great promise for achieving reliable and efficient energy harvesting systems. In this study, we focused on the receiving pattern reconfigurable antenna and its electronic steering capability. To the best of our knowledge, the proposed structure is the first to suggest a steerable pattern in energy harvesting applications for scavenging from various RF sources.

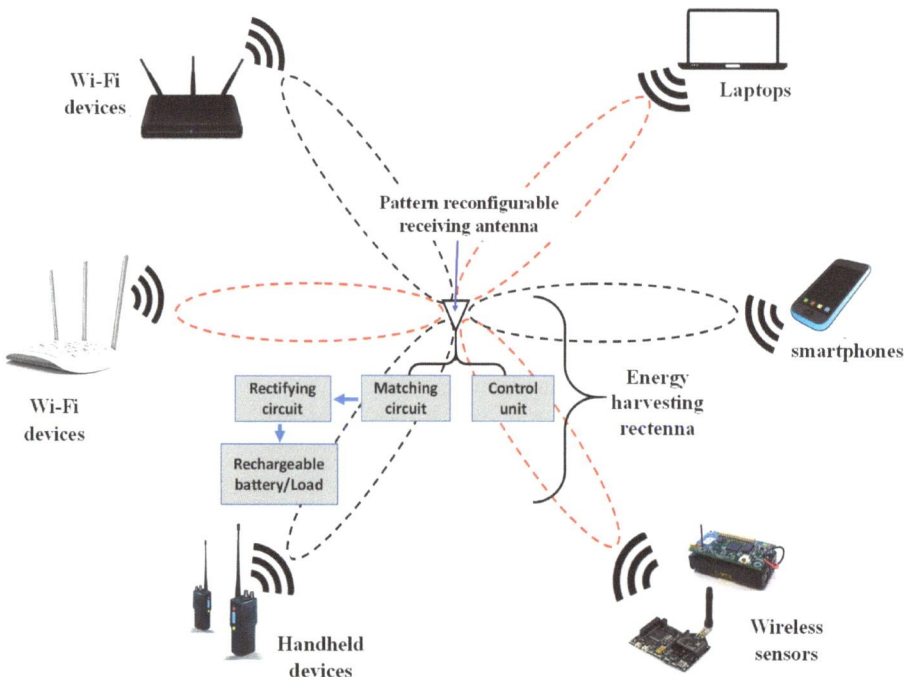

Figure 1. Configuration of energy harvesting from different RF sources.

2. Antenna Configuration

Figure 2a illustrates the configuration of the proposed antenna design. It comprises an eight-dipole configuration designed on the ground plane (bottom layer). The dipoles are connected to a common feeding point, and each dipole has a director to provide a better directional characteristic of the antenna. On the opposite side of the substrate, the top layer has eight open-circuited stubs positioned to excite the dipoles. Each stub excites a corresponding dipole through a coaxial probe feed. Photographs showing the top and bottom views of the fabricated prototype for the proposed antenna are displayed in Figure 2b,c. A balun, depicted in Figure 3a [24], is utilized for feeding the different dipoles. However, in the proposed design, we deviate from directly connecting the feed line to the other terminal of the dipole. Instead, we incorporate a $\lambda/4$ L-shaped open-circuited stub, which provides greater flexibility in adjusting the resonance frequency for the dipole structure. Consequently, the feeding structure comprises two parts, as illustrated in Figure 3b. Part (a) represents the feed line, with a width of 0.5 mm, utilized for exciting the dipole. Part (b) consists of an L-shaped open stub with a double width of the part (a), which connects the feed line to the other terminal of the dipole antenna. At the open-ended section of the L-shaped stub (at point o), the impedance is infinity. Thus, at a distance of $\lambda/4$ from this open-end point, the stub functions as a short circuit with the ground plane, connecting the feed line (at point f) with the other terminal of the dipole antenna on the ground plane.

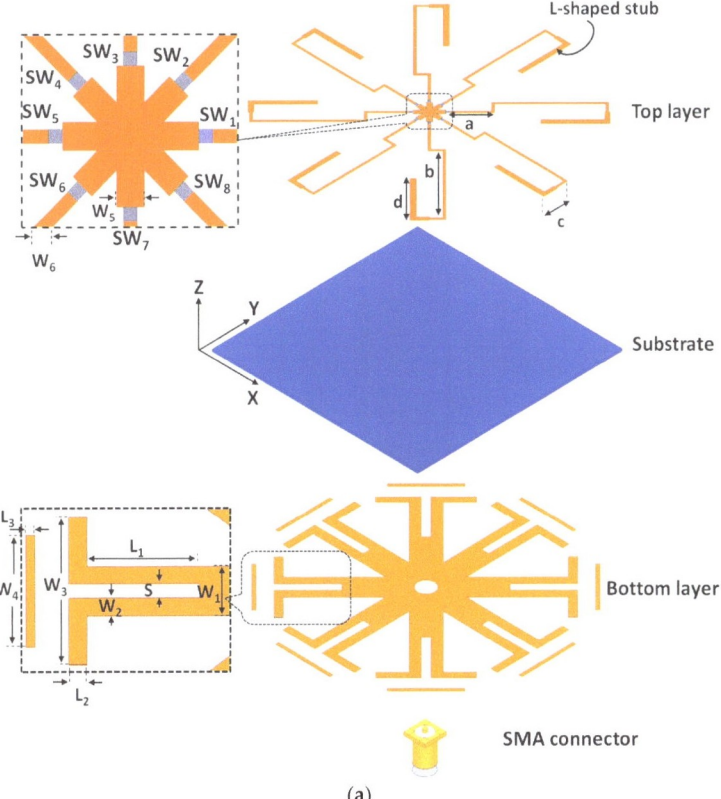

(a)

Figure 2. *Cont.*

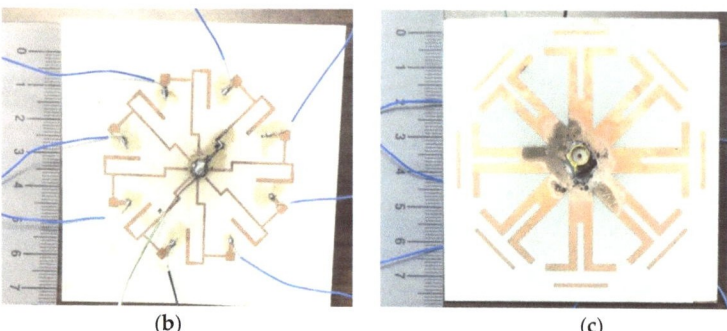

Figure 2. (**a**) The 3D assembly configuration of the proposed antenna, (**b**) top view of the antenna's fabricated prototype, and (**c**) bottom view of the antenna's fabricated prototype; a = 7 mm, b = 17 mm c = 4 mm, d = 10.5 mm, S = 2 mm, L_1 = 12 mm, L_2 = 2 mm, L_3 = 1 mm, W_1 = 7 mm, W_2 = 2.5 mm W_3 = 21 mm, W_4 = 16 mm, W_5 = 1 mm, and W_6 = 0.5 mm.

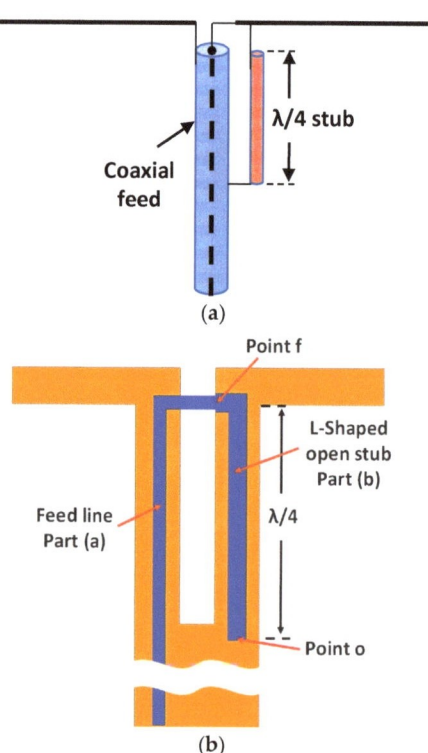

Figure 3. Configuration of (**a**) coaxial balun structure, and (**b**) balun used with the proposed antenna.

In Figure 2, eight switches (SW_1, SW_2, SW_3, SW_4, SW_5, SW_6, SW_7, and SW_8) are strategically placed along the feeding lines to enable pattern reconfigurability. Initially, during the simulation, the RF HPND-4005 diode series was employed to achieve the desired reconfigurability. To replicate the behavior of physical diodes, an equivalent circuit model consisting of a series R-L circuit in the ON state and a parallel R-C circuit in the OFF state was utilized. However, for the actual measurements, an RF switch matrix (Mini-Circuits USB-8SPDT-A18 [25]) with low insertion loss (0.2 dB) and high isolation (85 dB) was employed. Therefore, we substituted the equivalent circuit model with the S2P files of the switch matrix during the full-wave simulation analysis, yielding nearly identical results.

The proposed antenna was fabricated on a thin Rogers RT/duroid 6002™ substrate, featuring a thickness (h) of 0.76 mm, a dielectric constant (ε_r) of 2.94, a dielectric loss tangent (tanδ) of 0.0012, and a copper thickness (t) of 0.0035 mm. The optimized geometrical parameters of the proposed antenna are listed in the caption of Figure 2.

An equivalent circuit model is illustrated in Figure 4 to provide a more detailed explanation and is proposed to simplify the operation of the proposed antenna. Initially, the excitation is distributed among eight branches, each equipped with an RF switch. The feed line is symbolized with an inductor denoted as L_f. Additionally, we adopted a four-element equivalent circuit model [26] to represent the dipole antenna, which consists of a parallel connection involving R_d, L_d, and C_{d1} arranged in series with C_{d2}. Furthermore, each λ/4 open-circuit stub, linked to every feed line, can be effectively described as a series LC circuit (L_{st}, and C_{st}), as depicted in Figure 4.

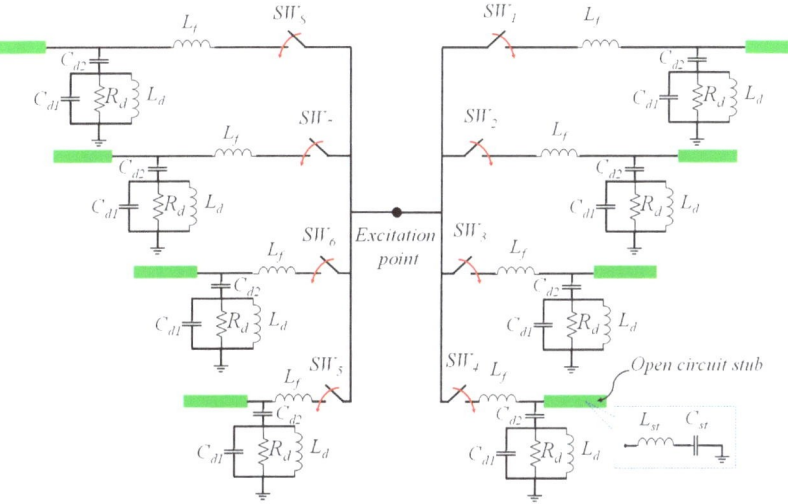

Figure 4. Equivalent circuit model of the proposed antenna structure.

3. Results and Discussion

3.1. Reflection Coefficient Results

To verify the reconfigurability property of the proposed antenna, the antenna was designed and optimized using a full-wave analysis simulation (ANSYS High Frequency Structure Simulator (HFSS)). Subsequently, the antenna was fabricated and measured to validate the simulation results. Figure 5 shows a comparison between the numerical and experimental results of the antenna's reflection coefficient when SW_1 is in the ON state, while the remaining switches are in the OFF state. Notably, the proposed antenna exhibited consistent reflection coefficient characteristics during pattern steering across various dedicated radiation directions. For this study, a frequency range of 4 to 4.5 GHz within the sub-6 GHz band was selected as the operational range. Nonetheless, the antenna parameters can be adjusted to accommodate different frequency bands. The proposed antenna operates within the range of 4.17 to 4.5 GHz, offering an impedance bandwidth of 7.6%. Parametric studies were performed to investigate the impact of various antenna parameters, such as the length of the λ/4 L-shaped open stub, dipole length, and feeding location, as depicted in Figure 6.

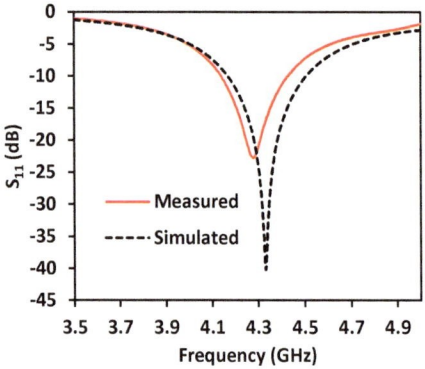

Figure 5. Numerical and experimental findings of the proposed pattern reconfigurable antenna when SW_1 is in ON state while the other switches are in OFF state.

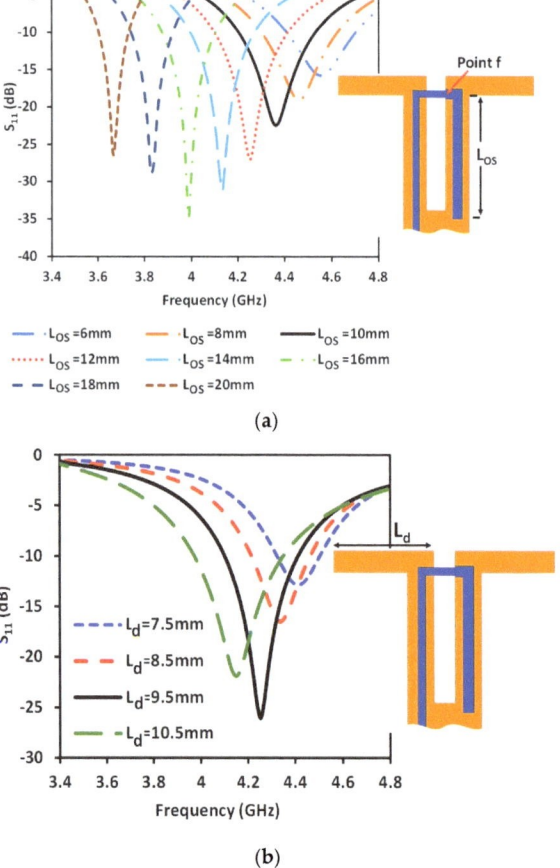

Figure 6. *Cont.*

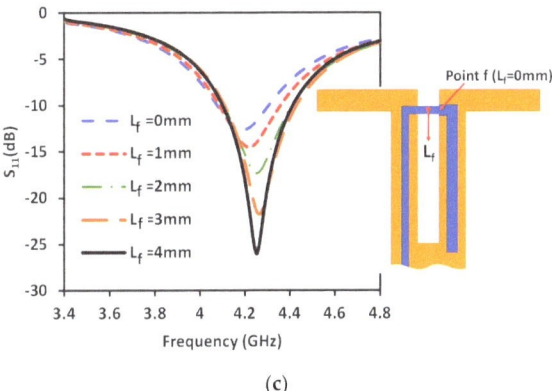

(c)

Figure 6. S_{11} parametric study for (**a**) different lengths of $\lambda/4$ open-circuit stub (L_{OS}), (**b**) different lengths of the dipole (L_d), and (**c**) different positions of the feed point.

By adjusting the length of the $\lambda/4$ open-circuit stub (L_{OS}), the matching can be achieved at different values of the operating frequency of the dipole, as shown in Figure 6a. When the stub length is shorter, it achieves the matching at a higher frequency for the dipole antenna. Conversely, increasing the stub length allows for matching adjustment at a lower frequency. Furthermore, the operating frequency can be adjusted by controlling the length of the dipole (L_d), as displayed in Figure 6b. Additionally, altering the feed position has an impact on the input impedance and consequently the antenna matching. The study of the feed location started when the feed point was at dipole as a reference position (when the feeding is at point f, i.e., $L_f = 0$ mm), shown in Figure 6c. Subsequently, the feeding position shifted away from the dipole towards the excitation position. This shift has a slight effect on the resonance frequency, as demonstrated in Figure 6c, where the currents in the two parallel lines of the ground plane counterbalance each other. However, it primarily influences the matching, as the input impedance changes with the alteration of the feed location. Therefore, the operating frequency and antenna matching can be adjusted by manipulating these parameters (L_{OS}, L_d, and L_f). As shown in Figure 6a, by adjusting the length of the open stub we can tune the frequency of the proposed antenna from approximately 3.7 GHz to 4.6 GHz. This frequency range falls within the proposed sub-6 5G spectrum for several countries, such as the US (3.7–3.98 GHz), Canada (3.65–4 GHz), South Korea (3.7–4 GHz), and Japan (3.6–4.1 GHz) [27]. We conducted our study around 4.3 GHz; however, the frequency can be easily adjusted to different bands. Figure 7 presents the surface current distribution for different states of the switches, revealing the concentration of current at each branch when its corresponding switch is in the ON state.

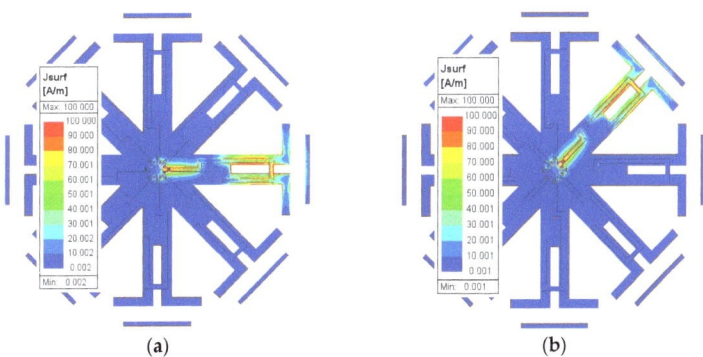

(a)   (b)

Figure 7. *Cont.*

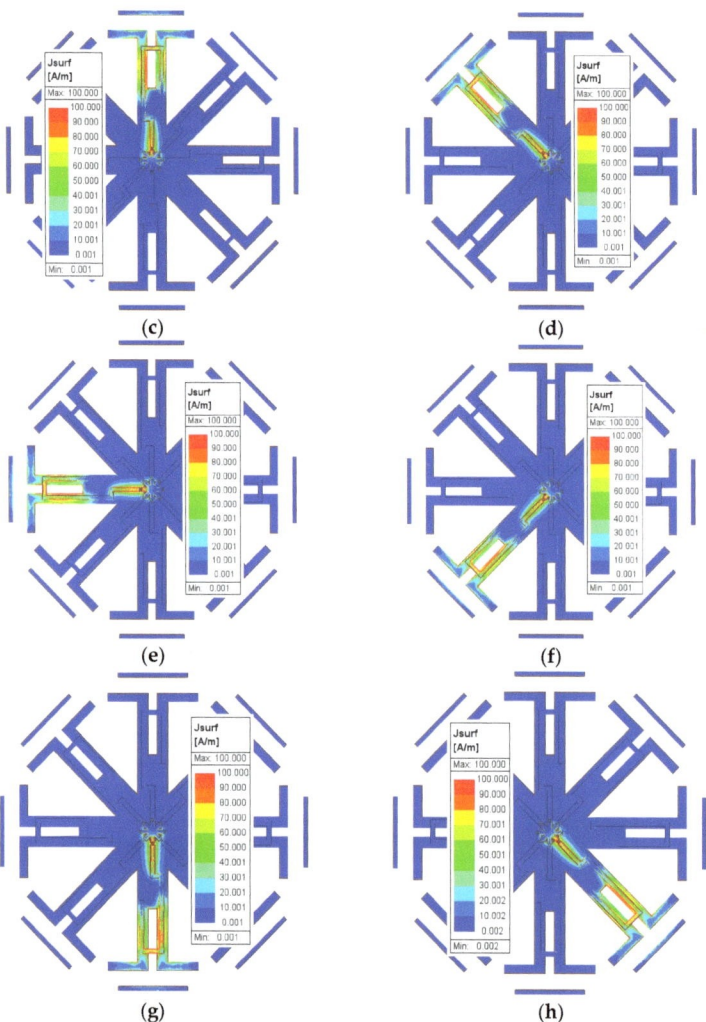

Figure 7. Surface current distribution when (**a**) SW$_1$ is ON, (**b**) SW$_2$ is ON, (**c**) SW$_3$ is ON, (**d**) SW$_4$ is ON, (**e**) SW$_5$ is ON, (**f**) SW$_6$ is ON, (**g**) SW$_7$ is ON, and (**h**) SW$_8$ is ON.

During the measurements, we utilized the Mini-Circuits USB-8SPDT-A18 RF switch matrix, as depicted in Figure 8. This switch matrix is capable of operating from DC to 18 GHz and offers low insertion loss (0.2 dB) and high isolation (85 dB). It served a dual function, both enabling ON/OFF switching operation and functioning as the control unit for the electronic steering process. With this switch matrix, we can electronically switch between different directions of the main beam of the proposed reconfigurable antenna, providing flexibility and adaptability to the system. Below is a summary of the sequence for performing the simulation and measurements:

- We initiated the simulation by employing an equivalent circuit model of a PIN diode (HPND4005) [6].

- During the measurements, we utilized a switch matrix instead of soldering the PIN diodes. In this new scheme, the switch matrix replaced the use of PIN diodes, and the antenna was connected to the switch matrix using only connecting wires. Figure 9 shows the circuit diagram of the switch matrix, illustrating how the switches were connected.
- In order to consider the effect of using the switch matrix during the simulation, we utilized the S2P files provided by the manufacturing company of the switch matrix. These files characterize both the ON and OFF states of the switch matrix. To achieve this, we removed the PIN diodes and substituted them with S2P blocks at the position between the two terminals of each switch. Each S2P block was used to read the S2P file for its respective switch.
- To incorporate the S2P blocks in the simulation model, we utilized the Keysight ADS simulator by conducting a co-simulation between the schematic and momentum, as illustrated in Figure 10. To switch between the different switching states in the simulation, two S2P files were loaded into the S2P block (one for the ON state and the other for the OFF state). For example, if only SW1 needed to be ON while the others were OFF, we loaded the S2P block assigned to SW1 with the S2P file for the ON state, while the other S2P blocks were loaded with the S2P file for the OFF state.
- On the other hand, to achieve the switching operation during the measurements, the matrix's software was used to control the states of the switches.

3.2. Radiation Characteristics

The numerical and measurement results for the 2D normalized radiation patterns (E-plane and H-plane) of the proposed reconfigurable antenna, under various switch configurations, are provided in Tables 1 and 2. In Table 1, the radiation results when switches SW_1, SW_2, SW_3, and SW_4 are in the ON position are displayed. Meanwhile, Table 2 showcases the results for the remaining four cases, where switches SW_5, SW_6, SW_7, and SW_8 are in the ON position. The graphs clearly demonstrate that by altering the states of the switches, a rotatable radiation pattern can be obtained. Through control of the switching matrix states, the radiation pattern can be steered at angles of $0°$, $45°$, $90°$, $135°$, $180°$, $225°$, $270°$, and $315°$. The antenna provides an average antenna gain of 4.2 dBi and a radiation efficiency of 80%.

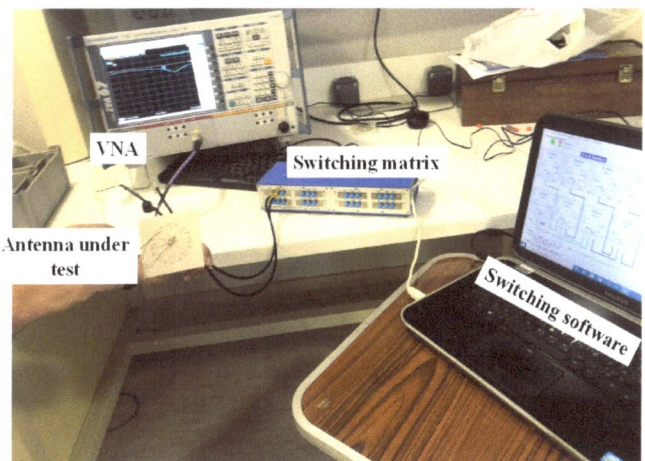

Figure 8. Measurement setup for antenna reflection parameters.

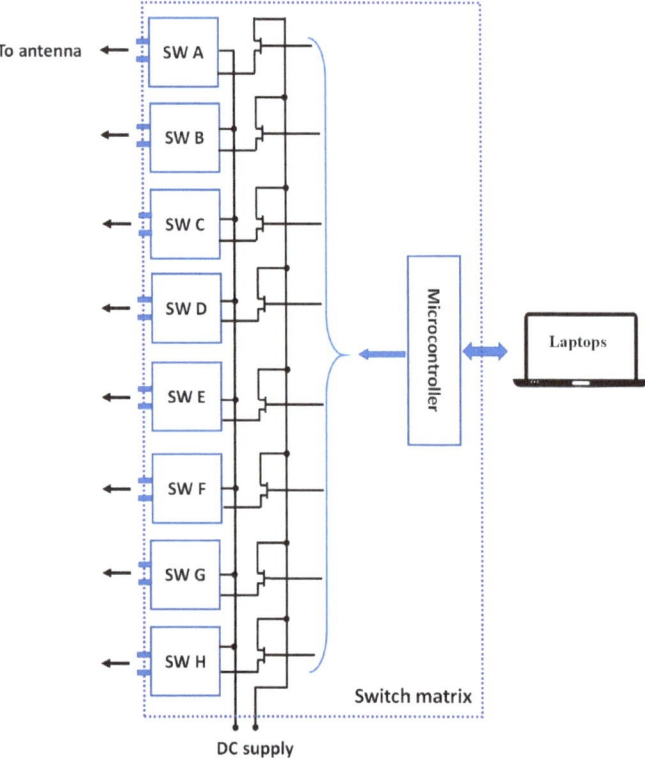

Figure 9. Circuit diagram of the switch matrix.

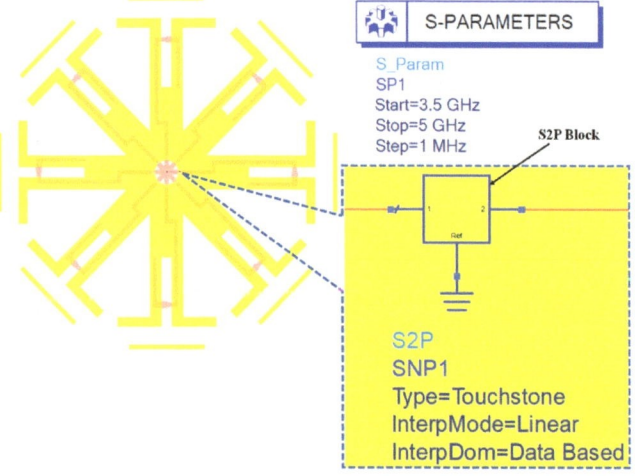

Figure 10. ADS model used for conducting a co-simulation between the schematic and momentum.

Table 1. The 2D normalized polar plots of E-plane and H-plane radiation patterns when switches SW_1, SW_2, SW_3, and SW_4 are in the ON position.

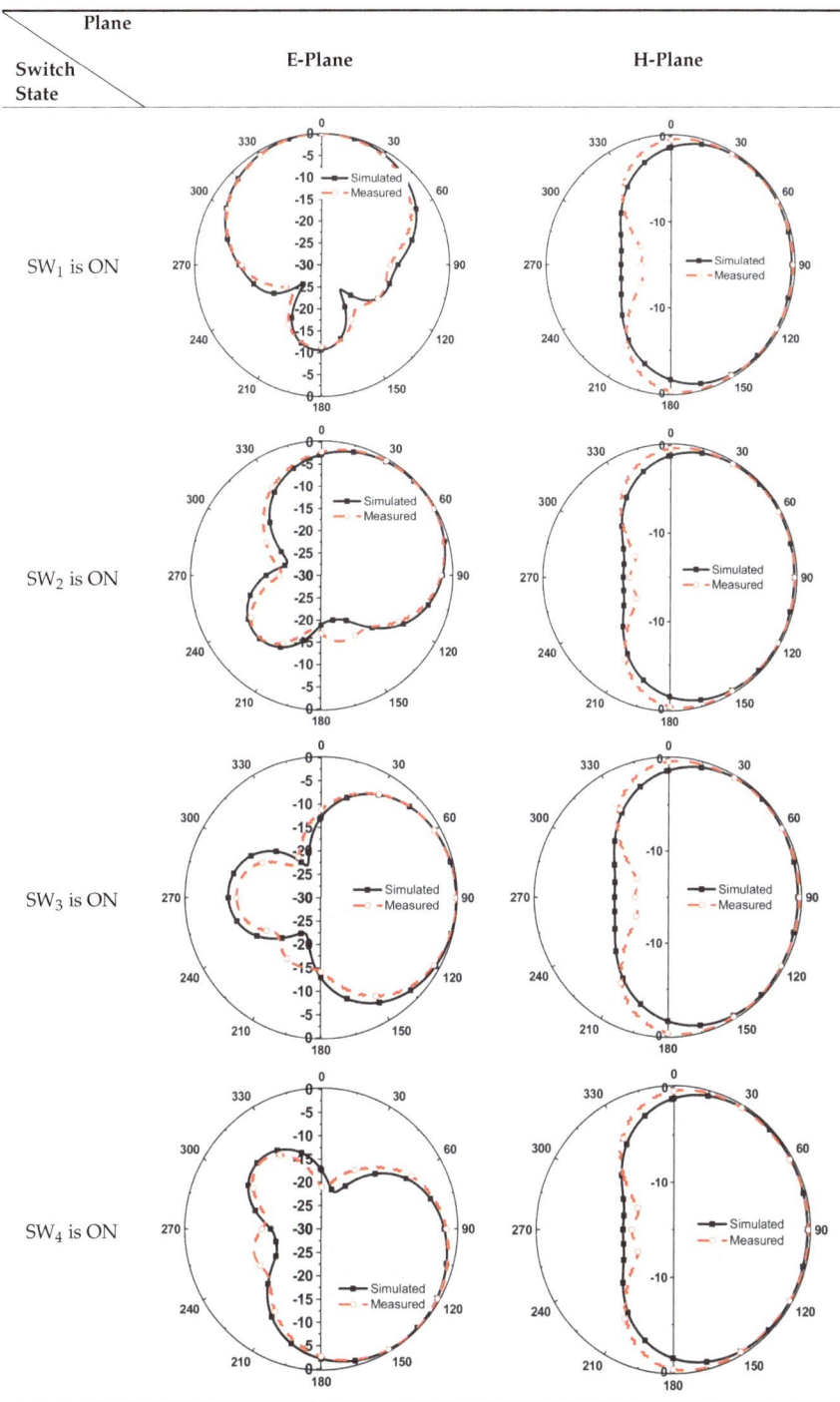

Table 2. The 2D normalized polar plots of E-plane and H-plane radiation patterns when switches SW_5, SW_6, SW_7, and SW_8 are in the ON position.

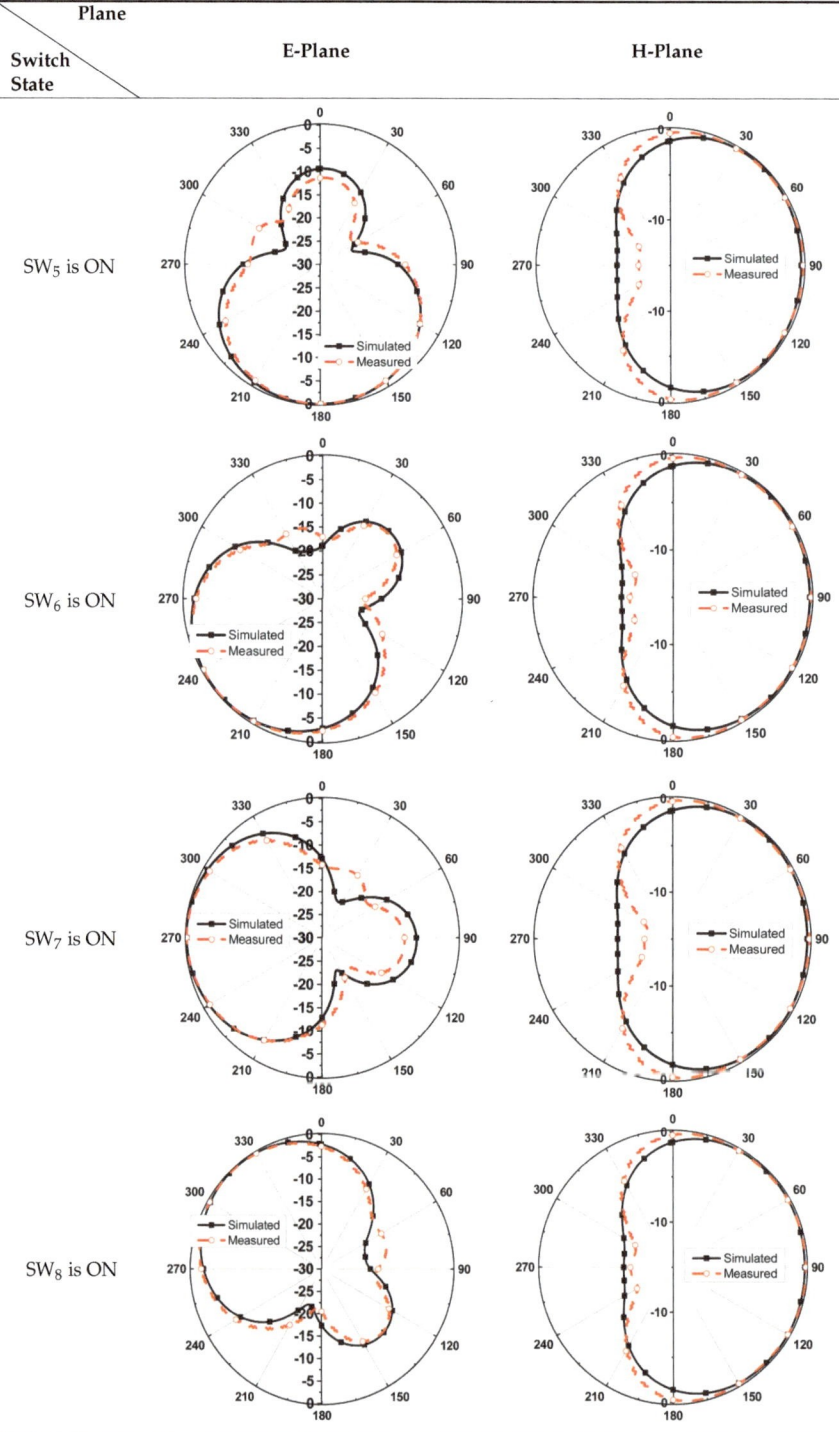

Furthermore, by activating more than one switch in the ON state, the radiation pattern can be steered with a smaller angular step. For instance, when both SW_1 and SW_2 are simultaneously in the ON state, the main radiation pattern can be directed towards the midpoint between the dipoles of these two switches (at $\phi = 22.5°$). A case study showcasing the connection of multiple switches is depicted in Figure 11. The figure displays the radiation patterns achieved by connecting SW_1 and SW_2, as well as SW_3 and SW_4, resulting in radiation patterns directed at $\phi = 22.5°$ and $\phi = 112.5°$, respectively. Moreover, when all switches are set to the same state, i.e., all switches are simultaneously in the ON state, the antenna operates in an omnidirectional mode, as demonstrated in Figure 12. The proposed dipole antenna exhibits linear polarization characteristics, as depicted in Figure 13. The antenna's axial ratio values in the main direction of the radiation pattern fall within the range of 30 to 40 dB. The radiation pattern measurement setup that was employed to assess the radiation characteristics of the proposed antenna is illustrated in Figure 14.

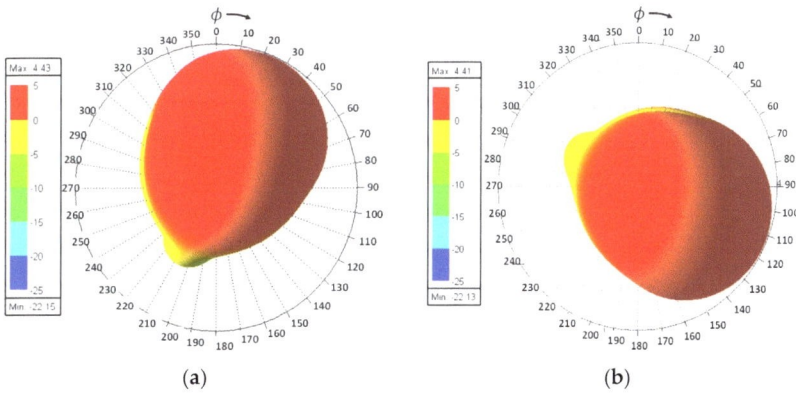

Figure 11. The 3D radiation patterns when: (**a**) SW_1 and SW_2 are ON, and (**b**) SW_3 and SW_4 are ON.

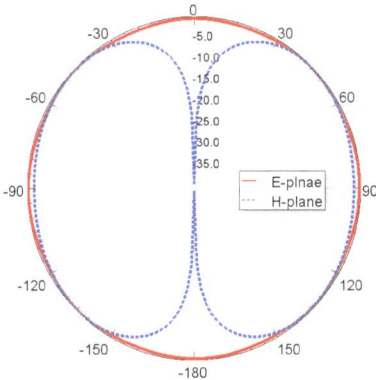

Figure 12. Omnidirectional propagation mode when all switches are in ON state.

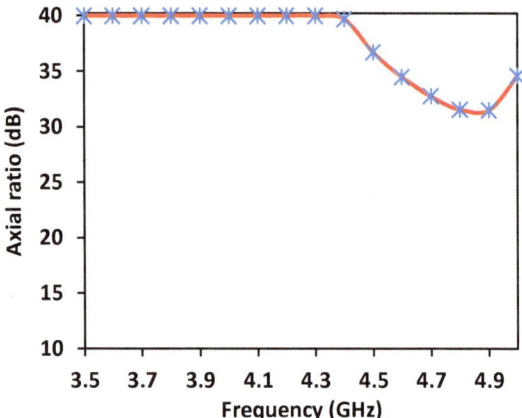

Figure 13. Axial ratio of the proposed antenna.

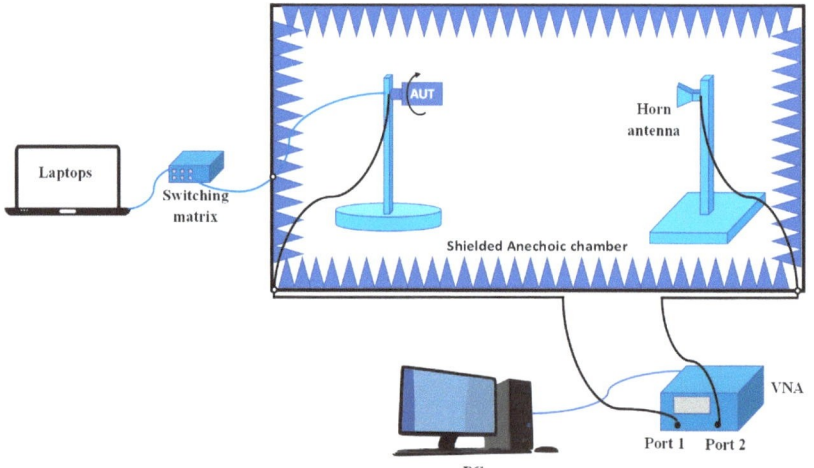

Figure 14. Radiation pattern measurement setup.

4. Conclusions

We propose a pattern reconfigurable antenna specifically designed for energy harvesting applications. This antenna possesses the ability to electronically steer at various rotational angles, covering a full 360°, enabling it to effectively scavenge energy from diverse sources located at different positions. To achieve matching, a balun comprising a λ/4 open-circuit stub was employed. The fabricated antenna was validated by verifying its performance at different directions using an RF switch matrix. The comparison between the numerical and measurement results demonstrates a high level of agreement, affirming the accuracy and reliability of the proposed antenna design.

Author Contributions: M.A. (Mohamed Aboualalaa) conceived the idea, performed the simulations and experiments, and contributed to writing the manuscript; H.A.M. wrote the main manuscript text and prepared the figures; T.A.H.A. and M.A. (Moath Alathbah) performed the literature research and reviewed the manuscript. All authors have read and agreed to the published version of the manuscript.

Funding: This research is supported by Researchers Supporting Project number (RSPD2023R868), King Saud University, Riyadh, Saudi Arabia.

Institutional Review Board Statement: Not applicable.

Informed Consent Statement: Not applicable.

Data Availability Statement: The data supporting the findings of this study are available upon request from the corresponding author.

Acknowledgments: The authors would like to acknowledge the support provided by Researchers Supporting Project at King Saud University, Riyadh, Saudi Arabia.

Conflicts of Interest: The authors declare no conflict of interest.

References

1. Al-Alaa, M.A.; Elsadek, H.A.; Abdallah, E.A.; Hashish, E.A. Pattern and frequency reconfigurable monopole disc antenna using PIN diodes and MEMS switches. *Microw. Opt. Technol. Lett.* **2014**, *56*, 187–195. [CrossRef]
2. Yahya, M.S.; Soeung, S.; Singh, N.S.S.; Yunusa, Z.; Chinda, F.E.; Rahim, S.K.A.; Musa, U.; Nor, N.B.M.; Sovuthy, C.; Abro, G.E.M. Triple-Band Reconfigurable Monopole Antenna for Long-Range IoT Applications. *Sensors* **2023**, *23*, 5359. [CrossRef] [PubMed]
3. Al-Alaa, M.A.; Elsadek, H.A.; Abdallah, E.A.; Hashish, E.A. PIFA frequency reconfigurable antenna. In Proceedings of the 2014 IEEE Antennas and Propagation Society International Symposium (APSURSI), Memphis, TN, USA, 6–12 July 2014; pp. 1256–1257. [CrossRef]
4. Jin, X.; Liu, S.; Yang, Y.; Zhou, Y. A Frequency-Reconfigurable Planar Slot Antenna Using S-PIN Diode. *IEEE Antennas Wirel. Propag. Lett.* **2022**, *21*, 1007–1011. [CrossRef]
5. Sailaja, B.V.S.; Naik, K.K. Design and analysis of reconfigurable fractal antenna with RF-switches on a flexible substrate for X-band applications. *Analog. Integr. Circuits Signal Process.* **2021**, *107*, 181–193. [CrossRef]
6. Al-Alaa, M.A.; Elsadek, H.; Abdallah, E. Compact multi-band frequency reconfigurable planar monopole antenna for several wireless communication applications. *J. Electr. Syst. Inf. Technol.* **2014**, *1*, 17–25. [CrossRef]
7. Zhang, Y.; Zhang, Y.; Huang, K.; Liu, S.-J.; Zhang, X.Y.; Liu, Q.H. A Reconfigurable Patch Antenna with Linear and Circular Polarizations Based on Double-Ring-Slot Feeding Structure. *IEEE Trans. Antennas Propag.* **2022**, *70*, 11389–11400. [CrossRef]
8. Kumar, J.; Basu, B.; Talukdar, F.A.; Nandi, A. Stable-multiband frequency reconfigurable antenna with improved radiation efficiency and increased number of multiband operations. *IET Microw. Antennas Propag.* **2019**, *13*, 642–648. [CrossRef]
9. Zhao, S.; Wang, Z.; Dong, Y. A Planar Pattern-Reconfigurable Antenna with Stable Radiation Performance. *IEEE Antennas Wirel. Propag. Lett.* **2022**, *21*, 784–788. [CrossRef]
10. Yuan, W.; Huang, J.; Zhang, X.; Cui, K.; Wu, W.; Yuan, N. Wideband Pattern-Reconfigurable Antenna with Switchable Monopole and Vivaldi Modes. *IEEE Antennas Wirel. Propag. Lett.* **2023**, *22*, 199–203. [CrossRef]
11. Morshed, K.M.; Karmokar, D.K.; Esselle, K.P.; Matekovits, L. Beam-Switching Antennas for 5G Millimeter-Wave Wireless Terminals. *Sensors* **2023**, *23*, 6285. [CrossRef]
12. Li, W.; Zhao, Y.; Ding, X.; Wu, L.; Nie, Z. A Wideband Pattern-Reconfigurable Loop Antenna Designed by Using Characteristic Mode Analysis. *IEEE Antennas Wirel. Propag. Lett.* **2022**, *21*, 396–400. [CrossRef]
13. Caselli, M.; Tonelli, M.; Boni, A. Analysis and design of an integrated RF energy harvester for ultra low-power environments. *Int. J. Circuit Theory Appl.* **2019**, *47*, 1086–1104. [CrossRef]
14. Mouapi, A. Radiofrequency Energy Harvesting Systems for Internet of Things Applications: A Comprehensive Overview of Design Issues. *Sensors* **2022**, *22*, 8088. [CrossRef] [PubMed]
15. Chang, H.-C.; Lin, H.-T.; Wang, P.-C. Wireless Energy Harvesting for Internet-of-Things Devices Using Directional Antennas. *Futur. Internet* **2023**, *15*, 301. [CrossRef]
16. Xie, Z.; Teng, L.; Wang, H.; Liu, Y.; Fu, M.; Liang, J. A Self-Powered Synchronous Switch Energy Extraction Circuit for Electromagnetic Energy Harvesting Enhancement. *IEEE Trans. Power Electron.* **2023**, *38*, 9972–9982. [CrossRef]
17. Nguyen, D.-A.; Bui, G.T.; Nam, H.; Seo, C. Design of Dual-Band Inverse Class-F Rectifier for Wireless Power Transfer and Energy Harvesting. *IEEE Microw. Wirel. Technol. Lett.* **2023**, *33*, 355–358. [CrossRef]
18. Aboualalaa, M.; Mansour, I.; Abdelrahman, A.B.; Allam, A.; Elsadek, H.; Pokharel, R.K.; Abo-Zahhad, M. Dual-band CPW rectenna for low input power energy harvesting applications. *IET Circuits Devices Syst.* **2020**, *14*, 892–897. [CrossRef]
19. Eidaks, J.; Kusnins, R.; Babajans, R.; Cirjulina, D.; Semenjako, J.; Litvinenko, A. Efficient Multi-Hop Wireless Power Transfer for the Indoor Environment. *Sensors* **2023**, *23*, 7367. [CrossRef]
20. Mansour, M.; Mansour, I.; Zekry, A. A reconfigurable class-F radio frequency voltage doubler from 650 MHz to 900 MHz for energy harvesting applications. *Alex. Eng. J.* **2022**, *61*, 8277–8287. [CrossRef]
21. Lopez, O.L.A.; Clerckx, B.; Latva-Aho, M. Dynamic RF Combining for Multi-Antenna Ambient Energy Harvesting. *IEEE Wirel. Commun. Lett.* **2022**, *11*, 493–497. [CrossRef]
22. Olgun, U.; Chen, C.-C.; Volakis, J.L. Investigation of Rectenna Array Configurations for Enhanced RF Power Harvesting. *IEEE Antennas Wirel. Propag. Lett.* **2011**, *10*, 262–265. [CrossRef]
23. Sun, H.; Huang, J.; Wang, Y. An Omnidirectional Rectenna Array with an Enhanced RF Power Distributing Strategy for RF Energy Harvesting. *IEEE Trans. Antennas Propag.* **2022**, *70*, 4931–4936. [CrossRef]
24. Ali, M. *Reconfigurable Antenna Design and Analysis*; Artech House: Norwood, MA, USA, 2021.

25. Available online: www.minicircuits.com/pdfs/USB-8SPDT-A18.pdf (accessed on 1 June 2023).
26. Tang, T.; Tieng, Q.; Gunn, M. Equivalent circuit of a dipole antenna using frequency-independent lumped elements. *IEEE Trans. Antennas Propag.* **1993**, *41*, 100–103. [CrossRef]
27. Qualcomm Technologies Inc. Spectrum for 4G and 5G. 2020. Available online: https://www.qualcomm.com/media/documents/files/spectrum-for-4g-and-5g.pdf (accessed on 16 March 2021).

Disclaimer/Publisher's Note: The statements, opinions and data contained in all publications are solely those of the individual author(s) and contributor(s) and not of MDPI and/or the editor(s). MDPI and/or the editor(s) disclaim responsibility for any injury to people or property resulting from any ideas, methods, instructions or products referred to in the content.

Article

A High-Performance Circularly Polarized and Harmonic Rejection Rectenna for Electromagnetic Energy Harvesting

Zaed S. A. Abdulwali [1], Ali H. Alqahtani [2,*], Yosef T. Aladadi [1], Majeed A. S. Alkanhal [1], Yahya M. Al-Moliki [1], Khaled Aljaloud [2] and Mohammed Thamer Alresheedi [1]

[1] Department of Electrical Engineering, King Saud University, Riyadh 11421, Saudi Arabia; zabdulwali@ksu.edu.sa (Z.S.A.A.); yaladadi@ksu.edu.sa (Y.T.A.); majeed@ksu.edu.sa (M.A.S.A.); yalmoliki@ksu.edu.sa (Y.M.A.-M.); malresheedi@ksu.edu.sa (M.T.A.)
[2] Department of Applied Electrical Engineering, Al-Muzahimya Campus, College of Applied Engineering, King Saud University, Riyadh 11421, Saudi Arabia; kaljaloud@ksu.edu.sa
* Correspondence: ahqahtani@ksu.edu.sa

Abstract: This paper presents a novel circularly polarized rectenna designed for efficient electromagnetic energy harvesting at the 2.45 GHz ISM band. A compact antenna structure is designed to achieve high performance in terms of radiation efficiency, axial ratio, directivity, effective area, and harmonic rejection over the entire bandwidth of the ISM frequency band. The optimized rectifier circuit enhances the RF harvested energy efficiency, with an AC-to-DC conversion efficiency ranging from 36% to 70% for low-level input power ranging from −10 dBm to 0 dBm. The stable output of DC power confirms the suitability of this design for various practical applications, including wireless sensor networks, energy harvesting power supplies, medical implants, and environmental monitoring systems. Experimental validation, which includes both the reflection coefficient and radiation patterns of the designed antenna, confirms the accuracy of the simulation. The study found that the proposed energy harvesting system has a high total efficiency ranging from 53% to 63% and is well-suited for low-power energy harvesting (0 dBm) from ambient electromagnetic radiation. The proposed circularly polarized rectenna is a competitive option for efficient electromagnetic energy harvesting, both as a standalone unit and in an array, due to its high performance, feasibility, and versatility in meeting various energy harvesting requirements. This makes it a promising and cost-effective solution for various wireless communication applications, offering great potential for efficient energy harvesting from ambient electromagnetic radiation.

Keywords: RF energy harvesting; circularly polarized antenna; harmonic rejection; stable DC power

1. Introduction

Electromagnetic (EM) energy harvesting has received a lot of attention for a long time. In the early 1900s, Nikola Tesla experimented with wirelessly transmitting power through microwaves. However, his work was largely left unimplemented, as his experiments were vastly ahead of their time and the technology did not yet exist to make energy harvesting via microwaves feasible [1]. Advances in wireless technologies in the last few decades have made far-field energy harvesting a useful technology that has numerous applications and benefits, such as the Internet of Things (IoT), RF identification (RFID), wireless sensor networks (WSNs), and more [2–5].

Electromagnetic energy harvesting relies heavily on rectifying antennas (rectennas), which can convert RF energy to DC power. The low-pass filter (LPF), diodes, and DC pass capacitor are the main components of the standard rectenna, as shown in the block diagram of Figure 1 [3]. The task of RF power harvesting is performed by converting sent/ambient RF electromagnetic waves into an AC signal through the receiving antenna. This AC signal is then rectified into DC which can be stored in batteries or used directly to drive a specific circuit.

Figure 1. Block diagram of RF energy harvesting system.

The greatest challenge in this system is the nonlinearity of the rectifier circuit, which has some drawbacks. On the one hand, the rectifier conversion efficiency from AC to DC depends on the input power. In other words, the greater the AC input power available at the rectifier, the higher the conversion efficiency that can be achieved, up to a certain point [6]. This issue can be mitigated by improving the performance of the pre-rectifying stages (i.e., the receiving antenna and matching circuit). Therefore, if a proper lossless matching network is used to match the antenna impedance to the rectifier impedance, maximum power transfer is achieved [4]. Consequently, the received antenna is the main part that improves the system's efficiency. The antenna structure should have a high conversion efficiency to convert RF power into AC power [7,8]. Furthermore, the antenna should have a higher capturing efficiency that enables it to capture sufficient RF power to drive the rectifier circuit into the interested region. Increasing capturing efficiency means having a higher radiation efficiency, higher directivity, a lower load mismatching factor, and a lower polarization mismatching factor [9]. In [10], a dual-purpose radial-array rectenna was proposed for RF-energy harvesting IoT sensors, significantly enhancing RF energy capturing from a 360° region. It also enables precise orientation sensing using 5.8 GHz antennas, showcasing its versatility and potential for various applications.

On the other hand, nonlinear rectifying circuits, such as diodes, produce harmonics of the fundamental frequency. The unwanted harmonics impair system performance and result in harmonic interference that reduces the efficiency of the antenna and nearby circuit due to the coupling effect. As a result, a low-pass filter (LPF) is needed to suppress the harmonics to prevent harmonic interference, power re-radiation, and noise with antennas and nearby circuits [11]. However, the LPF will have insertion losses and increase the size and cost of the system. To elevate this effect, filters and microstrip antennas are usually fabricated together on the same substrate to improve cost and efficiency [12]. However, designing an antenna that does not resonate at the second and third resonance frequencies (i.e., also called a filtenna) would be the optimum solution [11]. Additionally, a wide-bandwidth antenna is also preferred, but it comes with decreasing efficiency. Furthermore, the effective bandwidth of the antenna will be limited by the lowest bandwidth of other energy-harvesting circuitry, such as rectifiers, that suffer from decreasing efficiency with increased bandwidth [7,13]. Moreover, improving the mismatching polarization factor makes circularly polarized antennas more preferable to linear polarization for some electromagnetic energy harvesting applications, especially due to the independence of the rotating angle and fading resistance [14].

Therefore, researchers focus on obtaining a competitive design that satisfies the aforementioned aspects or those required by different applications of ambient RF energy harvesting. The antenna used in such systems can be of any type, yet microstrip patch antennas are widely used due to their economical and electromechanical advantages [9,15–18]. However, lower bandwidth, directivity, and efficiency are the main drawbacks of such antennas [19]. Various techniques can be employed to overcome the disadvantages of antenna systems. For instance, using multiple antenna elements in an array can enhance the directivity of the system. However, this comes at the cost of increasing the size of the receiving antenna, which is dependent on the number of elements used. As a result, there is an increasing amount of research being dedicated to the development of compact and highly efficient antenna elements, which can enable the realization of high-performance antenna arrays in a small form factor [14,20–22]. Therefore, an antenna structure that combines harmonic rejection, higher directivity, and efficiency could be a good candidate for low-cost and compact energy harvesting.

In the existing literature, numerous designs have been proposed for electromagnetic energy harvesting applications. However, it is important to note that no single design can be considered perfect or universally suitable for all applications in this field. In [23], compact circularly polarized filtennas at 2.45 GHz are presented, but they have radiation efficiencies of less than 60%. In [24], the authors design a broadband antenna for energy harvesting, but its RF to DC efficiency is less than 55%, and the implementation complexity is high since the filter is printed on another layer. The authors of [25] proposed a highly efficient cross-dipole antenna with a reflector and filter integrated on the same substrate, achieving a high efficiency of 90%. However, the antenna size was very large, approaching the wavelength of the electromagnetic wave.

In [6], two-layer, two-port antennas were used to improve the collected power by obtaining dual linear polarization, but the size and cost of the antenna increased due to the need for two LPF filters, a matching circuit, and a rectifier. In [26], the focus was on minimizing the antenna structures (with maximum dimensions approaching a quarter wavelength) for energy harvesting applications. Although the design achieved a gain of 4.6 dBi and was very compact, it was a linearly polarized antenna with a very narrow bandwidth of 0.55%. In [27], a circularly polarized antenna with a gain of 6 dBi was presented, but the matching and axial ratio (AR) bandwidths were very small at 1.7% and 0.68%, respectively. Additionally, the antenna could not reject harmonics. In [28], a circularly polarized antenna was designed for use in RFID applications with a gain of 6.9 dB, but its bandwidth was limited to 1.5% at 2.45 GHz.

Metamaterial-based energy harvesting structures have also garnered significant attention from researchers due to their unique properties. The ability to engineer a metasurface and metamaterial surface with negative permeability and permittivity has opened up novel applications in various frequency bands, including energy harvesting [29], perfect lensing [30], and perfect absorption [31]. For instance, the metamaterial's integration in the antenna design in [32] leads to a significant enhancement in the antenna's gain across all frequency bands. This notable improvement solidifies the antenna as a promising solution for high-performance wireless communication systems. Moreover, a compact metamaterial-inspired antenna (MIA) enables efficient WiFi energy harvesting without complex networks, producing a rectified DC voltage for powering distributed microsystems [33]. In a related study [34], a dual-band metasurface simplifies design and facilitates high-efficiency electromagnetic energy harvesting at Wi-Fi frequencies, making it suitable for applications such as wireless power transfer. Additionally, an efficient miniaturized metasurface achieves over 78% conversion efficiency at 5.54 GHz, making it ideal for compact wireless sensor networks with wide angles [35]. However, a significant limitation of metamaterials is their narrow bandwidth of operation due to the resonance frequencies of their structures. This bandwidth limitation, coupled with the cost of fabrication, presents a challenge for the practical implementation of metamaterial-based energy harvesting structures. Nonetheless, ongoing research in this area holds great promise for the development of efficient and cost-effective energy harvesting solutions for a wide range of applications.

This paper presents a highly effective area rectenna for the 2.45 GHz ISM band, achieving compactness, circular polarization, high efficiency, harmonic rejection, sufficient improved bandwidth covering the ISM band, and high gain compared to traditional patch antennas. By effectively combining these features in a balanced tradeoff, the design enhances efficiency and expands the potential applications for energy harvesting. The antenna design structure has a bandwidth ratio of 4.08%, a directivity of 7.2 dBi, and a radiation efficiency of 92.5%. Moreover, the total size of the antenna is around 0.5×0.5 wavelength. The antenna can reject the second and third harmonics, which means it can be used without an LPF. The structure is one port and one layer, making it a simple and low-cost design. An optimized rectifier is also presented to evaluate the total rectenna performance and meet the antenna requirements.

2. Rectenna Design

To build a rectenna system, the most important parameters to evaluate the system should be considered. One of these parameters is the received power by the antenna used in this system, which is directly related to the receiving antenna's effective area as [9]

$$P_r = A_{eff} W_{inc} \tag{1}$$

where A_{eff} is the antenna effective area that can be found by Equation (2), and W_{inc} is the incident power density of the plane wave that can also be given by Equation (3).

$$A_{eff} = \tau \, \eta_{rad} \frac{\lambda^2 D}{4\pi} \, PLF \tag{2}$$

$$W_{inc} = \frac{|E_i|^2}{2\eta_0} \tag{3}$$

where τ is the matching transmission coefficient of the loaded antenna, which can be given by Equation (4), η_{rad} is the radiation efficiency of the antenna, λ is the wavelength of the incident wave, D is the antenna directivity, PLF is the polarization mismatching loss factor that is given by Equation (5), E_i is the incident electric field intensity, and η_0 is the surrounding medium's characteristic impedance, which is assumed to be air in this research, so $\eta_0 = 120\pi \, \Omega$.

$$\tau = \frac{4 R_A R_L}{|Z_A + Z_L|^2} \tag{4}$$

$$PLF = |a_i . a_r|^2, \tag{5}$$

where R_A and Z_A are the receiving antenna input resistance and impedance, respectively; R_L and Z_L are the antenna load resistance and impedance, respectively; a_i and a_r are the incident field's instantaneous direction vector and the antenna polarization direction vector, respectively. It must be emphasized that (W_{inc} times $\lambda^2 D/(4\pi)$) is the maximum effective area used to find P_r when there is no mismatching loss (i.e., $\tau = 1$) or polarization mismatching loss (i.e., $PLF = 1$).

2.1. Antenna Geometry

The structure of a slotted circular patch antenna and its main important parameters are shown in Figure 2. The proposed antenna is a single-layer antenna consisting of a slotted circular patch on the front side, which is etched on a 60 mm × 60 mm Roger RT/duroid 5880 dielectric, a fully grounded reflector on the backside, a centered short via connecting the front and back sides, and coaxial probe feeding. The antenna shows the advantages of small size, high efficiency, directivity, and good matching in the 2.45 ISM band with circular polarization. In addition, it has harmonic rejection properties at the second and third harmonics. Thus, it is more suitable for power transfer and energy harvesting applications. This was achieved by minimizing the loss (i.e., increasing the radiation efficiency) and minimizing the rectenna size by self-suppressing the second and third harmonics.

Coaxial probe feeding was chosen because it has low transmission line loss and minimizes the area, thereby reducing the aforementioned problems. The feed position is chosen to achieve right-hand circular polarization.

The antenna is etched on a 60 mm × 60 mm (i.e., ~0.5 λ × 0.5 λ) Roger RT/duroid 5880 substrate with a dielectric constant (ε_r = 2.2) and electric tangent loss (tan(δ) = 0.0009), which are more suitable for enhancing the radiation efficiency. Although a low dielectric constant is not good for radiation characteristics, especially with a thin substrate [16], the proposed antenna has demonstrated good performance, as presented in the next section. The slotted edge is useful for obtaining this advantage since it causes the input impedance of the antenna to be constructively matched with the port impedance.

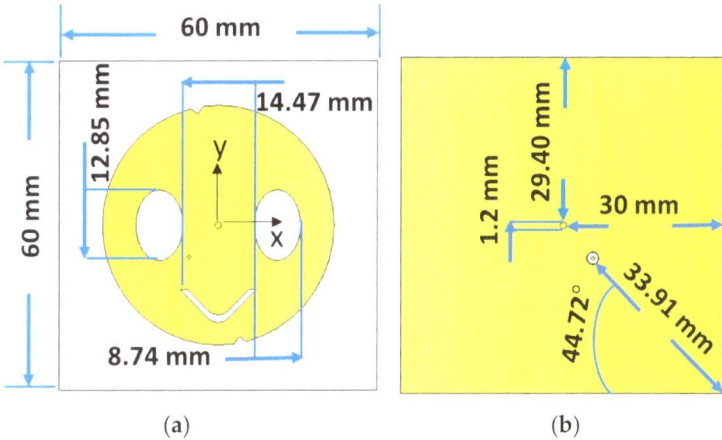

Figure 2. Structure of the proposed antenna, (**a**) the front side of the antenna, (**b**) the backside of the antenna.

2.2. Rectifier Design

To build a full rectenna system, a rectification circuit was designed based on the antenna's results. The rectifier circuit schematic built into ADS is shown in Figure 3. The goal was to maximize the DC output power at 1 KΩ, $P_{in} = -10$ to 10 dBm, and $f = 2.37 - 253$ GHz.

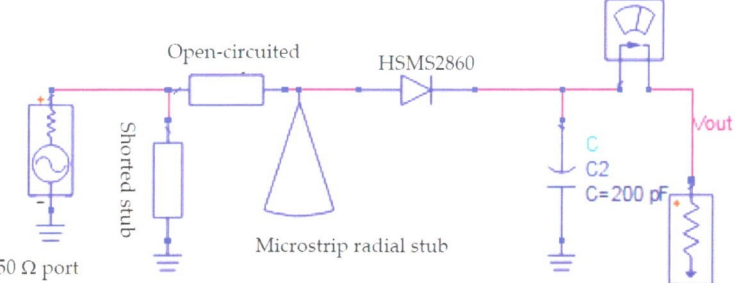

Figure 3. The rectifier circuit schematic with impedance matching circuit.

The design of the rectifier circuit was based on optimizing the microstrip transmission line (impedance matching circuit) with different lengths and widths to achieve the desired goal. To ensure proper impedance matching, a Roger RT5880 substrate with a dielectric constant ($\varepsilon_r = 2.2$) and electric tangent loss (tan(δ) = 0.0009) was used.

To analyze the rectifier in the proposed rectenna system, the Harmonic Balance (HB) simulator was used due to the presence of the nonlinear Schottky diode (HSMS2860). HSMS2860 belongs to the HSMS-286x family of DC-biased detector diodes, which have been designed and optimized for use from 915 MHz to 5.8 GHz. The HSMS-286x family is ideal for RF/ID and RF Tag applications as well as large signal detection, modulation, RF to DC conversion, or voltage doubling. The electrical specification and spice parameter elements of this circuit are defined by the datasheet of the HSMS286x series in [36]. The circuit contains a series of transmission lines, a shorted stub, an open-circuited stub, and a microstrip radial stub between the 50 Ω port and the HSMS2860 diode.

To achieve optimal performance, a series HSMS 2860 Schottky diode was selected for its low turn-on voltage and fast switching speed at a frequency of operation of 2.45 GHz.

Additionally, a shunt capacitor of 200 pF was chosen to appear as a short circuit for GHz frequencies and an open circuit for the rectified DC power.

Table 1 shows the optimized parameters of the impedance matching circuits, which were critical in achieving high AC-to-DC conversion efficiency ranging from 36% to 70% for low input power levels.

Table 1. Simulation results.

Shorted Stub L_1/W_1	Open-Circuited Stub L_2/W_2	Microstrip Radial Stub $L_3/W_3/\theta_3$
7.71/8.67	4/0.1	8.73/0.707/78°

3. Simulation Results

The antenna port is excited by a Gaussian pulse with a 50 Ω probe feeding, propagating from the back of the antenna towards a positive z-axis direction, as depicted in Figure 2. To solve the electromagnetic problem with open boundaries (far-field problem), the finite integration technique (FIT) solver in Computer Simulation Technology (CST) is utilized. Furthermore, practical measurements are employed to validate the designed antenna and ensure its performance in real-world scenarios. In parallel, the performance evaluation of the proposed rectifier is carried out using the Harmonic Balance (HB) technique within the ADS full-wave simulator. This approach allows for a comprehensive analysis of the rectifier's behavior and efficiency.

3.1. Antenna Reflection Coefficient and Input Impedance

The reflection coefficient (S_{11}) of a 50 Ω port as well as the input impedance of the antenna are shown in Figure 4. The antenna has a bandwidth of around 100 MHz with $S_{11} < -10$ dB, which covers the whole ISM bandwidth.

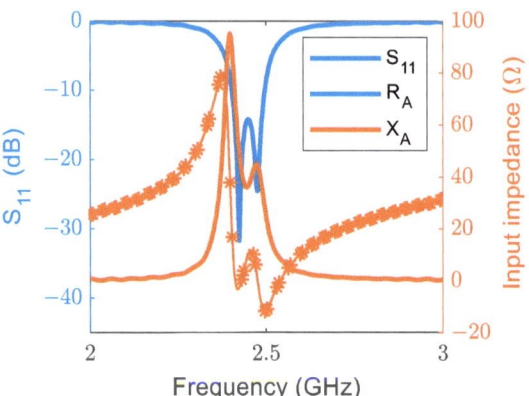

Figure 4. Return loss and input impedance of the final-designed antenna.

The antenna design process is arranged into five stages. The first stage started by selecting the structure that would minimize the size and achieve good performance at the required bandwidth of 2.45 GHz in the ISM band. Thus, a probe-fed circular-patch microstrip antenna with Roger RT/duroid 5880 low permittivity and loss substrate material is first chosen based on recommendations in [17]. Then, using the design formula Equation (6) in [9], the radius of the circular patch, r, is calculated at the center resonance frequency, 2.45 GHz, using Roger RT/duroid 5880 material of a 2.2 permittivity (ε_r) and 1.6 mm thickness (h). The radius calculation result of 23.13 mm is used to simulate the circular patch antenna using the full-wave simulator, where the feeding position and ground size are also chosen according to the recommendations of [9,17].

$$r = \frac{F}{\left\{1 + \frac{2h}{\pi \varepsilon_r F}\left[\ln\left(\frac{\pi F}{2h}\right) + 1.7726\right]\right\}^{1/2}}, \quad F = \frac{8.791 \times 10^9}{f_r \sqrt{\varepsilon_r}} \tag{6}$$

Figure 5a shows the circular structure with the S_{11} of the five design process stages. The first stage (stage #1) is to adjust the antenna size for the targeted ISM band without any perturbations. In this stage, the antenna has a resonating mode of around 2.45 GHz and also resonates around the second and third harmonics. Furthermore, the radiation in this case is linear, and the antenna bandwidth is only 42 MHz. Therefore, the following design process stages are to solve these challenges by allowing suitable current perturbations to change the linear polarization to circular, increase the bandwidth, and reject the second and third harmonics. To have circular polarization, two orthogonal modes are generated in the second stage by loading the antenna with two optimum symmetric oval slots, as shown in Figure 2, and locating the feeding at 45° from the slot axis. The symmetric slots also improve the bandwidth and help minimize the radius of the circular patch from 23.13 mm to 21.86 mm. The size and position of these slots are adjusted and optimized to achieve the desired bandwidth and polarization axial ratio. The achieved bandwidth is 100 MHz, as shown in Figure 5a (see stage #2 line). The third stage (stage #3) aims to reject the harmonic where the patch antenna is loaded with a V-slot on its center bottom side to reject the second and third harmonics. The symmetric position with optimized dimensions of the V-slot improves the harmonic rejection property. To finely adjust the axial ratio and improve the matching, two triangular slits were introduced in the fourth stage (stage #4). In the last stage (stage #5), a centered via is used to further suppress the second and third harmonic frequencies at the second and third resonances. Therefore, this structure works as a filtenna since it works as a stop-band filter for harmonics, as shown in Figure 5b. Harmonics can cause rectenna performance degradation as well as nearby circuitry degradation, as explained in the introduction. Accordingly, the antenna is suitable for energy harvesting applications. The most important performance parameters that affect the effective area of the antenna are discussed in the next section.

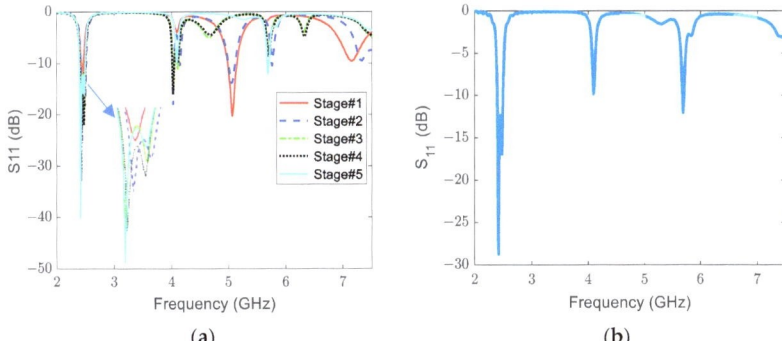

Figure 5. The reflection coefficient of (**a**) five design process stages and (**b**) final optimized stage.

Figure 6 shows the current distribution of the proposed right-hand circularly polarized antenna at 2.45 GHz, which varies at different phases, indicating the counterclockwise movement of the surface current distribution at the edges. At all phases, the current distribution exhibits maximum amplitude at the symmetric oval slots, V-slot, and edges.

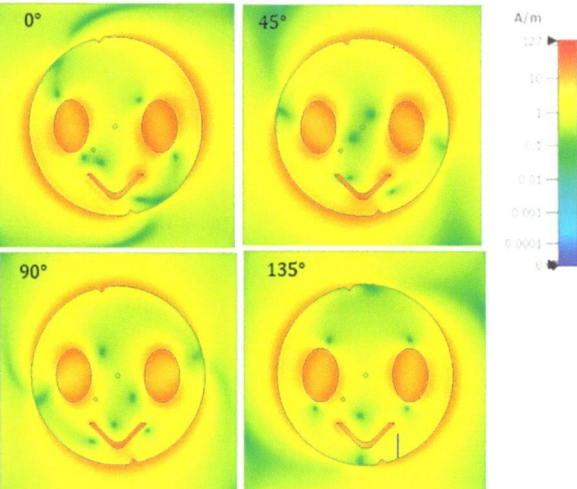

Figure 6. Current distribution of the proposed antenna at 2.45 GHz and different phases.

3.2. Antenna Radiation Efficiency and Axia Ratio

The proposed antenna structure exhibits a radiation efficiency that varies between 80% and 91% over the frequency bandwidth of the ISM band, as shown in Figure 7. This is a significant achievement, as high radiation efficiency is critical for converting RF power into AC power for energy harvesting applications.

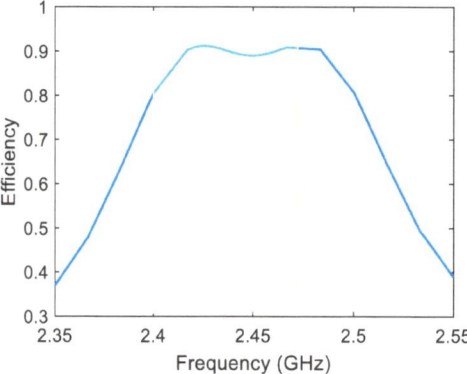

Figure 7. Antenna radiation efficiency.

Figure 8 shows the axial ratio performance of a circularly polarized antenna over a specific frequency range from 2.35 GHz to 2.55 GHz. The axial ratio, which is a measure of the polarization purity of an antenna, is plotted as a function of frequency, with the relevant frequency range highlighted for clarity. The results show that the proposed antenna is circularly polarized around the 2.45 GHz frequencies. The antenna exhibits outstanding circular polarization purity with an axial ratio consistently less than 3 dB. The fact that the axial ratio remains very small over the entire frequency range of interest is a testament to the antenna's exceptional performance and its ability to maintain the desired polarization state with minimal distortion. Consequently, the amount of captured power will be increased to match the circular polarization and be half the amount for linear polarization ambient EM waves.

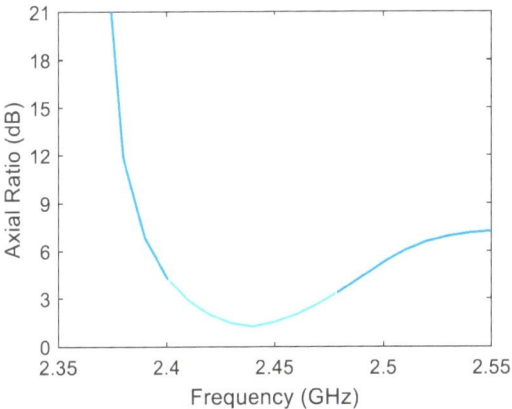

Figure 8. The axial ratio performance of a circularly polarized antenna.

3.3. Antenna Directivity and Effective Area

The co-polarized directivity pattern of the proposed antenna at the studied frequency is shown in Figure 9. The maximum directivity is around 7.2 dBi in the main lobe of the center operating frequency, 2.45 GHz. The radiation pattern is unidirectional and symmetric in both the azimuth and elevation planes, with no significant side or back lobes. The plot of directivity with frequency variation in the same main lobe direction is shown in Figure 10a. The directivity of the antenna indicates its ability to concentrate radiation in a specific direction. However, directivity alone is not sufficient for a complete understanding of the antenna's performance. To assess its effectiveness, it is important to consider the gain and realized gain as well. Gain refers to the ratio of the power radiated by the antenna in a specific direction to the power that an ideal isotropic antenna would radiate, assuming the same input power. It quantifies the antenna's ability to efficiently radiate energy in a specific direction.

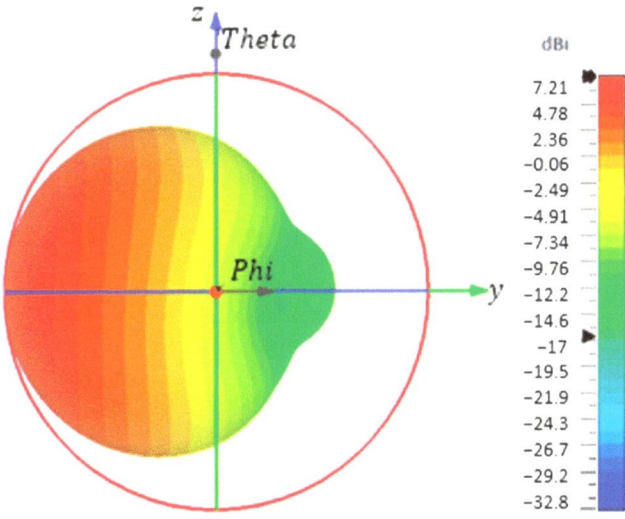

Figure 9. The antenna directivity pattern at 2.45 GHz.

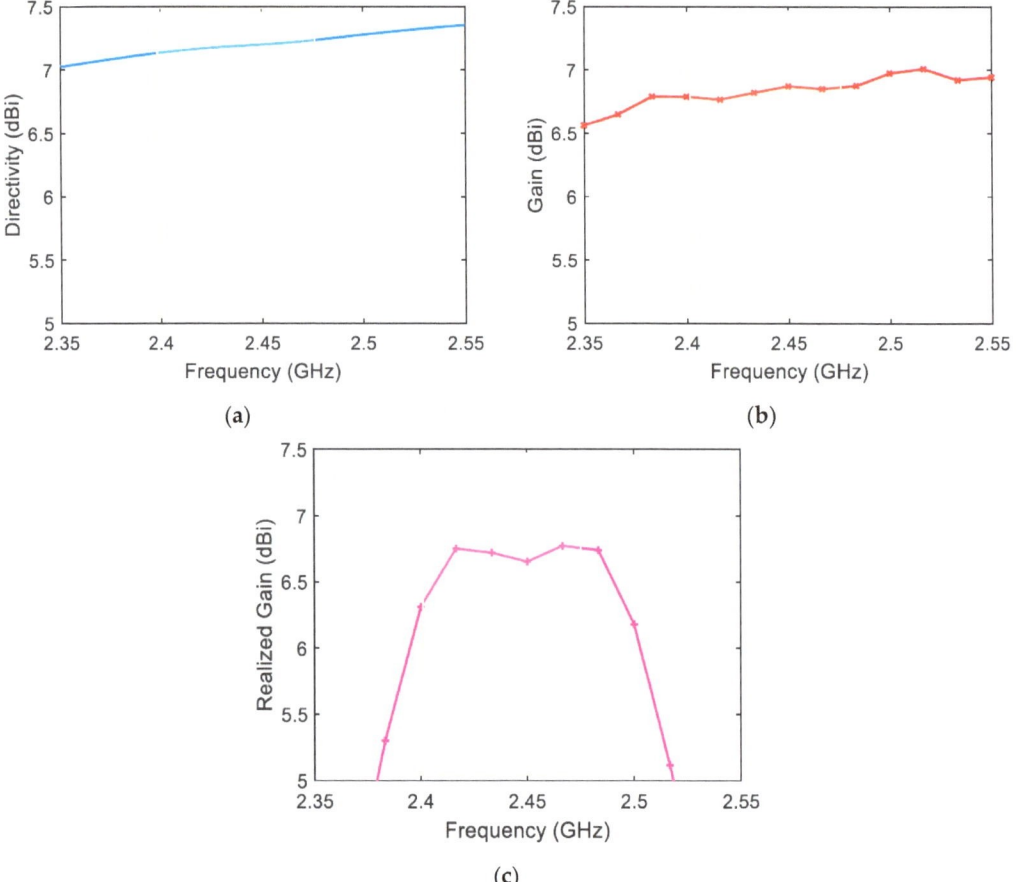

Figure 10. (a) Antenna directivity, (b) gain, and (c) realized gain vs. frequency.

On the other hand, realized gain takes into account various factors such as losses, impedance matching, and efficiency. It provides a more realistic measure of the antenna's performance in practical applications. Figure 10b,c present the antenna's gain and realized gain, respectively.

In the context of energy harvesting, the effective area is a critical parameter, as it determines how much energy can be harvested from the ambient electromagnetic waves. Once the incident field is co-polarized with the antenna structure (i.e., PLF $\approx$ 1) and the connected load circuit is conjugate-matched ($\tau \approx 1$), the maximum absorbed power is obtained. According to Equation (2), the effective area of the proposed antenna varies between 36 cm^2 and 55 cm^2 over the frequency range of 2.4 to 2.5 GHz.

A performance comparison between various published works and the proposed antenna is listed in Table 2. Among the listed designs, the proposed antenna exhibits a better trade-off between minimization, bandwidth, efficiency, and self-filtering for harmonic rejection (i.e., performing the LPF task). This makes it sufficient to drive the rectifier circuit for the targeted application with a low level of incident EM power.

Table 2. Comparison of antenna performance with the literature.

Ref.	Dimensions (λ)	Radiation Efficiency	Relative Bandwidth	Directivity	LPF Feature
[13]	$0.49 \times 0.49 \times 0.0136$	45%	5.3%	6.7 dBi	No
[24]	$0.49 \times 0.49 \times 0.0136$	50%	5.3%	6.9 dBi	Yes
[27]	$0.46 \times 0.46 \times 0.03$	89%	1.5%	6.98 dBi	-
[26]	$0.474 \times 0.474 \times 0.0158$	86%	1.7%	6.8 dBi	-
This work	$0.49 \times 0.49 \times 0.0136$	92%	4.08%	7.2 dBi	Yes

3.4. AC–DC Efficiency

The efficiency of converting the received AC power to DC power is critical for energy harvesting applications. The AC-to-DC power conversion efficiency is calculated using Equation (7).

$$\eta_{Ac-DC} = \frac{V_o \times I_o}{P_{in}} \quad (7)$$

where P_{in} is the total time-average power coupled to the 50 Ω input port of the impedance matching network, and V_o and I_o are the output voltage and current on the load of the DC filter.

Figure 11 illustrates the AC-to-DC conversion efficiency at 2.45 GHz with a 1 KΩ rectifier resistive load. The junction resistance in the equivalent circuit model of an HSMS 2860 Schottky diode [36] is dependent on the externally applied bias current, which makes AC-to-DC radiation conversion efficiency depend on the input power levels. The results show that for low input power levels ranging from -10 dBm to 0 dBm, the efficiency gradually increases from around 36% to 70%. The maximum DC power inversion occurs at an input AC power of 10 dBm. However, it is important to note that the performance of conversion efficiency deteriorates for higher input power levels (>11 dBm).

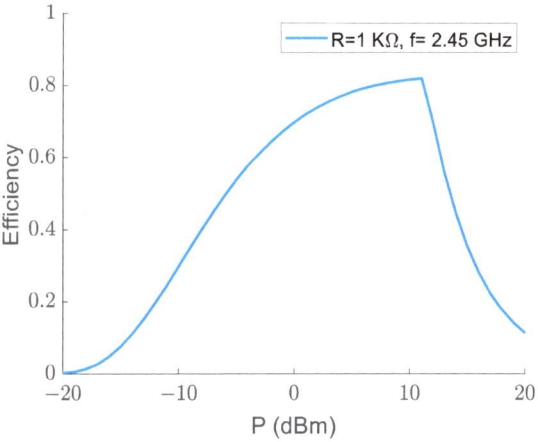

Figure 11. AC-to-DC radiation conversion efficiency.

For practical wireless communication applications, the received input power levels are typically between -10 dBm and 0 dBm (an interesting region). Therefore, the proposed antenna structure with a rectifier circuit with AC-to-DC radiation conversion efficiency ranging from 36% to 70% for low input power levels is suitable for energy harvesting from wireless communication signals. The results demonstrate the advantages of the proposed rectenna for efficient energy harvesting from ambient electromagnetic radiation, which could be useful in various applications.

The overall efficiency of the energy harvesting system is affected by various factors, including the resistive load of the rectifier. Figure 12 shows the effect of changing the resistive load of the rectifier on the output DC power. Electromagnetic energy harvesting is utilized in various applications, including wireless sensor networks, RFID systems, energy harvesting power supplies, medical implants, and environmental monitoring systems. The input impedance of these applications varies, with wireless sensor networks typically ranging from 1 kΩ to 10 kΩ, RFID systems ranging from 50 Ω to 100 Ω, and others typically ranging from 1 kΩ to 100 kΩ [37]. The results of this work indicate that the DC power is stable in the range of 1 KΩ to 4 KΩ resistive load rectifiers for practical input AC power levels (-10 dB to 0 dB), which makes it a suitable candidate to support wireless sensor networks, energy harvesting power supplies, medical implants, and environmental monitoring systems.

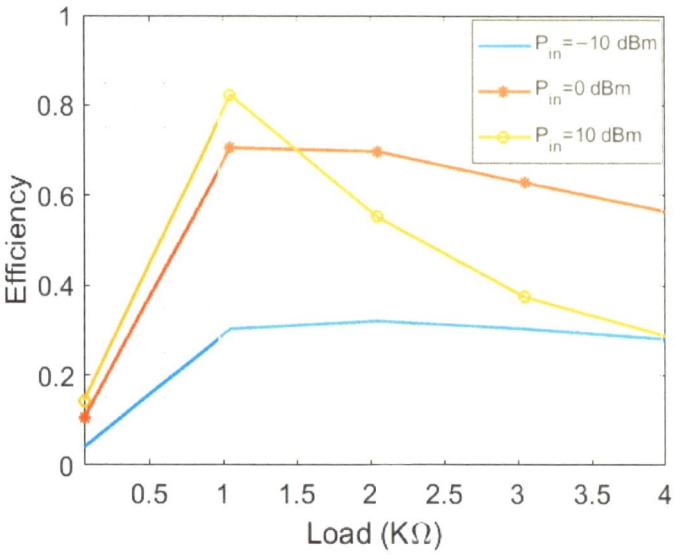

Figure 12. AC-to-DC radiation conversion efficiency at 2.45 GHz.

In addition to the resistive load, the energy harvesting performance also varies with the input AC power level and operating frequency. Figure 13 shows the AC-to-DC conversion efficiency for frequencies in the ISM band at three different input AC power levels. The results indicate that the efficiency is within the range of 20% to 75% for the targeted bandwidth and different input power levels ranging from -10 dBm to 0 dBm.

Figure 14 presents the total efficiency of the energy harvesting system, which includes both AC-to-AC and AC-to-DC conversion efficiencies, plotted against frequency for the 0 dBm input power level and 1 KΩ resistive load. The results demonstrate that the proposed energy harvesting system offers a high total efficiency, ranging from 53% to 63%.

The high total efficiency of the energy harvesting system has important implications for various applications where low-power energy harvesting is essential. The results highlight the potential of the proposed antenna structure and energy harvesting system to provide a cost-effective and efficient energy harvesting solution for various wireless communication applications.

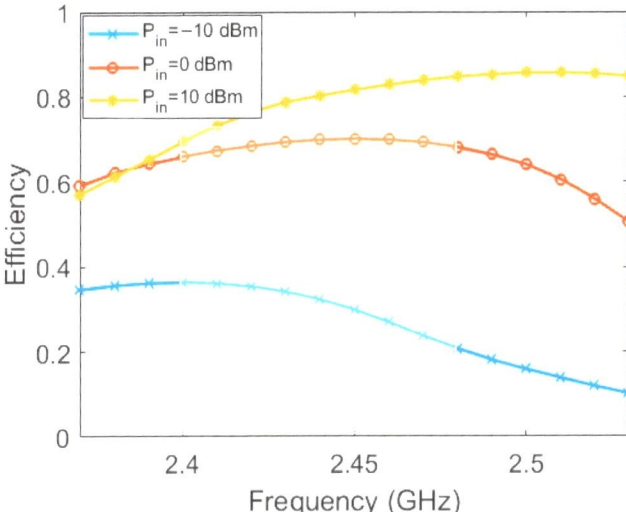

Figure 13. AC-to-DC radiation conversion efficiency at 1 KΩ resistive load.

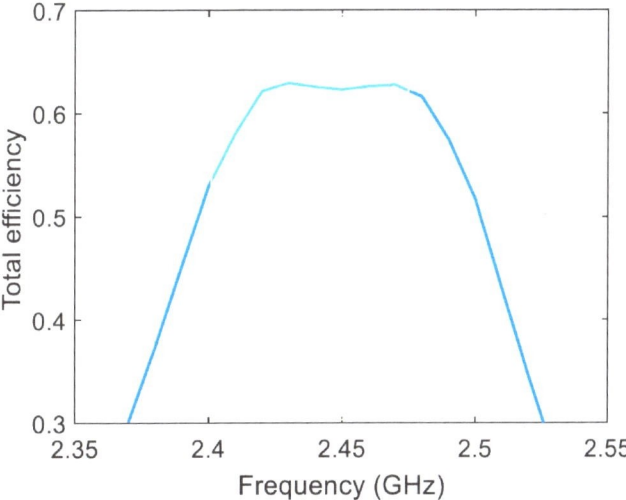

Figure 14. Total efficiency at 1 KΩ resistive load and 0 dBm input power.

3.5. Experimental Validation

This section provides experimental validation of the antenna simulation results. A comparison is performed by analyzing the reflection coefficients and radiation patterns of both the simulated and measured data.

The fabricated antenna and vector network analyzer system are shown in Figure 15. Figure 16 presents a comparison between the measured and simulated S_{11} values of the designed antenna, which indicates good agreement between the simulation and measurement results. This agreement confirms that the designed antenna is highly accurate and reliable.

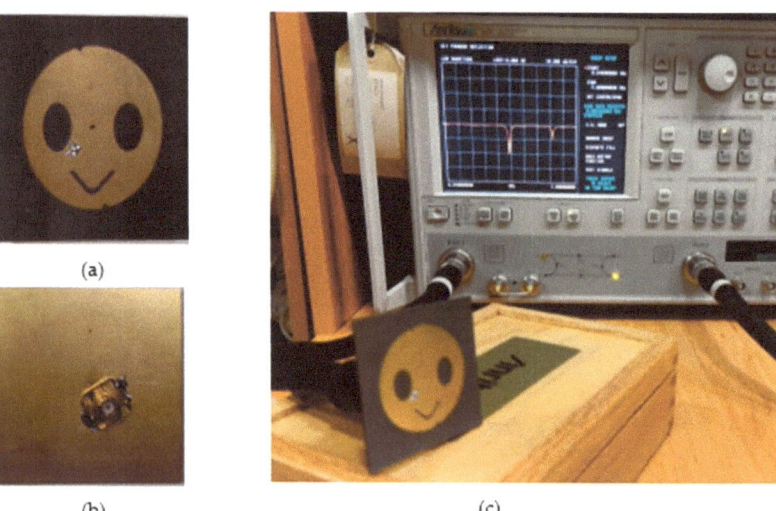

Figure 15. The fabricated antenna with vector network analyzer system. (**a**) The front side of the antenna, (**b**) the backside of the antenna, and (**c**) the vector network analyzer measurement system with the proposed antenna.

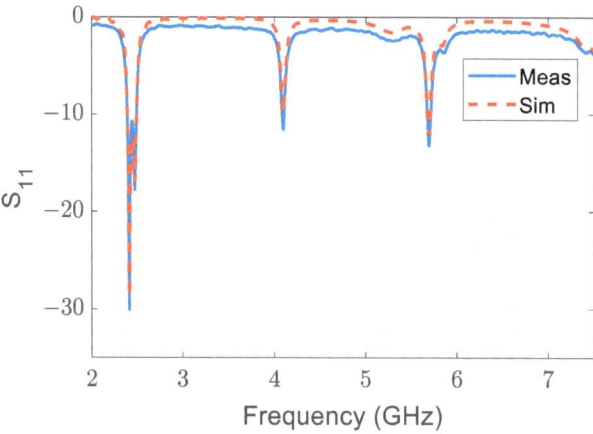

Figure 16. The measured and simulated reflection coefficients, S_{11}.

Figure 17 shows the Geozondas [38] time-domain antenna measurement setup used for measuring the radiation pattern of the designed antenna in the time domain. This setup consists of a pulsed signal generator, a digital sampling converter, a transmitting antenna, a receiving antenna, and an oscilloscope. The pulsed signal generator sends a short pulse to the transmitting antenna, which then radiates the pulse into space. The receiving antenna captures the signal, which is then analyzed by the oscilloscope to determine the receiving antenna's radiation pattern. Figure 18a,b illustrates the measured and simulated polar radiation patterns of the designed antenna in the azimuth and elevation planes, respectively. The agreement between the simulation and measurement results demonstrates the accuracy and reliability of the designed antenna. The close match between the two sets of data in both the reflection coefficients and radiation patterns provides strong evidence to support the validity of the simulation approach and its ability to predict the behavior of the whole rectenna (antenna and rectifier) system.

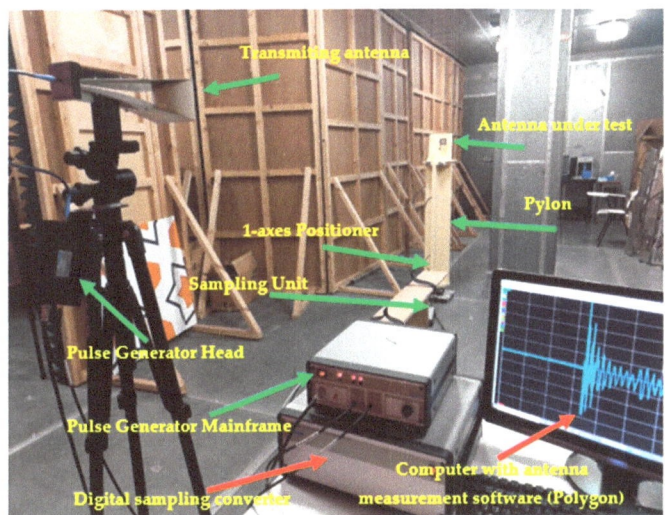

Figure 17. Geozondas time-domain antenna measurement setup.

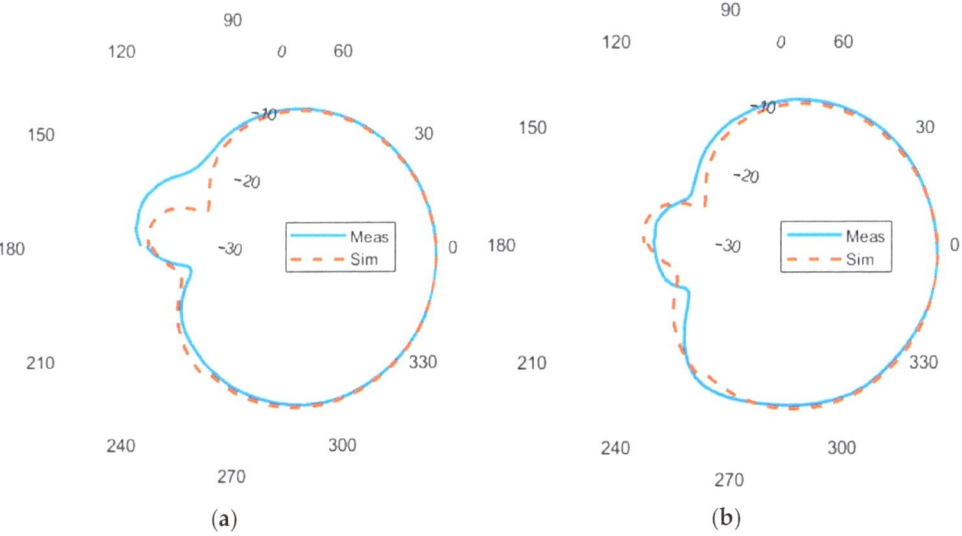

Figure 18. Normalized circularly polarized radiation patterns at 2.45 GHz: (**a**) azimuth and (**b**) elevation plane.

The gain is measured by calculating the performance of the proposed antenna in comparison to a reference antenna with a known gain. The two antennas are connected alternately to the same transmitter, and the received power is measured for each antenna. Both used antennas, the reference and the transmitter, are double-ridged horn antennas with lens GZ0126 DRH (see Figure 17). The gain of the proposed antenna is then determined by comparing the power levels. The measured gain of the proposed antenna is depicted in Figure 19, demonstrating an accepted agreement with the simulated gain.

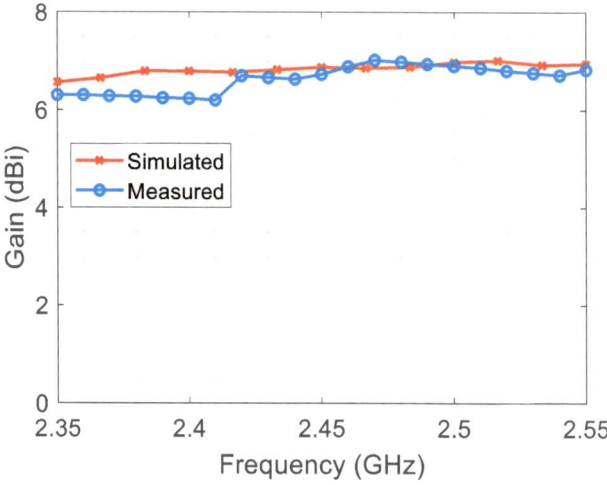

Figure 19. The measured gain compared to the simulation gain.

4. Conclusions

This paper presents a circularly polarized rectenna designed for high-performance electromagnetic energy harvesting in the 2.45 GHz ISM frequency band. The proposed antenna structure achieves a radiation efficiency ranging from 80% to 91%, an effective area between 36 cm^2 and 55 cm^2, an axial ratio consistently less than 3 dB, and harmonic rejection over the bandwidth of the ISM band. Furthermore, it achieves a directivity of 7.2 dBi in the main lobe direction with negligible side/back lobes at the center frequency. The optimized rectifier enhances the RF harvested energy efficiency, with an AC-to-DC conversion efficiency ranging from 36% to 70% for input power levels ranging from −10 dBm to 0 dBm. The proposed design is feasible, versatile, and a competitive option for electromagnetic energy harvesting applications as a single element or in an array to meet various energy harvesting requirements. The stable output DC power in the range of 1 KΩ to 4 KΩ resistive load rectifiers for practical input AC power levels (−10 dB to 0 dB) confirms the suitability of this design for various applications, including wireless sensor networks, energy harvesting power supplies, medical implants, and environmental monitoring systems.

Additionally, the study finds that the proposed energy harvesting system has a high total efficiency that ranges from 53% to 63% at 0 dBm ambient electromagnetic power. Therefore, it is well suited for low-power energy harvesting, which has significant implications for various wireless communication applications.

The manufactured novel circularly polarized antenna was tested for its radiation pattern and S$_{11}$ characteristics, with simulation and measurement results exhibiting remarkable agreement.

Author Contributions: Conceptualization, Z.S.A.A., Y.T.A. and M.A.S.A.; software, Z.S.A.A. and Y.T.A.; validation, Z.S.A.A., Y.T.A., Y.M.A.-M. and M.A.S.A.; formal analysis, Z.S.A.A., Y.T.A., Y.M.A.-M. and K.A.; investigation, Z.S.A.A. and Y.T.A.; resources, A.H.A. and M.T.A.; data curation, Z.S.A.A. and Y.T.A.; writing—original draft preparation, Z.S.A.A., Y.T.A. and Y.M.A.-M.; writing—review and editing, Z.S.A.A., A.H.A., Y.T.A., Y.M.A.-M. and M.A.S.A.; visualization, Z.S.A.A., A.H.A. and K.A.; supervision, A.H.A. and M.A.S.A.; project administration, A.H.A., K.A. and M.T.A.; funding acquisition, A.H.A., K.A. and M.T.A. All authors have read and agreed to the published version of the manuscript.

Funding: This research was funded by the Deputyship for Research and Innovation, the "Ministry of Education" in Saudi Arabia, grant number IFKSUDR_E169.

Institutional Review Board Statement: Not applicable.

Informed Consent Statement: Not applicable.

Data Availability Statement: Data sharing not applicable.

Acknowledgments: The authors extend their appreciation to the Deputyship for Research and Innovation, the "Ministry of Education" in Saudi Arabia, for funding this research work through the project number (IFKSUDR_E169).

Conflicts of Interest: The authors declare no conflict of interest.

References

1. Brown, W.C. The History of Power Transmission by Radio Waves. *IEEE Trans. Microw. Theory Tech.* **1984**, *32*, 1230–1242. [CrossRef]
2. Schlesak, J.J.; Alden, A.; Ohno, T. Microwave powered high altitude platform. *IEEE MTT-S Int. Microw. Symp. Dig.* **1988**, *1*, 283–286.
3. Srinivasu, G.; Sharma, V.K.; Nella, A. A Survey on Conceptualization of RF Energy Harvesting. *J. Appl. Anal. Comput.* **2019**, *6*, 791–800.
4. Valenta, C.R.; Durgin, G.D. Harvesting wireless power: Survey of energy-harvester conversion efficiency in far-field, wireless power transfer systems. *IEEE Microw. Mag.* **2014**, *15*, 108–120.
5. Zhang, Z.; Pang, H.; Georgiadis, A.; Cecati, C. Wireless Power Transfer—An Overview. *IEEE Trans. Ind. Inform.* **2019**, *66*, 1044–1058. [CrossRef]
6. Sun, H.; Geyi, W. A New Rectenna with All-Polarization-Receiving Capability for Wireless Power Transmission. *IEEE Antennas Wirel. Propag. Lett.* **2016**, *15*, 814–817. [CrossRef]
7. Wagih, M.; Weddell, A.S.; Beeby, S. Rectennas for radio-frequency energy harvesting and wireless power transfer: A review of antenna design [Antenna Applications Corner]. *IEEE Antennas Propag. Mag.* **2020**, *62*, 95–107. [CrossRef]
8. Abdulhasan, R.A.; Mumin, A.O.; Jawhar, Y.A.; Ahmed, M.S.; Alias, R.; Ramli, K.N.; Homam, M.J.; Audah, L.H.M. Antenna Performance Improvement Techniques for Energy Harvesting: A Review Study. *Artic. Int. J. Adv. Comput. Sci. Appl.* **2017**, *8*, 97–102.
9. Balanis, C.A. *Antenna Theory: Analysis and Design*, 4th ed.; John Wiley & Sons: Hoboken, NJ, USA, 2016.
10. Kumar, M.; Kumar, S.; Sharma, A. Dual-purpose planar radial-array of rectenna sensors for orientation estimation and RF-energy harvesting at IoT nodes. *IEEE Microw. Wirel. Compon. Lett.* **2022**, *32*, 245–248. [CrossRef]
11. Divakaran, S.K.; Das Krishna, D.; Nasimuddin. RF energy harvesting systems: An overview and design issues. *Int. J. RF Microw. Comput.-Aided Eng.* **2019**, *29*, e21633. [CrossRef]
12. Suh, Y.H.; Chang, K. A high-efficiency dual-frequency rectenna for 2.45- and 5.8-GHz wireless power transmission. *IEEE Trans. Microw. Theory Tech.* **2002**, *50*, 1784–1789. [CrossRef]
13. Lin, Y.L.; Zhang, X.Y.; Du, Z.X.; Lin, Q.W. High-efficiency microwave rectifier with extended operating bandwidth. *IEEE Trans. Circuits Syst. II Express Briefs* **2018**, *65*, 819–823. [CrossRef]
14. Yo, T.C.; Lee, C.M.; Hsu, C.M.; Luo, C.H. Compact circularly polarized rectenna with unbalanced circular slots. *IEEE Trans. Antennas Propag.* **2008**, *56*, 882–886. [CrossRef]
15. Bakkali, A.; Pelegri-Sebastia, J.; Sogorb, T.; Llario, V.; Bou-Escriva, A. A Dual-Band Antenna for RF Energy Harvesting Systems in Wireless Sensor Networks. *J. Sens.* **2016**, *2016*, 5725836. [CrossRef]
16. Shrestha, S.; Noh, S.K.; Choi, D.Y. Comparative study of antenna designs for RF energy harvesting. *Int. J. Antennas Propag.* **2013**, *2013*, 385260. [CrossRef]
17. Kraus, J.D.; Marhefka, R.J. *Antennas for All Applications*, 3rd ed.; McGraw-Hill: New York, NY, USA, 2002.
18. Garg, A.I.R.; Bhartia, P.; Bahl, I.J. *Microstrip Antenna Design Handbook*; Artech House: Norwood, MA, USA, 2001.
19. James, J.R.; Hall, P.S.; Wood, C. *Microstrip Antenna Theory and Design*; Peter Peregrinus Ltd.: London, UK, 1981.
20. Chandrasekaran, K.T.; Nasimuddin, N.; Alphonse, A.; Karim, M.F. Compact circularly polarized beam-switching wireless power transfer system for ambient energy harvesting applications. *Int. J. RF Microw. Comput. Eng.* **2019**, *29*, e21642. [CrossRef]
21. Hu, Y.Y.; Sun, S.; Xu, H. Compact Collinear Quasi-Yagi Antenna Array for Wireless Energy Harvesting. *IEEE Access* **2020**, *8*, 35308–35317. [CrossRef]
22. Collado, A.; Member, S.; Georgiadis, A. A Compact Dual-Band Rectenna Using Slot-Loaded. *IEEE Antennas Wirel. Propag. Lett.* **2013**, *12*, 1634–1637.
23. Huang, F.J.; Yo, T.C.; Lee, C.M.; Luo, C.H. Design of circular polarization antenna with harmonic suppression for rectenna application. *IEEE Antennas Wirel. Propag. Lett.* **2012**, *11*, 592–595. [CrossRef]
24. Song, C.; Huang, Y.; Zhou, J.; Zhang, J.; Yuan, S.; Carter, P. A high-efficiency broadband rectenna for ambient wireless energy harvesting. *IEEE Trans. Antennas Propag.* **2015**, *63*, 3486–3495. [CrossRef]
25. Sun, H.; Guo, Y.X.; He, M.; Zhong, Z. Design of a high-efficiency 2.45-GHz rectenna for low-input-power energy harvesting. *IEEE Antennas Wirel. Propag. Lett.* **2012**, *11*, 929–932.
26. Lin, W.; Ziolkowski, R.W. Electrically Small, Single-Substrate Huygens Dipole Rectenna for Ultra-Compact Wireless Power Transfer Applications. *IEEE Trans. Antennas Propag.* **2020**, *69*, 1130–1134. [CrossRef]

27. Wang, M.S.; Zhu, X.Q.; Guo, Y.X.; Wu, W. Compact Circularly Polarized Patch Antenna with Wide Axial-Ratio Beamwidth. *IEEE Antennas Wirel. Propag. Lett.* **2018**, *17*, 714–718. [CrossRef]
28. Das, T.K.; Dwivedy, B.; Behera, D.; Behera, S.K.; Karmakar, N.C. Design and modelling of a compact circularly polarized antenna for RFID applications. *AEU-Int. J. Electron. Commun.* **2020**, *123*, 153313. [CrossRef]
29. Chandrasekaran, K.T.; Agarwal, K.; Nasimuddin; Alphones, A.; Mittra, R.; Karim, M.F. Compact Dual-Band Metamaterial-Based High-Efficiency Rectenna: An Application for Ambient Electromagnetic Energy Harvesting. *IEEE Antennas Propag. Mag.* **2020**, *62*, 18–29. [CrossRef]
30. Rosenblatt, G.; Orenstein, M. Perfect lensing with lossy metamaterials: Maintaining a singular focus by avoiding feedback. In Proceedings of the 2016 Progress in Electromagnetic Research Symposium (PIERS), Shanghai, China, 8–11 August 2016; p. 4946.
31. Upender, P.; Kumar, A. THz Dielectric Metamaterial Sensor With High Q for Biosensing Applications. *IEEE Sens. J.* **2023**, *23*, 5737–5744. [CrossRef]
32. Singh, H.; Mittal, N.; Gupta, A.; Kumar, Y.; Woźniak, M.; Waheed, A. Metamaterial integrated folded dipole antenna with low SAR for 4G, 5G and NB-IoT applications. *Electronics* **2021**, *10*, 2612. [CrossRef]
33. Sun, Z.; Zhao, X.; Zhang, L.; Mei, Z.; Zhong, H.; You, R.; Lu, W.; You, Z.; Zhao, J. WiFi energy-harvesting antenna inspired by the resonant magnetic dipole metamaterial. *Sensors* **2022**, *22*, 6523. [CrossRef] [PubMed]
34. Amer, A.A.G.; Othman, N.; Sapuan, S.Z.; Alphones, A.; Hassan, M.F.; Al-Gburi, A.J.A.; Zakaria, Z. Dual-Band, Wide-Angle, and High-Capture Efficiency Metasurface for Electromagnetic Energy Harvesting. *Nanomaterials* **2023**, *13*, 2015. [CrossRef]
35. Amer, A.A.G.; Sapuan, S.Z.; Ashyap, A.Y. Efficient metasurface for electromagnetic energy harvesting with high capture efficiency and a wide range of incident angles. *J. Electromagn. Waves Appl.* **2023**, *37*, 245–256. [CrossRef]
36. Packard, H. Technical Data, "Surface Mount Microwave Schottky Detector Diodes". 1999. Available online: https://www.google.com.hk/url?sa=t&rct=j&q=&esrc=s&source=web&cd=&ved=2ahUKEwi3ueX52JCBAxWWilYBHXeQD6wQFnoECA4QAQ&url=http%3A%2F%2Fwww.hp.woodshot.com%2Fhprfhelp%2F4_downld%2Fproducts%2Fdiodes%2Fhsms2850.pdf&usg=AOvVaw3Q6hCE3Xm0w2WwkArNd2m_&opi=89978449 (accessed on 16 August 2023).
37. Song, C.; Huang, Y.; Carter, P.; Zhou, J.; Yuan, S.; Xu, Q.; Kod, M. A novel six-band dual CP rectenna using improved impedance matching technique for ambient RF energy harvesting. *IEEE Trans. Antennas Propag.* **2016**, *64*, 3160–3171. [CrossRef]
38. Geozandas.com. Time Domain Antenna Measurement Systems. [Online]. Available online: https://www.geozondas.com/main_page.php?pusl=12 (accessed on 20 July 2023).

Disclaimer/Publisher's Note: The statements, opinions and data contained in all publications are solely those of the individual author(s) and contributor(s) and not of MDPI and/or the editor(s). MDPI and/or the editor(s) disclaim responsibility for any injury to people or property resulting from any ideas, methods, instructions or products referred to in the content.

Article

Statistical Characterization of Wireless Power Transfer via Unmodulated Emission

Sebastià Galmés [1,2]

[1] Departament de Ciències Matemàtiques i Informàtica, Universitat de les Illes Balears, 07122 Palma, Spain; sebastia.galmes@uib.es
[2] Institut d'Investigació Sanitària Illes Balears, 07120 Palma, Spain

Abstract: In the past few years, the ability to transfer power wirelessly has experienced growing interest from the research community. Because the wireless channel is subject to a large number of random phenomena, a crucial aspect is the statistical characterization of the energy that can be harvested by a given device. For this characterization to be reliable, a powerful model of the propagation channel is necessary. The recently proposed generalized-K model has proven to be very useful, as it encompasses the effects of path loss, shadowing, and fast fading for a broad set of wireless scenarios, and because it is analytically tractable. Accordingly, the purpose of this paper is to characterize, from a statistical point of view, the energy harvested by a static device from an unmodulated carrier signal generated by a dedicated source, assuming that the wireless channel obeys the generalized-K propagation model. Specifically, by using simulation-validated analytical methods, this paper provides exact closed-form expressions for the average and variance of the energy harvested over an arbitrary time period. The derived formulation can be used to determine a power transfer plan that allows multiple or even massive numbers of low-power devices to operate continuously, as expected from future network scenarios such as the Internet of things or 5G/6G.

Keywords: radio-frequency energy harvesting; wireless power transfer; path loss; shadowing; multipath fading; unmodulated carrier; additive white Gaussian noise; mean; variance; correlation

Citation: Galmés, S. Statistical Characterization of Wireless Power Transfer via Unmodulated Emission. *Sensors* **2022**, *22*, 7828. https://doi.org/10.3390/s22207828

Academic Editors: Onel Luis Alcaraz López and Katsuya Suto

Received: 5 September 2022
Accepted: 10 October 2022
Published: 14 October 2022

Publisher's Note: MDPI stays neutral with regard to jurisdictional claims in published maps and institutional affiliations.

Copyright: © 2022 by the authors. Licensee MDPI, Basel, Switzerland. This article is an open access article distributed under the terms and conditions of the Creative Commons Attribution (CC BY) license (https://creativecommons.org/licenses/by/4.0/).

1. Introduction

Recent developments in low-power integrated circuits and wireless technologies, the emergence of new application paradigms in the context of the Internet of things (IoT), and a better understanding of propagation phenomena, have led the scientific community to revise Tesla's initial idea of the wireless power transfer. This idea is now seen as a promising and achievable solution to overcome the limitations of conventional power supply methods, such as batteries or wired connections to fixed power grids. Given the large number of nodes expected to be interconnected in IoT applications and other 5G/6G scenarios, the benefits of radio frequency-based energy harvesting (RF-EH) in terms of operating cost savings and self-sustainability are undoubted. In addition, whether based solely on RF-EH or combined with other primary energy sources (solar radiation, mechanical vibration, air flow, etc.), the panacea of perpetual operation of wireless networks seems somewhat closer today.

Research in RF-EH has already produced significant results, as indicated in recent survey papers. Examples are [1–6]. Moreover, relatively new books like [7] or [8] contain good compilations of the main contributions and results. Until now, however, less attention has been paid to the role of the propagation model, either because current work has focused on design aspects or applications of the energy-harvesting technology, or because it has relied on assumptions that neglect the importance of the wireless environment. For example, many papers assume the presence of a dedicated energy source, which performs channel estimation to adjust the transmitted power accordingly. Therefore, the specific propagation model plays a secondary role in these works. However, an important application area for RF-EH is IoT, which is expected to comprise thousands or even millions of extremely simple and low-cost wireless devices. Thus, it is quite possible that these devices will not be able to participate in channel state information (CSI) estimation procedures, nor will the power transmitter be able to cope with the excessive workload involved in keeping track of each connection.

Given the erratic behavior of the wireless channel, the main motivation of this paper is to analyze in detail the impact of the variability of the received power on the amount of electromagnetic energy that can be captured by a device. To fully capture this variability, a key aspect is the selection of a powerful channel propagation model that encompasses all sources of signal variation in a flexible manner. Thus, the main contributions regarding the use of channel propagation models in the characterization of the energy-harvesting process are reviewed below.

First, the propagation models used in RF-EH are outlined in some of the above survey papers. This is the case of [1,6], which includes a brief description of wireless propagation through manageable models such as free-space, two-ray, or Rayleigh fading [4], which recalls the free-space and two-ray models and [5], which only makes reference to the Friis path-loss formula. As for regular papers, most also adopt simplified propagation models. For example, the work presented in [9] assumes free-space path loss and additive white Gaussian noise, which is consistent with the use of a power transmitter mounted on an unmanned autonomous vehicle (UAV) flying along a circular path over the wireless nodes. The objective is to optimize the trajectory radius of the UAV so that a fair allocation of energy between the participating devices is achieved. In [10], a two-ray model is used to account for both line of sight (LOS) and non-line of sight (NLOS) propagation. The analysis focuses on the impact of the NLOS component and other factors (radiation pattern of the transmit and receive antennas, losses associated with different polarization of transmitting field, and efficiency of the power harvester circuit) on the average energy harvested. The use of the two-ray model applies to situations in which there is a direct LOS component and a clear NLOS component reflected by a uniform ground plane. This model is further extended to include path loss, lognormal shadowing, and Rician fading. The latter characterizes the aggregation of many weak scattered rays rather than a single dominant one. The paper then focuses on the estimation of the path loss and shadowing parameters as well as the Rician factor on the basis of experimental measurements. The paper concludes with the assertion that the Rayleigh fading model is not well suited to characterize practical scenarios of RF-EH due to the presence of a strong LOS component. Reference [11] discusses a power beamforming strategy for distributed power transfer from multiple transmitters to a single receiver, based on a relatively simple path-loss model combined with Rayleigh fading and Gaussian noise. Another example is [12], which characterizes the channel by a Nakagami-m fading model. Its purpose is to determine an optimal transmission policy for the RF energy-harvesting device, which switches between two modes: on and off. In the on mode, the device is operational and powered by a battery; in the off mode, the device turns off, and the battery feeding process stops. In [13], various empirical path-loss models are used to determine the usability and fundamental limits of joint RF and photovoltaic harvesting-based M2M communications. Essentially, the theoretical bounds derived in this paper are based on the well-known Shannon's capacity theorem. In [14], several empirical path-loss models are used with the objective of proposing design guidelines for all stages of the end-to-end RF-to-DC energy-conversion process. In [15], both the Nakagami-m and generalized-K fading models are considered in the statistical characterization of the battery recharging time. In [16], the HATA model, the Ericsson model, and the ITU-R model are used to provide outdoor RF spectral survey results in suburban areas. Basically, the main objective of this paper is to determine the frequency bands most suitable for energy harvesting from RF ambient sources in populated environments.

Among the reviewed literature, the closest contributions to the work presented in this paper are made by [17,18]. The first characterizes the average and variance of the energy collected by a node from the RF signal generated by multiple transmitters distributed according to a spatial Poisson process. The channel between each transmitter and the collector node is initially assumed to be generalized-K, but it is then approximated by a gamma distribution and finally mixed with the rest of the channel distributions to produce a general Gaussian process according to the central limit theorem. Very general propagation models are also considered in [18], such as the generalized η-μ and κ-μ models, which are used to obtain exact closed-form expressions for the distribution, mean, variance, and higher-order moments of the recharge time. However, no relationship is provided between the mathematical parameters of these models and the physical parameters of the system.

This paper fills the previous gaps by fully considering the generalized-K propagation model for the statistical characterization of the energy collected by a single device over an arbitrary time interval. The generalized K-model is doubly advantageous in that it is analytically tractable (in contrast to what is stated in [17]) while covering a wide variety of scenarios with respect to all perturbations introduced by wireless channels, namely path loss, shadowing, and fast fading. Specifically, the main contributions of this paper can be listed as follows.

- Exact closed-form expressions are obtained for the expectation and variance of the energy harvested by a static device, which is assumed to be illuminated by a dedicated power source emitting an unmodulated carrier (a WPT system is thus considered). To the author's knowledge, this is the first work that adopts such a very general propagation model to accurately characterize the statistics of the RF energy-harvesting process.
- A detailed evaluation is performed showing the sensitivity of the statistical parameters with respect to the physical parameters of the propagation environment (transmission distance, path-loss exponent, shadowing spread, and Nakagami parameter).

Table 1 summarizes the current state-of-the art work in RF-EH related to the specification and use of channel propagation models.

Table 1. Current state-of-the-art work in RF-EH involving channel propagation models.

Reference	Propagation Model	Main Focus
[9]	Free space path-loss and additive Gaussian noise	Fair energy allocation
[10]	Two-ray/Path-loss, lognormal shadowing and Rician fading	Average energy harvested/Estimation of model parameters
[11]	Path-loss, Rayleigh fading and additive Gaussian noise	Beamforming strategy
[12]	Nakagami-m fading	Optimal transmission policy
[13]	Empirical	Theoretical bounds on transmission rate
[14]	Empirical	Design guidelines for RF-to-DC circuitry
[15]	Nakagsami-m/Generalized-K	Battery recharging time
[16]	Empirical (HATA/Ericsson/ITU-R)	Spectral behavior of RF-EH in suburban environments
[17]	Gaussian	Mean and variance of harvested energy (from multiple transmitters)
[18]	Generalized η-μ/κ-μ	Battery recharging time
Current contribution	Generalized-K	Mean and variance of harvested energy

In the context of RF energy harvesting, one of the paradigms with the greatest projection is the wireless energy network (WEN) [19]. The results obtained in this paper are useful in several aspects related to the planning and performance of this type of network.

- To define areas with different levels of energy coverage, once a potential location of the primary and secondary power sources is defined. Here, a grade of service metric such as the probability of energy outage will be of the utmost importance.
- To analyze the queuing time experienced by the energy requests directed to the same power source.

The rest of the paper is organized as follows. In Section 2, the basics of RF-EH are reviewed. In Section 3, the problem is formulated by stating the system model and assumptions, the energy harvesting equations and the generalized-K distribution. Exact closed-form expressions for the average and variance of the energy harvested by a static device are respectively obtained in Sections 4 and 5. In Section 6, the analytical expressions are validated by simulation and then numerical results are obtained. Finally, in Section 7, the main conclusions and suggestions for further research are drawn.

2. Fundamentals of RF-EH

RF-EH has recently emerged as a disruptive technology that allows low-power portable devices and energy-constrained wireless networks to convert the electromagnetic energy present in the environment into DC current. This idea is not new, as it dates back to the early years of the last century, when Nikola Tesla designed and built an experimental station (Wardenclyffe Tower) for the wireless transfer of information and energy to remote devices. However, the project did not receive enough funds and was quickly abandoned before it became operational due to several reasons: the low efficiency of the electric-to-electromagnetic-to-electric conversion process as well as health concerns related to high-power transmitters. Fortunately, in the past few years we have witnessed a resurgence of the concept, as a result of its reformulation for low-power wireless devices.

Wireless power transfer (WPT) techniques fall into one of two major categories, namely near field and far field. The distinction is made because electromagnetic waves behave very differently in these two regions, and correspondingly the techniques to collect energy from them are also quite different. Near-field propagation takes place within an area of about one wavelength of the transmitting antenna, which typically corresponds to distances of at most several meters. As detailed in [3], propagation in the near-field region is essentially non-radiative, meaning that power leaves the transmitter only when there is a receiver to couple to within such a region. Accordingly, power can be transferred in the near-field region by employing inductive coupling, capacitive coupling or their enhanced versions, which consist of adding resonant circuits in order to increase the power-coupling coefficient and, consequently, the transmission range. However, for distances of hundreds of meters or even several kilometers, far field is the only possible region of operation. In contrast to near-field propagation, far-field propagation is radiative, and it obeys the well-known Friis equation when no obstacles are present between transmitter and receiver. RF-EH encompasses systems and techniques devoted to far-field WPT via electromagnetic signals like radio waves, microwaves, or light waves.

Various architectures have been proposed for energy-harvesting systems that coexist with traditional data receivers. One solution is to have independent segments for WPT and wireless information transfer (WIT), as depicted in Figure 1. Such a global scenario of wireless information and power transfer (WIPT) is also referred to as separated receiver architecture. As can also be seen, power sources can be classified into two classes: dedicated power sources and ambient power sources. Dedicated power sources are specifically deployed to transfer RF energy to one or several nodes. These sources can use the license-free ISM frequency bands, though subject to restrictive upper bounds on transmission power. On the other hand, ambient power sources, that is, transmitters that are not intended for RF energy transfer, are always available at no cost, but the collected energy can be very small. Consequently, for applications that require stable and predictable energy supply, dedicated power sources are preferable.

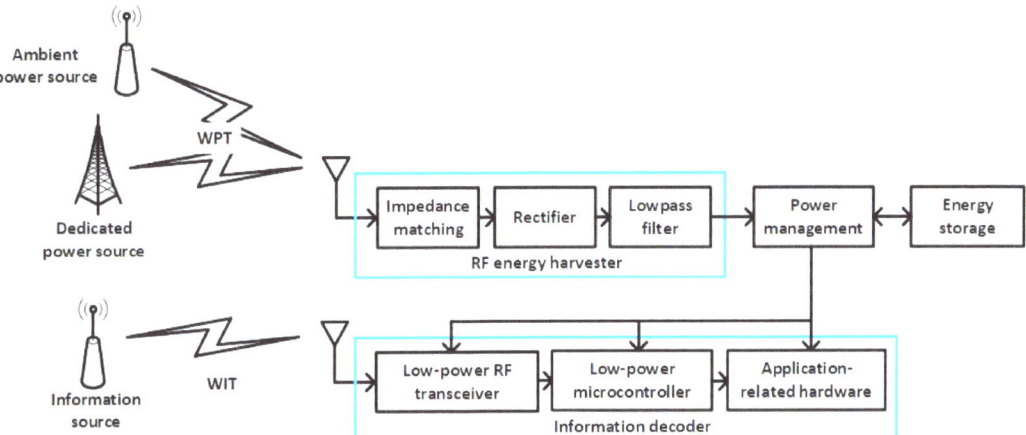

Figure 1. WIPT scenario, with separated WIT and WPT segments. The WIT segment contains conventional hardware to receive information, whereas the WPT segment is essentially a voltage rectifier followed by a low-pass filter devoted to extracting the DC component of the received signal.

The architecture represented in Figure 1 is also known as out-of-band RF energy harvesting, because the node collects energy from an RF signal different from that used to receive information. On the other hand, because the information signal also carries energy, a new modality called simultaneous wireless information and power transfer (SWIPT), or in-band energy harvesting, was devised. The new architecture is represented in Figure 2. As can be seen, SWIPT allows for use of a single antenna (or antenna array) to obtain both information and energy. However, a splitting architecture is required in order to distribute the received signal among the two processes. The reason is that a serial implementation would not be feasible regardless of which process was implemented first. Either the energy harvester would destroy information, or the information decoder would consume all signal power. References [1,3] provide very complete descriptions of the WIPT and SWIPT architectures.

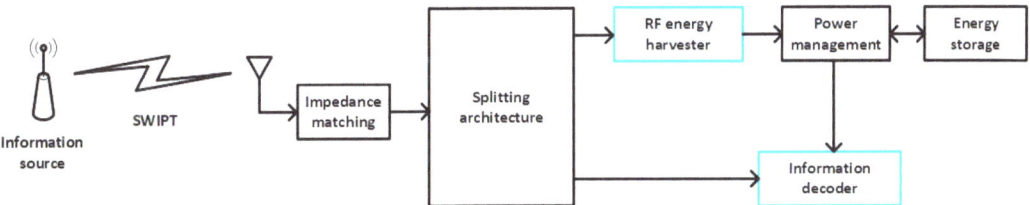

Figure 2. SWIPT scenario. Power is obtained from the same information signal, by using a splitting architecture that generates an input for every independent process.

Finally, another architecture is frequency splitting, the diagram of which is shown in Figure 3. In this architecture, the input signal transports both information and power by using separated frequencies. In fact, this signal consists of the information-bearing modulated component plus an unmodulated sine wave that essentially results from shifting the carrier frequency used in the modulation process, from a value f_c to a different value f_p. The figure also highlights the fact that though the transmitter sends a pure sinusoidal to transfer power, the received signal exhibits some spread due to variations in the channel coefficient. Consequently, there must be a sufficiently high guard band between f_c and f_p.

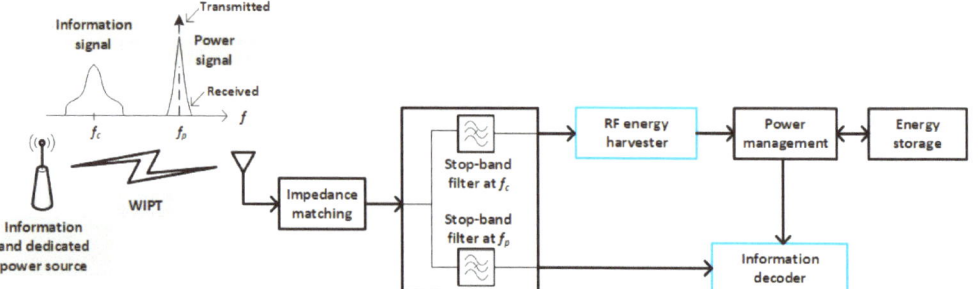

Figure 3. Frequency-splitting architecture. This architecture is really a variant of WIPT, because information and power are transferred via separated signals in the frequency domain, but just a single antenna is required at each side of the link (as in SWIPT).

3. Problem Formulation

In this section, the system model and assumptions, the fundamental energy harvesting equation and the generalized-K distribution are formulated.

3.1. System Model and Assumptions

As stated in [1], the SWIPT architectures always impose a trade-off between information rate and amount of RF energy harvested. To bypass this trade-off, especially in situations where a stable and predictable energy supply is required, WPT from a dedicated source is preferable. Accordingly, this paper focuses on scenarios like those shown in Figure 1 or Figure 3. More specifically, the system under analysis is shown in Figure 4, where a dedicated power source generates an unmodulated carrier at frequency f_p in the RF, microwave, or visible light bands. Both the power source and the receivers are located at fixed positions and equipped with directional antennas of gains G_t and G_r, respectively. In particular, the power source uses a phased array that allows for tuning the main beam in the direction of current receiver, whereas receiver antennas are always aligned with the power source. An additive white Gaussian noise (AWGN) channel is assumed, with spectral noise density N_0 and a channel coefficient $h(t)$ that obeys the abovementioned generalized-K distribution. Moreover, the energy-harvesting device is assumed to operate linearly, in spite of the presence of non-linear components like diodes. For the sake of generality, the analysis will start from an arbitrary waveform $x(t)$ for the power-bearing signal, and only in the end will it be particularized to the special case of an unmodulated carrier. The corresponding power transfer bandwidth is B, meaning that the impedance matching circuit (or the combined effect of this circuit and the stop-band filter at f_c if the scenario of Figure 3 is considered) has an equivalent bandwidth B centered at frequency f_p.

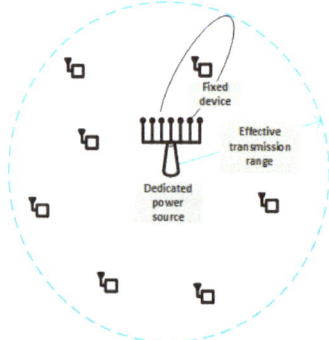

Figure 4. System model.

3.2. Energy-Harvesting Equation

Using complex notation, the power-bearing signal $x(t)$ can be expressed in terms of its equivalent low-pass signal or complex envelope $\tilde{x}(t)$ as follows:

$$x(t) = \Re\left\{\tilde{x}(t) \cdot e^{j2\pi f_p t}\right\}. \tag{1}$$

Here, $\Re$ stands for the real part. As pointed out in [20], the received signal can be generally formulated in this way:

$$r(t) = \Re\left\{\sum_{k=0}^{K(t)} \alpha_k(t) \cdot \tilde{x}(t - \tau_k(t)) \cdot e^{j\left(2\pi f_p(t-\tau_k(t))+\varphi_k\right)}\right\} + n(t). \tag{2}$$

In this expression, $n(t)$ is the AWGN component, $K(t)$ is the total number of resolvable multi-path components and $\alpha_k(t)$, $\tau_k(t)$, and ϕ_k are, respectively, the time-dependent amplitude and delay parameters and a constant phase offset associated with the k:th resolvable multi-path component. If the delay spread of the channel, T_m, verifies $T_m \ll \frac{1}{B}$, then $\tilde{x}(t - \tau_k(t)) \cong \tilde{x}(t)$. This is the so-called narrow-band fading condition. Note that for $x(t)$ consisting of an unmodulated carrier, this condition holds for any T_m. Consequently, under narrow-band fading, we have [20]

$$r(t) = \Re\left\{\left(\sum_{k=0}^{K(t)} \alpha_k(t) \cdot e^{-j\left(2\pi f_p \tau_k(t)-\varphi_k\right)}\right) \cdot \tilde{x}(t) \cdot e^{j2\pi f_p t}\right\} + n(t). \tag{3}$$

The summation term between parentheses is independent of the complex envelope $\tilde{x}(t)$, and hence it simply introduces a multiplicative effect on the signal. It is the so-called channel coefficient, typically denoted by $h(t)$. Accordingly, Equation (3) can be rewritten in the following way:

$$r(t) = \Re\left\{\tilde{x}(t) \cdot h(t) \cdot e^{j2\pi f_p t}\right\} + n(t). \tag{4}$$

The noise component can also be reformulated in terms of its complex envelope $\tilde{n}(t)$ as $n(t) = \Re\{\tilde{n}(t) \cdot e^{j2\pi f_p t}\}$, and therefore the channel output signal can be expressed as follows:

$$r(t) = \Re\left\{(\tilde{x}(t) \cdot h(t) + \tilde{n}(t)) \cdot e^{j2\pi f_p t}\right\}. \tag{5}$$

The term multiplying the exponential is nothing more than the complex envelope of the received signal $r(t)$, namely $\tilde{r}(t)$:

$$\tilde{r}(t) = \tilde{x}(t) \cdot h(t) + \tilde{n}(t). \tag{6}$$

Figure 5 shows the equivalent low-pass representation of the channel effect, as well as the subsequent energy-harvesting and power-management units. The energy-harvesting unit is capable of producing a certain amount of energy at the end of an exposition time period T. This energy, denoted as $\mathcal{E}(T)$, can be mathematically formulated as follows:

$$\mathcal{E}(T) = \eta \int_T r^2(t) \cdot dt. \tag{7}$$

Here, the integral represents the RF energy captured by the receiving antenna along the period T, whereas η denotes the RF-to-DC conversion efficiency. Equation (7) constitutes the starting point of the analysis performed in this paper.

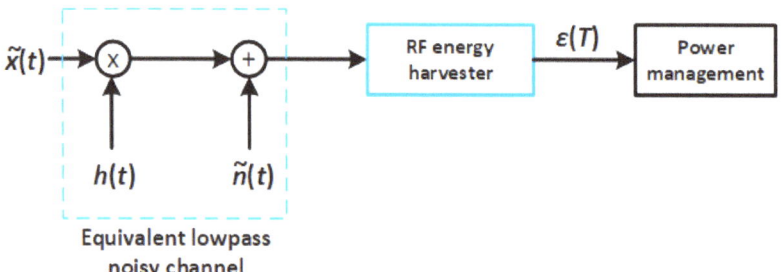

Figure 5. Equivalent low-pass model of the channel.

3.3. The Generalized-K Distribution

The generalized-K distribution is a powerful description of the perturbation effects experienced by signals in wireless propagation environments. It was proposed in the early 2000s as a compound probability density function (PDF) encompassing all sources of wireless signal degradation, namely path loss, shadowing, and fast fading. In addition, it enjoys the property of analytical tractability, which helps to obtain closed-form expressions for the performance measures of interest. An example is the statistical characterization of the energy harvesting performed in this paper. The availability of closed-form expressions for magnitudes such as the mean and variance of the energy harvested is very useful for network planning, especially in scenarios like IoT, where the energy provider cannot rely on CSI to adjust its transmission power because of the expected massive number of participating devices.

The generalized-K model combines the Nakagami-m distribution for fast fading and the gamma distribution for path loss and shadowing (Nakagami gamma). In addition, it can be particularized to numerous well-known models, such as Rayleigh lognormal (Suzuki model), Nakagami lognormal, Rayleigh gamma (K model) and, in an approximate way, Rician gamma, and Rician lognormal.

Let $z = z(t) = \|h(t)\|$ be the envelope of the channel coefficient. Assuming that the channel is stationary, z can be treated as a single random variable whose statistical characterization is independent of time. If this characterization obeys the generalized-K model, the probability density function (PDF) of z can be expressed as follows [21]:

$$f_Z(z) = \frac{4\sqrt{\frac{m}{b}}}{\Gamma(a)\Gamma(m)} \left(\sqrt{\frac{m}{b}}z\right)^{a+m-1} K_{a-m}\left(2\sqrt{\frac{m}{b}}z\right). \tag{8}$$

In this expression, a and b are respectively the shape and scale parameters of the gamma distribution ($a > 0, b > 0$), whereas $m > 0$ is the so-called Nakagami parameter. K_{a-m} stands for the modified Bessel function of order $(a - m)$. The shape and scale parameters can be formulated in terms of physical parameters [22]:

$$a = \frac{1}{e^{\frac{\sigma^2_{\Psi dB}}{\zeta^2}} - 1}; \tag{9}$$

$$b = e^{\frac{\mu_{\Psi dB}}{\zeta} + \frac{\sigma^2_{\Psi dB}}{2\zeta^2}}\left(e^{\frac{\sigma^2_{\Psi dB}}{\zeta^2}} - 1\right). \tag{10}$$

In these expressions, $\zeta = \frac{10}{\ln 10}$, and $\mu_{\Psi dB}$ and $\sigma_{\Psi dB}$ are, respectively, the average and standard deviation of $\Psi dB = 10\log_{10}\Psi = 10\log_{10}\frac{P_r}{P_t}$, that is, the ratio of power received to power transmitted expressed in dB. The expected value $\mu_{\Psi dB}$ is given by the following expression [20]:

$$\mu_{\Psi dB} = 10\log_{10}\alpha - 10\beta\log_{10}\frac{d}{d_0}. \tag{11}$$

Here, d is the distance between the transmitter and the receiver (transmission distance), d_0 is the reference distance, β is the path-loss exponent, and α is a constant that depends on multiple parameters (antenna gains, reference distance, path-loss exponent, average blockage, and carrier frequency). There is no mathematical expression for $\sigma_{\Psi dB}$ (shadowing spread), but its value has been experimentally set up within the range [4, 13] dB [20]. The generic path-loss model given in (11) can be replaced by more specific models (free-space, two-ray, Okumura, Hata, COST 231 models, etc.) when further details about the scenario are provided [20].

Equations (8)–(11) define a versatile propagation model that is entirely formulated in terms of physical (and measurable) parameters. To summarize, the inputs to this model are the constant α, the path-loss exponent, both the transmission and reference distances, the delay spread and the Nakagami parameter. This parameter can take on any value above 0, thus providing high flexibility to capture quite different small-scale multipath fading conditions:

- If $m > 1$, the channel is Rician, meaning that there is a dominant LOS propagation component over the scattered non-LOS component. This is the so-called non-isotropic propagation environment. In this case, the degree of fading is low, becoming less severe with increasing m. In particular, for $m \to \infty$ there is no fading.
- If $m \leq 1$, there is no dominant LOS component in the received signal and the degree of fading is high, increasing as m decreases. Such scenario is referred to as isotropic propagation environment. Particular cases are $m = 1$ (Rayleigh channel), $m = 0.5$ (one-sided Gaussian channel) and $m \ll 1$ (very severe fading channel).

4. Average Energy Harvested

The first step in the characterization of the amount of energy harvested as a random variable is to obtain its average. Recalling Equation (7), the expected energy harvested can be formulated in this way:

$$E\{\mathcal{E}(T)\} = E\left\{\eta \int_T r^2(t) \cdot dt\right\} = \eta \int_T E\{r^2(t)\} \cdot dt = \frac{\eta}{2} \int_T E\{\|\tilde{r}(t)\|^2\} \cdot dt. \quad (12)$$

On the other hand, from Equation (7), we have

$$\tilde{r}(t) = (\tilde{x}_I(t) + j \cdot \tilde{x}_Q(t)) \cdot (h_{re}(t) + j \cdot h_{im}(t)) + \tilde{n}_I(t) + j \cdot \tilde{n}_Q(t). \quad (13)$$

In this expression, both $\tilde{x}(t)$ and $\tilde{n}(t)$ have been decomposed into their in-phase and quadrature components, respectively $(\tilde{x}_I(t), \tilde{x}_Q(t))$ and $(\tilde{n}_I(t), \tilde{n}_Q(t))$, and the channel coefficient into its real and imaginary parts, namely $(h_{re}(t), h_{im}(t))$. Further manipulation allows to separate the real and imaginary components of $\tilde{r}(t)$ as follows:

$$\tilde{r}(t) = \tilde{x}_I(t) \cdot h_{re}(t) - \tilde{x}_Q(t) \cdot h_{im}(t) + \tilde{n}_I(t) + j \cdot (\tilde{x}_I(t) \cdot h_{im}(t) + \tilde{x}_Q(t) \cdot h_{re}(t) + \tilde{n}_Q(t)). \quad (14)$$

Then, the squared module of $\tilde{r}(t)$ is nothing else but the sum of the squared real and imaginary parts:

$$\|\tilde{r}(t)\|^2 = (\tilde{x}_I(t) \cdot h_{re}(t) - \tilde{x}_Q(t) \cdot h_{im}(t) + \tilde{n}_I(t))^2 + (\tilde{x}_I(t) \cdot h_{im}(t) + \tilde{x}_Q(t) \cdot h_{re}(t) + \tilde{n}_Q(t))^2. \quad (15)$$

Proceeding through standard calculations, we can end up with the following exact result for $\|\tilde{r}(t)\|^2$:

$$\|\tilde{r}(t)\|^2 = \|\tilde{x}(t)\|^2 \cdot \|h(t)\|^2 + \tilde{n}_I^2(t) + \tilde{n}_Q^2(t) + 2(\tilde{x}_I(t) \cdot h_{re}(t) - \tilde{x}_Q(t) \cdot h_{im}(t)) \cdot \tilde{n}_I(t) \\ + 2(\tilde{x}_I(t) \cdot h_{im}(t) + \tilde{x}_Q(t) \cdot h_{re}(t)) \cdot \tilde{n}_Q(t). \quad (16)$$

Next, we can obtain the expectation of $\|\tilde{r}(t)\|^2$. The analysis can be simplified by recalling some valuable properties of AWGN ([23]): (i) $E\{\tilde{n}_I^2(t)\} = E\{\tilde{n}_Q^2(t)\} = E\{n^2(t)\} = N_R = \eta_0 \cdot B$, where N_R stands for the received noise power, η_0 for the spectral density power and B for the power transfer bandwidth, (ii) $E\{\tilde{n}_I(t)\} = E\{\tilde{n}_Q(t)\} = 0$, and (iii) $\tilde{n}_I(t)$

and $\tilde{n}_Q(t)$ are mutually independent Gaussian random variables. In addition, because noise is independent of both, input signal and channel coefficient, and $E\{\tilde{n}_I(t)\} = 0$, we also have $E\{(\tilde{x}_I(t) \cdot h_{re}(t) - \tilde{x}_Q(t) \cdot h_{im}(t)) \cdot \tilde{n}_I(t)\} = E\{(\tilde{x}_I(t) \cdot h_{re}(t) - \tilde{x}_Q(t) \cdot h_{im}(t))\} \cdot E\{\tilde{n}_I(t)\} = 0$. Similarly, $E\{(\tilde{x}_I(t) \cdot h_{im}(t) + \tilde{x}_Q(t) \cdot h_{re}(t)) \cdot \tilde{n}_Q(t)\} = 0$. Because the input signal and the channel coefficient are also mutually independent random variables, the squared module of the complex envelope can be expressed in this way:

$$E\{\|\tilde{r}(t)\|^2\} = E\{\|\tilde{x}(t)\|^2\} E\{\|h(t)\|^2\} + 2E\{n^2(t)\}. \quad (17)$$

Accordingly, the average energy harvested formulated in (12) obeys the following expression:

$$E\{\mathcal{E}(T)\} = \eta \int_T \left(\frac{E\{\|\tilde{x}(t)\|^2\}}{2} E\{\|h(t)\|^2\} + E\{n^2(t)\} \right) \cdot dt. \quad (18)$$

The equivalent low-pass input signal $\tilde{x}(t)$ is defined by the modulation. Consequently, in general, this signal is non-stationary and can be expressed as $\tilde{x}(t) = A(t)e^{j\theta(t)}$, where $A(t)$ and $\theta(t)$ denote, respectively, the time-varying amplitude and phase of the modulating signal. On the other hand, the channel is assumed to be stationary, and hence its effect represented by $E\{\|h(t)\|^2\}$ can be taken out of the integral in the previous equation. In [22], exact closed-form expressions are provided for the moments of the generalized-K distribution, among which the second moment is given by $E\{\|h(t)\|^2\} = a \cdot b$, where a and b obey, respectively, expressions (9) and (10). In agreement with [24], this second moment will be renamed from now on as Ω_p. Accordingly, we have

$$E\{\mathcal{E}(T)\} = \eta \left(\Omega_p \int_T \frac{E\{A^2(t)\}}{2} \cdot dt + N_R \cdot T \right). \quad (19)$$

In this equation, the relevant term regarding how efficient the energy-transfer process can be is Ω_p, because it represents the global channel effect due to path loss, shadowing and multi-path fading on the received signal. Instead, the integral in Equation (19) evaluates the energy contained in an interval T of the transmitted signal, whatever its shape. Thus, without loss of generality, we can assume the simplest case of $A(t) = A$, which corresponds to an unmodulated carrier. Accordingly, the previous equation can be reformulated in this way:

$$E\{\mathcal{E}(T)\} = \eta \cdot T \left(\frac{A^2}{2} \cdot \Omega_p + N_R \right). \quad (20)$$

The term $\frac{A^2}{2}$ is nothing else but the transmitted power (normalized to a 1Ω-load). If we denote this power as P_t, the expected energy harvested is as follows:

$$E\{\mathcal{E}(T)\} = \eta \cdot T (P_t \cdot \Omega_p + N_R). \quad (21)$$

Note that the average energy harvested depends on the shape and scale parameters of the gamma distribution (via Ω_p), but not on the Nakagami parameter that characterizes multi-path fading. Note also that it has two components, the main one due to the power-bearing signal, and the thermal noise.

5. Variance of Energy Harvested

To obtain the variance of the energy harvested, first we can analyze the second moment about zero:

$$E\{\mathcal{E}^2(T)\} = \eta^2 E\left\{\left(\int_T r^2(t)dt\right)^2\right\} = \eta^2 E\left\{\left(\int_T r^2(s)ds\right)\left(\int_T r^2(t)dt\right)\right\}$$
$$= \eta^2 E\left\{\int_T\int_T r^2(s)r^2(t)ds\cdot dt\right\} = \eta^2 \int_T\int_T E\{r^2(s)r^2(t)\}ds\cdot dt \quad (22)$$
$$= \frac{\eta^2}{4}\int_T\int_T E\{\|\tilde{r}(s)\|^2\|\tilde{r}(t)\|^2\}ds\cdot dt.$$

The next step is to evaluate the expectation inside the integral, which is nothing else but a correlation. However, the analytical procedure that yields an exact closed-form expression for this correlation is very complex, and thus the details have been relegated to the Appendix A. For the setting $s = t + \tau$, the result is Equation (A31):

$$E\{\|\tilde{r}(t+\tau)\|^2\|\tilde{r}(t)\|^2\} = \phi_{\|\tilde{r}\|^2}(t,\tau) = \phi_{\|\tilde{x}\|^2}(t,\tau)\phi_{\|h\|^2}(\tau) + 4N_R\cdot E\{\|\tilde{x}(t)\|^2\}E\{\|h(t)\|^2\} \\ +16\phi_{\tilde{n}}(\tau)\cdot\Re\{\phi_{\tilde{x}}(t,\tau)\phi_h(\tau)\} + 4N_R^2 + 4\phi_{\tilde{n}}^2(\tau). \quad (23)$$

Next, the terms that appear in this equation are analyzed individually.

If we assume that the input signal is an unmodulated carrier, that is, $\tilde{x}(t) = Ae^{j\theta}$, we have

$$E\{\|\tilde{x}(t)\|^2\} = A^2 = 2P_t; \quad (24)$$

$$\phi_{\|\tilde{x}\|^2}(t,\tau) = E\{\|\tilde{x}(t+\tau)\|^2\|\tilde{x}(t)\|^2\} = A^4 = 4P_t^2; \quad (25)$$

$$\phi_{\tilde{x}}(t,\tau) = \frac{1}{2}E\{\tilde{x}(t+\tau)\tilde{x}^*(t)\} = \frac{1}{2}E\{Ae^{j\theta}\cdot Ae^{-j\theta}\} = \frac{A^2}{2} = P_t. \quad (26)$$

Note that, for the case of an unmodulated carrier, the input signal does not only become a stationary process, but its associated correlations are constant. Moreover, despite the fact that it has been assumed that θ is a constant phase offset, the analysis that follows would also be valid, with some minor modifications, for a phase-modulated signal, that is, $\theta = \theta(t)$.

Another auto-correlation involved in (23) is $\phi_{\tilde{n}}(\tau)$. For the case of AWGN, the following expression is provided in [23]:

$$\phi_{\tilde{n}}(\tau) = \eta_0\frac{\sin(\pi B\tau)}{\pi\tau} = \eta_0 B\frac{\sin(\pi B\tau)}{\pi B\tau} = N_R\cdot\text{sinc}(\pi B\tau). \quad (27)$$

The remaining terms are $\phi_h(\tau)$ and $\phi_{\|h\|^2}(\tau)$, which are channel auto-correlations. They capture the variations, in the statistical sense, perceived by the user as it moves at a certain speed v over the combined path loss, shadowing, and multi-path fading scenario under consideration (in the present case, the scenario that leads to the generalized-K distribution). These temporal correlations can always be transformed into spatial correlations, because the dependence on τ is, in fact, on the product $v\cdot\tau$. However, because the focus of this paper is on static users, that is, $v = 0$, which has the same effect as $\tau = 0$, we are really interested in $\phi_h(0)$ and $\phi_{\|h\|^2}(0)$. Regarding the first one, we can write

$$\phi_h(0) = \frac{1}{2}E\{h(t)\cdot h^*(t)\} = \frac{1}{2}E\{\|h(t)\|^2\} = \frac{\Omega_p}{2}. \quad (28)$$

To analyze the second term, it is useful to introduce the auto-covariance $\mu_{\|h\|^2}(\tau)$, which is related to the auto-correlation as follows:

$$\mu_{\|h\|^2}(\tau) = \phi_{\|h\|^2}(\tau) - E\{\|h(t+\tau)\|^2\}\cdot E\{\|h(t)\|^2\}. \quad (29)$$

Accordingly,

$$\phi_{\|h\|^2}(0) = \mu_{\|h\|^2}(0) + E^2\{\|h(t)\|^2\} = \mu_{\|h\|^2}(0) + \Omega_p^2. \quad (30)$$

Because the auto-covariance at the origin is nothing else but the variance, we have

$$\phi_{\|h\|^2}(0) = Var\{\|h(t)\|^2\} + \Omega_p^2. \tag{31}$$

Moreover, $Var\{\|h(t)\|^2\}$ can be expressed in terms of the second and fourth moments of the distribution of $\|h(t)\|$:

$$Var\{\|h(t)\|^2\} = E\{\|h(t)\|^4\} - E^2\{\|h(t)\|^2\}. \tag{32}$$

Note that the second term in the right-hand side of this equation is Ω_p^2. In [22], a generic closed-form expression is provided for the moments of the generalized-K distribution. This expression is exact and can be particularized for the fourth moment as follows:

$$E\{\|h(t)\|^4\} = \frac{\Gamma(a+2)\Gamma(m+2)}{\Gamma(a)\Gamma(m)}\left(\frac{b}{m}\right)^2. \tag{33}$$

Recall that a and b are, respectively, the scale and shape parameters of the gamma distribution that describes path loss and shadowing, and m is the Nakagami parameter that characterizes multi-path fading. Next, considering Equations (32) and (33), Equation (31) can be rewritten in terms of the parameters of the generalized-K distribution:

$$\phi_{\|h\|^2}(0) = \frac{\Gamma(a+2)\Gamma(m+2)}{\Gamma(a)\Gamma(m)}\left(\frac{b}{m}\right)^2. \tag{34}$$

Now, introducing (24)–(28) and (34) into Equation (23), and rearranging terms, we can obtain the definite result for $E\{\|\tilde{r}(t+\tau)\|^2\|\tilde{r}(t)\|^2\}$:

$$E\{\|\tilde{r}(t+\tau)\|^2\|\tilde{r}(t)\|^2\} = 4P_t^2\frac{\Gamma(a+2)\Gamma(m+2)}{\Gamma(a)\Gamma(m)}\left(\frac{b}{m}\right)^2 \\ + 8N_R P_t \Omega_p(1+\text{sinc}(\pi B\tau)) \\ + 4N_R^2(1+\text{sinc}^2(\pi B\tau)). \tag{35}$$

As can be seen, this expression depends exclusively on system parameters. In particular, $P_t\Omega_p$ is the average received power, because it is the product of the average transmitted power and the channel effect represented by Ω_p. Finally, the variance of the energy harvested can be obtained by integrating expression (35) according to (22), and then subtracting the square of the expected energy harvested given by (21). An exact closed-form expression is obtained:

$$Var\{\mathcal{E}(T)\} = (\eta P_t T)^2 \left(\frac{\Gamma(a+2)\Gamma(m+2)}{\Gamma(a)\Gamma(m)}\left(\frac{b}{m}\right)^2 - \Omega_p^2\right) \\ + \frac{4\eta^2 N_R P_t \Omega_p}{\pi^2 B^2}(-1+\cos(\pi BT)+\pi BT \cdot Si(\pi BT)) \\ + \frac{\eta^2 N_R^2}{\pi^2 B^2}\Big(-1-\gamma+\cos(2\pi BT)+Ci(2\pi BT) \\ -\ln(2\pi BT)+2\pi BT \cdot Si(2\pi BT)\Big). \tag{36}$$

Here, γ is the Euler's constant ($\gamma \simeq 0.577216$), and $Si()$ and $Ci()$ stand, respectively, for the sine integral and cosine integral functions, which are defined as follows:

$$Si(x) = \int_0^x \frac{\sin(u)}{u}du; \tag{37}$$

$$Ci(x) = -\int_x^\infty \frac{\cos(u)}{u} du. \tag{38}$$

As can be seen from Equation (36), the variance of the energy harvested has three components: one that depends exclusively on the power-bearing signal, a cross-term that depends on both the power-bearing signal and the thermal noise, and finally a third component that only depends on noise.

Even more meaningful than the variance is the squared coefficient of variation, which is nothing more than the ratio of the variance to the squared expectation. In essence, it represents the relative variability of the distribution around its expected value. In particular, the squared coefficient of variation of the energy harvested, namely $SCV\{\mathcal{E}(T)\}$, is given by

$$\begin{aligned}
SCV\{\mathcal{E}(T)\} = {} & P_t^2 \frac{\left(\frac{\Gamma(a+2)\Gamma(m+2)}{\Gamma(a)\Gamma(m)}\left(\frac{b}{m}\right)^2 - \Omega_p^2\right)}{(P_t\Omega_p + N_R)^2} \\
& + \frac{4N_R P_t \Omega_p(-1 + \cos(\pi BT) + \pi BT \cdot Si(\pi BT))}{\pi^2 B^2 T^2 (P_t\Omega_p + N_R)^2} \\
& + \frac{N_R^2}{\pi^2 B^2 T^2 (P_t\Omega_p + N_R)^2}\Big(-1 - \gamma + \cos(2\pi BT) \\
& + Ci(2\pi BT) - \ln(2\pi BT) + 2\pi BT \cdot Si(2\pi BT)\Big).
\end{aligned} \tag{39}$$

If the signal-to-noise ratio is very high, we can approximate the squared coefficient of variation of the energy harvested ($SCV\{\mathcal{E}(T)\}$) by the next limit:

$$\lim_{N_R \to 0} SCV\{\mathcal{E}(T)\} = \frac{\Gamma(a+2)\Gamma(m+2)}{\Gamma(a)\Gamma(m)\Omega_p^2}\left(\frac{b}{m}\right)^2 - 1. \tag{40}$$

Recall that $\Omega_p = a \cdot b$. Equation (40) is nothing else but the squared coefficient of variation of $\|h(t)\|^2$.

6. Validation and Performance Assessment

In this section, the analysis performed in this paper is validated via simulation, and then the evolution of $E\{\mathcal{E}(T)\}$ and $SCV\{\mathcal{E}(T)\}$ in terms of multiple input variables is studied. With no loss of generality, an exposition time of 1 minute is assumed. To highlight the possibilities of the RF energy harvesting technology, a long-range scenario is considered, which is based on a real case: the KING-TV tower located at Seattle (Washington, DC, USA). This telecommunications tower transmits several analog and digital TV channels in the VHF and UHF bands, respectively. In particular, it uses a source power of 960 kW to broadcast a 6-MHz digital TV signal at the frequency of 0.677 GHz. The evaluation that follows assumes that all transmit power is concentrated on this frequency ($f_p = 0.677$ GHz), though the receiver bandwidth is kept to 6 MHz ($B = 6$ MHz). Other fixed parameters are the reference distance ($d_0 = 1$ m), the energy conversion efficiency ($\eta = 0.5$), the ambient temperature (290 K), and the receiver noise figure (9 dB).

For the simulation and performance assessment, four input variables were taken into consideration: the transmission distance, the path-loss exponent, the shadowing spread, and the Nakagami parameter. Accordingly, Table 2 shows the analytical and simulation results for both the expected and squared coefficient of variation of the energy harvested. For each set of values, at least 30,000 runs were executed in order to achieve relative errors within 10% at a 90% confidence level. The relative errors between the analytical and simulation results have also been added to the table. As can be seen, there is a high agreement between the two sets of results.

Table 2. Comparison between analytical and simulation results for different parameter sets. The relative error (in absolute value) incurred by the simulation results has also been included.

$d(m), \beta, \sigma_{\Psi dB}, m$	$E\{\mathcal{E}(T)\}$ Analytical $SCV\{\mathcal{E}(T)\}$ Analytical	$E\{\mathcal{E}(T)\}$ Simulation $SCV\{\mathcal{E}(T)\}$ Simulation
10,000, 3.0, 8.5, 2.0	24.3135	25.1643 (3.5%)
	68.1367	68.2532 (0.2%)
10,000, 2.0, 5.5, 0.3	79,855.3	77,096.3 (3.5%)
	20.5454	21.0171 (2.3%)
20,000, 2.0, 6.5, 4.0	27,440.9	26,991.4 (1.6%)
	10.7423	10.0688 (6.3%)
20,000, 3.0, 8.5, 5.0	3.03919	2.9721 (2.2%)
	54.3092	54.1187 (0.4%)
30,000, 3.5, 6.5, 7.0	0.00235284	0.00233108 (0.9%)
	9.6885	9.19882 (5.1%)
30,000, 2.5, 4.5, 9.0	39.2981	39.7297 (1.1%)
	2.2511	2.24226 (0.4%)
40,000, 4.0, 8.5, 6.0	0.0000152195	0.0000151148 (0.7%)
	20.5507	19.1669 (6.7%)
40,000, 2.0, 6.5, 0.5	6860.24	6720.51 (2.0%)
	27.1815	24.6008 (9.5%)
50,000, 2.5, 4.5, 3.0	10.9585	11.0875 (1.2%)
	2.90132	2.7808 (4.2%)
50,000, 3.5, 10.5, 8.0	0.00238773	0.00235364 (1.4%)
	385.967	361.542 (6.3%)
60,000, 3.0, 8.0, 5.0	0.0904563	0.0897801 (0.7%)
	34.7094	33.9685 (2.1%)
60,000, 2.0, 4.0, 10.0	1520.35	1529.88 (0.6%)
	1.56925	1.551 (1.2%)
70,000, 2.5, 6.0, 1.0	7.17395	7.23829 (0.9%)
	12.4884	12.2202 (2.1%)
70,000, 2.5, 6.0, 0.1	7.17395	6.77048 (5.6%)
	73.1861	77.2072 (5.5%)
80,000, 4.0, 10.0, 3.0	6.96074×10^{-6}	6.95204×10^{-6} (0.1%)
	8.44362	8.39472 (0.6%)
80,000, 3.0, 7.0, 7.0	0.0256447	0.0253276 (1.2%)
	14.3489	13.7613 (4.1%)
90,000, 2.5, 10.5, 6.0	27.3988	27.1551 (0.9%)
	402.224	407.723 (1.4%)
90,000, 3.5, 8.0, 1.0	0.0000950559	0.0000937424 (1.4%)
	51.6893	50.6754 (2.0%)
100,000, 2.0, 7.5, 4.0	1590.89	1523.16 (4.3%)
	23.6669	23.0866 (2.5%)
100,000, 3.0, 4.5, 0.7	0.00613169	0.0061943 (1.0%)
	6.0946	6.17654 (1.3%)

To assess the performance of the energy harvesting process, several input–output relations were explored, the results of which are reflected in subsequent figures. For instance, Figure 6 plots the evolution of the average energy harvested in terms of distance for different values of the path-loss exponent. As expected, the average energy harvested increases as the distance and the path-loss exponent decrease. A similar plot is shown in Figure 7, but parameterized by the shadowing spread instead of the path-loss exponent. The figure reveals that the average energy harvested increases with the shadowing spread. The interpretation is less intuitive, but we can think of shadowing as a low-frequency "noise" superimposed on the signal, the power of which is directly proportional to its variability (as occurs with thermal noise). Figures 8–10 describe the behavior of the squared coefficient of variation. In particular, Figure 8 shows the dependence of this coefficient on distance, for different path-loss exponents. We can observe that the squared coefficient of variation decreases as the distance and/or the path-loss exponent increase, that is, as the expected energy harvested decreases. Such a reduction of variability with the decrease of the average is typical of non-negative random variables, like the energy harvested considered here. Figure 8 does not allow us to distinguish between the curves obtained for the lowest path-loss exponents. However, these differences can be better highlighted by exchanging the roles of distance and path-loss exponent in the representation. This is shown in Figure 9, which confirms that beyond $\beta \cong 3.5$ the decay profiles begin to distinguish. Finally, Figure 10 shows how the squared coefficient of variation varies with the shadowing spread and the Nakagami parameter. As can be seen, the influence of the shadowing spread is much higher than that of the Nakagami parameter. The figure also highlights the fact that the squared coefficient of variation of the energy harvested can vary within a very large range, consistent with the relatively shorter variability of the shadowing spread.

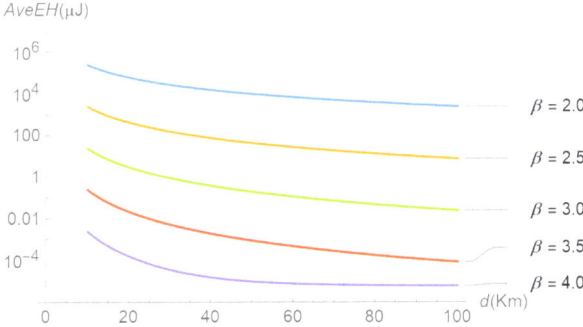

Figure 6. Evolution of the average energy harvested as a function of distance, for different path-loss exponents and a shadowing spread of 8.5 dB.

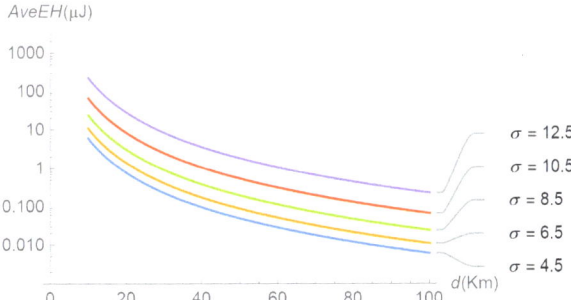

Figure 7. Evolution of the average energy harvested as a function of distance, for different levels of shadowing spread and a path-loss exponent equal to 3.0.

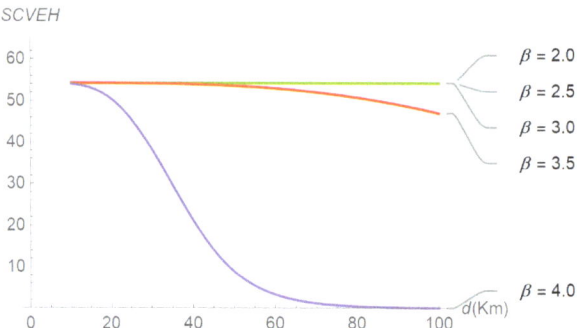

Figure 8. Evolution of the squared coefficient of variation in terms of distance, for different path-loss exponents. The shadowing spread and the Nakagami parameter have been set to 8.5 and 5.0, respectively.

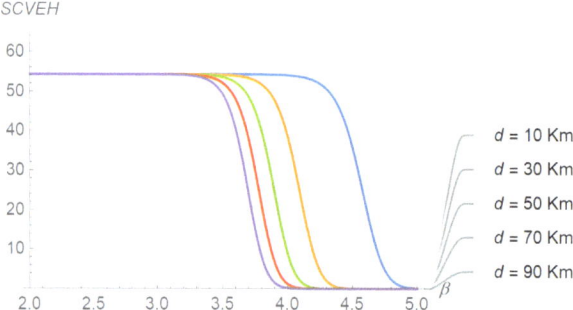

Figure 9. Evolution of the squared coefficient of variation in terms of the path-loss exponent, for different distances. The shadowing spread and the Nakagami parameter have been set to 8.5 and 5.0, respectively. The outermost curve corresponds to $d = 10{,}000$ m.

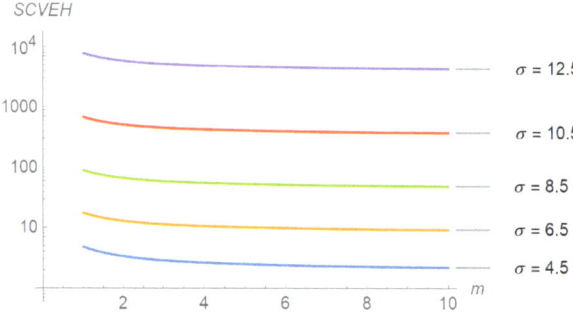

Figure 10. Evolution of the squared coefficient of variation as a function of the Nakagami parameter, for different values of the shadowing spread. The transmission distance and the path-loss exponent have been set to 50,000 m. and 3.0, respectively.

7. Discussion

In this paper, exact closed-form expressions for the mean and variance (and squared coefficient of variation) of the energy harvested by a static device have been obtained and validated via simulation. To model the propagation scenario, the generalized-K model has been adopted, as it encompasses the effects of path-loss, shadowing and multi-path fading for a wide set of wireless scenarios. It has been assumed that the device is illuminated for an arbitrary exposure time by a dedicated source emitting an unmodulated carrier. A long-range scenario has also been assumed, in order to highlight the capabilities of the RF energy-harvesting technology.

The results obtained in this paper reveal that the path-loss and shadowing components of the propagation model have a much greater influence on the amount of energy harvested than the multi-path component. They also reveal that the squared coefficient of variation of the harvested energy can be very large. This is in agreement with the high variability that the signal level can experience in a generalized-K propagation environment. Consequently, the RF-EH device must be designed to operate under wide dynamic ranges at its input, which in turn means a very low sensitivity threshold and high saturation point.

Future work can go in several directions. The following list includes some suggestions.

- As stated in Section 1, the results obtained in this paper can be used to determine relevant metrics in the context of future WENs. Examples of these metrics are the probability of energy outage and the mean waiting time experienced by energy requests.
- Also, this work can be extended to the case of mobile devices and to the evaluation of higher-order moments of the harvested energy.
- The RF energy harvester considered in this paper is ideal. However, a more realistic model of such device should include undesirable phenomena, such as limited sensitivity, saturation effects and non-linearities. Therefore, another extension could consist of evaluating the impact of these perturbations on the energy-harvesting process.
- Finally, to reinforce the results derived from this work, an experimental validation is required.

Funding: This research received no external funding.

Institutional Review Board Statement: Not applicable.

Informed Consent Statement: Not applicable.

Data Availability Statement: Not applicable.

Acknowledgments: The author would like to thank the anonymous reviewers of this journal for their helpful comments.

Conflicts of Interest: The author declares no conflict of interest.

Abbreviations

The following abbreviations are used in this manuscript:

5G/6G	Fifth Generation/Sixth Generation
AWGN	Additive White Gaussian Noise
CSI	Channel State Information
EH	Energy Harvesting
IoT	Internet of Things
LOS	Line of Sight
NLOS	Non Line of Sight
RF-EH	Radio Frequency-based Energy Harvesting
SWIPT	Simultaneous Wireless Information and Power Transfer
UAV	Unmanned Autonomous Vehicle
WEN	Wireless Energy Network
WIPT	Wireless Information and Power Transfer
WIT	Wireless Information Transfer
WPT	Wireless Power Transfer

Appendix A

As stated in Section 5, the variance of the energy harvested is based on the expectation $E\{\|\tilde{r}(s)\|^2\|\tilde{r}(t)\|^2\}$. By recalling Equation (16), the argument of such expectation can be formulated as follows:

$$\|\tilde{r}(s)\|^2\|\tilde{r}(t)\|^2 = \Big(\|\tilde{x}(s)\|^2\|h(s)\|^2 + \tilde{n}_I^2(s) + \tilde{n}_Q^2(s)$$
$$+ 2\Big(\tilde{x}_I(s)\cdot h_{re}(s) - \tilde{x}_Q(s)\cdot h_{im}(s)\Big)\cdot \tilde{n}_I(s) + 2\Big(\tilde{x}_I(s)\cdot h_{im}(s) + \tilde{x}_Q(s)\cdot h_{re}(s)\Big)\cdot \tilde{n}_Q(s)\Big)$$
$$\cdot \Big(\|\tilde{x}(t)\|^2\|h(t)\|^2 + \tilde{n}_I^2(t) + \tilde{n}_Q^2(t)$$
$$+ 2\Big(\tilde{x}_I(t)\cdot h_{re}(t) - \tilde{x}_Q(t)\cdot h_{im}(t)\Big)\cdot \tilde{n}_I(t) + 2\Big(\tilde{x}_I(t)\cdot h_{im}(t) + \tilde{x}_Q(t)\cdot h_{re}(t)\Big)\cdot \tilde{n}_Q(t)\Big). \quad \text{(A1)}$$

To evaluate the expected value of such a long expression, we can first identify the following terms:

- $I_1 : \|\tilde{x}(s)\|^2\|h(s)\|^2$.
- $I_2 : 2\Big(\tilde{x}_I(s)\cdot h_{re}(s) - \tilde{x}_Q(s)\cdot h_{im}(s)\Big)\cdot \tilde{n}_I(s) + 2\Big(\tilde{x}_I(s)\cdot h_{im}(s) + \tilde{x}_Q(s)\cdot h_{re}(s)\Big)\cdot \tilde{n}_Q(s)$.
- $I_3 : \tilde{n}_I^2(s) + \tilde{n}_Q^2(s)$.
- $I_4 : \|\tilde{x}(t)\|^2\|h(t)\|^2$.
- $I_5 : 2\Big(\tilde{x}_I(t)\cdot h_{re}(t) - \tilde{x}_Q(t)\cdot h_{im}(t)\Big)\cdot \tilde{n}_I(t) + 2\Big(\tilde{x}_I(t)\cdot h_{im}(t) + \tilde{x}_Q(t)\cdot h_{re}(t)\Big)\cdot \tilde{n}_Q(t)$.
- $I_6 : \tilde{n}_I^2(t) + \tilde{n}_Q^2(t)$.

Note that I_1, I_2 and I_3 are formally equivalent to I_4, I_5 and I_6, respectively. According to these new variables, Equation (A1) can be simply reformulated in the following way:

$$\|\tilde{r}(s)\|^2\|\tilde{r}(t)\|^2 = (I_1 + I_2 + I_3)\cdot (I_4 + I_5 + I_6). \quad \text{(A2)}$$

The subsequent procedure consists of developing all terms generated by the product in Equation (A2), and then take their respective expectations. In fact, only six out of the nine terms lead to different results:

Product $I_1 \cdot I_4$

$$I_1 \cdot I_4 = \|\tilde{x}(s)\|^2\|h(s)\|^2\|\tilde{x}(t)\|^2\|h(t)\|^2. \quad \text{(A3)}$$

Then, the expected value can be obtained after several manipulations:

$$E\{I_1 \cdot I_4\} = E\Big\{\|\tilde{x}(s)\|^2\|h(s)\|^2\|\tilde{x}(t)\|^2\|h(t)\|^2\Big\}$$
$$= E\Big\{\|\tilde{x}(s)\|^2\|\tilde{x}(t)\|^2\Big\}E\Big\{\|h(s)\|^2\|h(t)\|^2\Big\}$$
$$= E\Big\{\|\tilde{x}(t+\tau)\|^2\|\tilde{x}(t)\|^2\Big\}E\Big\{\|h(t+\tau)\|^2\|h(t)\|^2\Big\} \quad \text{(A4)}$$
$$= \phi_{\|\tilde{x}\|^2}(t,\tau)\cdot \phi_{\|h\|^2}(\tau).$$

Note that the total expectation has been expressed as a product of expectations thanks to the independence between $\tilde{x}(t)$ and $h(t)$. Moreover, the temporal variable s has been replaced, with no loss of generality, by $t + \tau$, where τ is a time lag. The first expectation is nothing else but the auto-correlation of the squared module of $\tilde{x}$ in terms of the absolute time and the time lag τ, because no assumption has yet been made about the input signal being stationary (it depends on the type of modulation). Instead, because the channel is assumed to be stationary, the second auto-correlation depends exclusively on the time lag.

Product $I_1 \cdot I_5$

$$I_1 \cdot I_5 = 2\|\tilde{x}(s)\|^2 \|h(s)\|^2 \cdot \Big(\big(\tilde{x}_I(t) \cdot h_{re}(t) - \tilde{x}_Q(t) \cdot h_{im}(t)\big) \cdot \tilde{n}_I(t) \\ + \big(\tilde{x}_I(t) \cdot h_{im}(t) + \tilde{x}_Q(t) \cdot h_{re}(t)\big) \cdot \tilde{n}_Q(t) \Big).$$
(A5)

Because noise is independent of both input signal and channel coefficient, the expected value of the previous expression depends directly on the expected values of the in-phase and quadrature components of noise, which are identically zero: $E\{\tilde{n}_I(t)\} = E\{\tilde{n}_Q(t)\} = 0$. Accordingly,

$$E\{I_1 \cdot I_5\} = 0.$$
(A6)

Product $I_1 \cdot I_6$

$$I_1 \cdot I_6 = \|\tilde{x}(s)\|^2 \|h(s)\|^2 \big(\tilde{n}_I^2(t) + \tilde{n}_Q^2(t)\big).$$
(A7)

Again because of the mutual independence between input signal and channel coefficient, we have

$$E\{I_1 \cdot I_6\} = E\big\{\|\tilde{x}(s)\|^2\big\} E\big\{\|h(s)\|^2\big\} E\big\{\big(\tilde{n}_I^2(t) + \tilde{n}_Q^2(t)\big)\big\} \\ = E\big\{\|\tilde{x}(s)\|^2\big\} E\big\{\|h(s)\|^2\big\} \big(E\big\{\tilde{n}_I^2(t)\big\} + E\big\{\tilde{n}_Q^2(t)\big\}\big) \\ = 2N_R \cdot E\big\{\|\tilde{x}(s)\|^2\big\} E\big\{\|h(s)\|^2\big\}.$$
(A8)

Here, the first property of AWGN stated in Section 4 has been recalled.

Product $I_2 \cdot I_4$

This product is formally equivalent to $I_1 \cdot I_5$; thus, we have

$$E\{I_2 \cdot I_4\} = 0.$$
(A9)

Product $I_2 \cdot I_5$

$$I_2 \cdot I_5 = 4 \Big(\big(\tilde{x}_I(s) \cdot h_{re}(s) - \tilde{x}_Q(s) \cdot h_{im}(s)\big) \cdot \tilde{n}_I(s) \\ \cdot \big(\tilde{x}_I(t) \cdot h_{re}(t) - \tilde{x}_Q(t) \cdot h_{im}(t)\big) \cdot \tilde{n}_I(t) \\ + \big(\tilde{x}_I(s) \cdot h_{re}(s) - \tilde{x}_Q(s) \cdot h_{im}(s)\big) \cdot \tilde{n}_I(s) \\ \cdot \big(\tilde{x}_I(t) \cdot h_{im}(t) + \tilde{x}_Q(t) \cdot h_{re}(t)\big) \cdot \tilde{n}_Q(t) \\ + \big(\tilde{x}_I(s) \cdot h_{im}(s) + \tilde{x}_Q(s) \cdot h_{re}(s)\big) \cdot \tilde{n}_Q(s) \\ \cdot \big(\tilde{x}_I(t) \cdot h_{re}(t) - \tilde{x}_Q(t) \cdot h_{im}(t)\big) \cdot \tilde{n}_I(t) \\ + \big(\tilde{x}_I(s) \cdot h_{im}(s) + \tilde{x}_Q(s) \cdot h_{re}(s)\big) \cdot \tilde{n}_Q(s) \\ \cdot \big(\tilde{x}_I(t) \cdot h_{im}(t) + \tilde{x}_Q(t) \cdot h_{re}(t)\big) \cdot \tilde{n}_Q(t) \Big).$$
(A10)

This expression involves the following expectations:

$$\begin{aligned} E\{\tilde{n}_I(s) \cdot \tilde{n}_I(t)\} &= E\{\tilde{n}_I(t+\tau) \cdot \tilde{n}_I(t)\} = \phi_I^{noise}(\tau) \\ E\{\tilde{n}_I(s) \cdot \tilde{n}_Q(t)\} &= E\{\tilde{n}_I(s)\} \cdot E\{\tilde{n}_Q(t)\} = 0 \cdot 0 = 0 \\ E\{\tilde{n}_Q(s) \cdot \tilde{n}_I(t)\} &= E\{\tilde{n}_Q(s)\} \cdot E\{\tilde{n}_I(t)\} = 0 \cdot 0 = 0 \\ E\{\tilde{n}_Q(s) \cdot \tilde{n}_Q(t)\} &= E\{\tilde{n}_Q(t+\tau) \cdot \tilde{n}_Q(t)\} = \phi_Q^{noise}(\tau). \end{aligned} \quad (A11)$$

Note that the fact that noise is stationary and that its in-phase and quadrature components are independent (Section 4) has been taken into account. On the other hand, from [23] we have $\phi_I^{noise}(\tau) = \phi_Q^{noise}(\tau) = \phi_{\tilde{n}}(\tau)$. Then, based on these preliminary results, the expected value of $I_2 \cdot I_5$ can be initially written as follows:

$$\begin{aligned} E\{I_2 \cdot I_5\} = 4E\Big\{ &\Big(\tilde{x}_I(s) \cdot h_{re}(s) - \tilde{x}_Q(s) \cdot h_{im}(s)\Big) \\ &\cdot \Big(\tilde{x}_I(t) \cdot h_{re}(t) - \tilde{x}_Q(t) \cdot h_{im}(t)\Big)\Big\} \cdot \phi_{\tilde{n}}(\tau) \\ + 4E\Big\{ &\Big(\tilde{x}_I(s) \cdot h_{im}(s) + \tilde{x}_Q(s) \cdot h_{re}(s)\Big) \\ &\cdot \Big(\tilde{x}_I(t) \cdot h_{im}(t) + \tilde{x}_Q(t) \cdot h_{re}(t)\Big)\Big\} \cdot \phi_{\tilde{n}}(\tau). \end{aligned} \quad (A12)$$

Next, by performing standard calculations, the latter equation can be rewritten in this way:

$$\begin{aligned} E\{I_2 \cdot I_5\} = 4E\Big\{\Re\{\tilde{x}(s) \cdot h(s)\}\Re\{\tilde{x}(t) \cdot h(t)\}\Big\} \cdot \phi_{\tilde{n}}(\tau) \\ + 4E\Big\{\Im\{\tilde{x}(s) \cdot h(s)\}\Im\{\tilde{x}(t) \cdot h(t)\}\Big\} \cdot \phi_{\tilde{n}}(\tau). \end{aligned} \quad (A13)$$

Here, $\Im$ stands for the imaginary part. Given the linearity of the expectation operator, the latter result can be further simplified:

$$\begin{aligned} E\{I_2 \cdot I_5\} &= 4\phi_{\tilde{n}}(\tau) \\ &\cdot E\Big\{\Re\{\tilde{x}(s) \cdot h(s)\}\Re\{\tilde{x}(t) \cdot h(t)\} \\ &+ \Im\{\tilde{x}(s) \cdot h(s)\}\Im\{\tilde{x}(t) \cdot h(t)\}\Big\} \\ &= 4\phi_{\tilde{n}}(\tau) E\Big\{\Re\Big\{\tilde{x}(s) \cdot h(s)\big(\tilde{x}(t) \cdot h(t)\big)^*\Big\}\Big\}. \end{aligned} \quad (A14)$$

Recall that the asterisk denotes the complex conjugate operator. Now, we can replace s by $t + \tau$ and formulate $E\{I_2 \cdot I_5\}$ in terms of correlations

$$\begin{aligned} E\{I_2 \cdot I_5\} &= 4\phi_{\tilde{n}}(\tau) \\ &\cdot E\Big\{\Re\Big\{\tilde{x}(t+\tau) \cdot h(t+\tau)\big(\tilde{x}(t) \cdot h(t)\big)^*\Big\}\Big\} \\ &= 4\phi_{\tilde{n}}(\tau)\Re\Big\{E\{\tilde{x}(t+\tau) \cdot \tilde{x}^*(t) \cdot h(t+\tau) \cdot h(t)^*\}\Big\} \\ &= 4\phi_{\tilde{n}}(\tau)\Re\Big\{E\{\tilde{x}(t+\tau) \cdot \tilde{x}^*(t)\} \cdot E\{h(t+\tau) \cdot h(t)^*\}\Big\} \\ &= 4\phi_{\tilde{n}}(\tau)\Re\Big\{2\phi_{\tilde{x}}(t,\tau)2\phi_h(\tau)\Big\} \\ &= 16\phi_{\tilde{n}}(\tau)\Re\Big\{\phi_{\tilde{x}}(t,\tau)\phi_h(\tau)\Big\}. \end{aligned} \quad (A15)$$

Note from [25] that, in the complex field, $\phi_{U \cdot V} = \frac{1}{2} E\{U \cdot V^*\}$.

Product $I_2 \cdot I_6$

$$I_2 \cdot I_6 = 2\left(\tilde{n}_I^2(t) + \tilde{n}_Q^2(t)\right)$$
$$\cdot \left(\left(\tilde{x}_I(s) \cdot h_{re}(s) - \tilde{x}_Q(s) \cdot h_{im}(s)\right) \cdot \tilde{n}_I(s) \right. \quad \text{(A16)}$$
$$\left. + \left(\tilde{x}_I(s) \cdot h_{im}(s) + \tilde{x}_Q(s) \cdot h_{re}(s)\right) \cdot \tilde{n}_Q(s)\right).$$

Now, we can determine the corresponding expected value by making use of the independence properties already outlined:

$$E\{I_2 \cdot I_6\} = 2E\left\{\left(\tilde{x}_I(s) \cdot h_{re}(s) - \tilde{x}_Q(s) \cdot h_{im}(s)\right)\right\}$$
$$\cdot E\left\{\left(\tilde{n}_I^2(t) + \tilde{n}_Q^2(t)\right) \cdot \tilde{n}_I(s)\right\}$$
$$+ 2E\left\{\left(\tilde{x}_I(s) \cdot h_{im}(s) + \tilde{x}_Q(s) \cdot h_{re}(s)\right)\right\} \quad \text{(A17)}$$
$$\cdot E\left\{\left(\tilde{n}_I^2(t) + \tilde{n}_Q^2(t)\right) \cdot \tilde{n}_Q(s)\right\}.$$

Note that this formula relies on four expectations that only involve noise, namely $E\{\tilde{n}_I^2(t) \cdot \tilde{n}_I(s)\}$, $E\{\tilde{n}_Q^2(t) \cdot \tilde{n}_I(s)\}$, $E\{\tilde{n}_I^2(t) \cdot \tilde{n}_Q(s)\}$ and $E\{\tilde{n}_Q^2(t) \cdot \tilde{n}_Q(s)\}$. Two of them are zero as a result of the statistical independence between the in-phase and quadrature components of the equivalent low-pass noise signal:

$$E\left\{\tilde{n}_Q^2(t) \cdot \tilde{n}_I(s)\right\} = E\left\{\tilde{n}_Q^2(t)\right\} \cdot E\{\tilde{n}_I(s)\} = N_R \cdot 0 = 0; \quad \text{(A18)}$$

$$E\left\{\tilde{n}_I^2(t) \cdot \tilde{n}_Q(s)\right\} = E\left\{\tilde{n}_I^2(t)\right\} \cdot E\{\tilde{n}_Q(s)\} = N_R \cdot 0 = 0. \quad \text{(A19)}$$

On the other hand, because $\tilde{n}_I(t)$ is a zero-mean normal random process, any set of random variables obtained from it by selecting particular time instants constitute a zero-mean multivariate normal random vector. Hence, it satisfies the Isserlis' theorem [26], which states that any odd-order moment of a zero-mean multivariate normal random vector is zero. Accordingly,

$$E\left\{\tilde{n}_I^2(t) \cdot \tilde{n}_I(s)\right\} = 0. \quad \text{(A20)}$$

The same statements can be formulated about process $\tilde{n}_Q(t)$ to conclude that

$$E\left\{\tilde{n}_Q^2(t) \cdot \tilde{n}_Q(s)\right\} = 0. \quad \text{(A21)}$$

Thus, we definitely have:

$$E\{I_2 \cdot I_6\} = 0. \quad \text{(A22)}$$

Product $I_3 \cdot I_4$

As this product is formally equivalent to $I_1 \cdot I_6$, we have

$$E\{I_3 \cdot I_4\} = E\{I_1 \cdot I_6\}. \quad \text{(A23)}$$

Product $I_3 \cdot I_5$

This product is formally equivalent to $I_2 \cdot I_6$. Thus,

$$E\{I_3 \cdot I_5\} = 0. \tag{A24}$$

Product $I_3 \cdot I_6$

$$I_3 \cdot I_6 = \left(\tilde{n}_I^2(s) + \tilde{n}_Q^2(s)\right) \cdot \left(\tilde{n}_I^2(t) + \tilde{n}_Q^2(t)\right). \tag{A25}$$

Its expectation is as follows:

$$\begin{aligned}
E\{I_3 \cdot I_6\} &= E\left\{\left(\tilde{n}_I^2(s) + \tilde{n}_Q^2(s)\right) \cdot \left(\tilde{n}_I^2(t) + \tilde{n}_Q^2(t)\right)\right\} \\
&= E\left\{\tilde{n}_I^2(s) \cdot \tilde{n}_I^2(t)\right\} + E\left\{\tilde{n}_I^2(s) \cdot \tilde{n}_Q^2(t)\right\} \\
&\quad + E\left\{\tilde{n}_Q^2(s) \cdot \tilde{n}_I^2(t)\right\} + E\left\{\tilde{n}_Q^2(s) \cdot \tilde{n}_Q^2(t)\right\}.
\end{aligned} \tag{A26}$$

This expression contains two types of expectations—those that involve both the in-phase and quadrature components of noise, and those that only involve one of these components. Regarding the latter, we can substitute s by $t + \tau$ and formulate them as auto-correlations, whereas for the former we can make use of the properties highlighted in Section 4:

$$\begin{aligned}
E\{I_3 \cdot I_6\} &= E\left\{\tilde{n}_I^2(t+\tau) \cdot \tilde{n}_I^2(t)\right\} \\
&\quad + E\left\{\tilde{n}_I^2(s)\right\} \cdot E\left\{\tilde{n}_Q^2(t)\right\} \\
&\quad + E\left\{\tilde{n}_Q^2(s)\right\} \cdot E\left\{\tilde{n}_I^2(t)\right\} \\
&\quad + E\left\{\tilde{n}_Q^2(t+\tau) \cdot \tilde{n}_Q^2(t)\right\} \\
&= \phi_{\tilde{n}_I^2}(\tau) + N_R \cdot N_R + N_R \cdot N_R + \phi_{\tilde{n}_Q^2}(\tau) \\
&= 2N_R^2 + \phi_{\tilde{n}_I^2}(\tau) + \phi_{\tilde{n}_Q^2}(\tau).
\end{aligned} \tag{A27}$$

Next, by making use of the Isserlis' theorem for the even-order moments [26], we have

$$\begin{aligned}
\phi_{\tilde{n}_I^2}(\tau) &= E\left\{\tilde{n}_I^2(t+\tau) \cdot \tilde{n}_I^2(t)\right\} \\
&= E\{\tilde{n}_I(t+\tau) \cdot \tilde{n}_I(t+\tau) \cdot \tilde{n}_I(t) \cdot \tilde{n}_I(t)\} \\
&= E\{\tilde{n}_I(t+\tau) \cdot \tilde{n}_I(t+\tau)\} \cdot E\{\tilde{n}_I(t) \cdot \tilde{n}_I(t)\} \\
&\quad + E\{\tilde{n}_I(t+\tau) \cdot \tilde{n}_I(t)\} \cdot E\{\tilde{n}_I(t+\tau) \cdot \tilde{n}_I(t)\} \\
&\quad + E\{\tilde{n}_I(t+\tau) \cdot \tilde{n}_I(t)\} \cdot E\{\tilde{n}_I(t+\tau) \cdot \tilde{n}_I(t)\} \\
&= N_R \cdot N_R + 2\left(\phi_I^{noise}(\tau)\right)^2 = N_R^2 + 2\phi_{\tilde{n}}^2(\tau).
\end{aligned} \tag{A28}$$

Similarly:

$$\phi_{\tilde{n}_Q^2}(\tau) = N_R^2 + 2\phi_{\tilde{n}}^2(\tau). \tag{A29}$$

Next, by introducing Equations (A28) and (A29) into Equation (A27), and making some additional manipulations, we can obtain the following result:

$$E\{I_3 \cdot I_6\} = 4N_R^2 + 4\phi_{\tilde{n}}^2(\tau). \tag{A30}$$

Finally, from Equations (A4), (A6), (A8), (A9), (A15), (A22), (A23), (A24) and (A30), we can derive the following expression for $E\{\|\tilde{r}(s)\|^2\|\tilde{r}(t)\|^2\}$, with $s = t + \tau$:

$$E\{\|\tilde{r}(t+\tau)\|^2\|\tilde{r}(t)\|^2\} = \phi_{\|\tilde{x}\|^2}(t,\tau)\phi_{\|h\|^2}(\tau)$$
$$+ 4N_R \cdot E\{\|\tilde{x}(t)\|^2\}E\{\|h(t)\|^2\}$$
$$+ 16\phi_{\tilde{n}}(\tau) \cdot \Re\{\phi_{\tilde{x}}(t,\tau)\phi_h(\tau)\}$$
$$+ 4N_R^2 + 4\phi_{\tilde{n}}^2(\tau).$$

(A31)

References

1. Lu, X.; Wang, P.; Niyato, D.; Kim, D.I.; Han, Z. Wireless networks with RF energy harvesting: A contemporary survey. *IEEE Commun. Surv. Tutor.* **2015**, *17*, 757–789. [CrossRef]
2. Tran, L.; Cha, H.; Park, W. RF power harvesting: A review on designing methodologies and applications. *Micro Nano Syst. Lett.* **2017**, *5*, 14. [CrossRef]
3. Ponnimbaduge Perera, T.D.; Jayakody, D.N.K.; Sharma, S.K.; Chatzinotas, S.; Li, J. Simultaneous wireless information and power transfer (SWIPT): Recent advances and future challenges. *IEEE Commun. Surv. Tutor.* **2018**, *20*, 264–302. [CrossRef]
4. Srinivasu, G.; Sharma, V.K.; Anveshkumar, N. A survey on conceptualization of RF energy harvesting. *J. Appl. Sci. Comput.* **2019**, *VI*, 791–800.
5. Ibrahim, H.H.; Singh, M.J.; Al-Bawri, S.S.; Ibrahim, S.K.; Islam, M.T.; Alzamil, A.; Islam, M.S. Radio frequency energy harvesting technologies: A comprehensive review on designing, methodologies, and potential applications. *Sensors* **2022**, *22*, 4144. [CrossRef] [PubMed]
6. Sherazi, H.H.R.; Zorbas, D.; O'Flynn, B. A comprehensive survey on RF energy harvesting: Applications and performance determinants. *Sensors* **2022**, *22*, 2990. [CrossRef] [PubMed]
7. Jakayodi, D.N.K.; Thompson, J.; Chatzinotas, S.; Durrani, S. *Wireless Information and Power Transfer: A New Paradigm for Green Communications*; Springer: New York, NY, USA, 2017.
8. Ng, D.W.K.; Duong, T.Q.; Zhong, C.; Schober, R. *Wireless Information and Power Transfer: Theory and Practice*; Wiley-IEEE Press: Hoboken, NJ, USA, 2018.
9. Hadzi-Velkov, Z.; Pejoski, S.; Schober, R.; Zlatanov, N. Wireless powered ALOHA networks with UAV-mounted-base stations. *IEEE Wirel. Commun. Lett.* **2020**, *9*, 56–60. [CrossRef]
10. Kumar, S.; De, S.; Mishra, D. RF energy transfer channel models for sustainable IoT. *IEEE Internet Things J.* **2018**, *5*, 2817–2828. [CrossRef]
11. Luo, B.; Yeoh, P.L.; Schober, R.; Krongold, B.S. Optimal energy beamforming for distributed wireless power transfer over frequency-selective channels. In Proceedings of the IEEE International Conference on Communications, Shanghai, China, 20–24 May 2019.
12. Castro, I.T.; Landesa, L.; Serna, A. Modeling the energy harvested by an RF energy harvesting system using gamma processes. *Math. Probl. Eng.* **2019**, *2019*, 8763580. [CrossRef]
13. Rinne, J.; Keskinen, J.; Berger, P.R.; Lup, D.; Valkama, M. M2M communication assessment in energy-harvesting and wake-up radio assisted scenarios using practical components. *Sensors* **2018**, *18*, 3992. [CrossRef] [PubMed]
14. Mouapi, A.; Hakem, N.; Kandil, N. Radiofrequency energy harvesting for wireless sensor node: Design guidelines and current circuits performance. In *IOT Applications Computing*; Singh, I., Gao, Z., Massarelli, C., Eds.; IntechOpen: London, UK, 2022.
15. Altinel, D.; Kurt, G.K. Statistical models for battery recharging time in RF energy harvesting systems. In Proceedings of the IEEE Wireless Communications and Networking Conference, Istanbul, Turkey, 6–9 April 2014.
16. AbdelGhany, O.M.; Sobih, A.G.; El-Tager, A.M. Outdoor RF spectral study available from cell-phone towers in sub-urban areas for ambient RF energy harvesting. In Proceedings of the 18th International Conference on Aerospace Sciences & Aviation Technology, Cairo, Egypt, 9–11 April 2019.
17. Oliveira, D.; Oliveira, R. Characterization of energy availability in RF energy harvesting networks. *Math. Probl. Eng.* **2016**, *2016*, 1–9. [CrossRef]
18. Salahat, E.; Yang, N. Statistical models for battery recharge time from RF energy scavengers in generalized wireless fading channels. In Proceedings of the IEEE Globecom Workshops, Singapore, 4–8 December 2017.
19. Lladó, J.; Galmés, S. *Wireless Energy Networks—How Cooperation Extends to Energy*; Lecture Notes in Computer Science; Springer: Berlin/Heidelberg, Germany, 2022; Volume 13492.
20. Goldsmith, A. *Wireless Communications*; Cambridge University Press: Cambridge, UK, 2005.
21. Bithas, P.S.; Sagias, N.C.; Mathiopoulos, P.T.; Rontogiannis, A.A. On the performance analysis of digital communications over generalized-K fading channels. *IEEE Commun. Lett.* **2006**, *5*, 353–355. [CrossRef]
22. Shankar, P.M. Outage probabilities in shadowed fading channels using a compound statistical model. *IEE Proc. Commun.* **2005**, *152*, 828–832. [CrossRef]
23. Proakis, J.; Manolakis, D. *Digital Signal Processing*, 4th ed.; Pearson: London, UK, 2013.

24. Stüber, G.L. *Principles of Mobile Communication*, 2nd ed.; Kluwer Academic Publishers: Dordrecht, The Netherlands, 2001.
25. Papoulis, A.; Pillai, S.U. *Probability, Random Variables and Stochastic Processes*, 4th ed.; McGraw-Hill Europe: London, UK, 2002.
26. Isserlis, L. On a formula for the product-moment coefficient of any order of a normal frequency distribution in any number of variables. *Biometrika* **1918**, *12*, 134–139. [CrossRef]

Article

Performance Analysis of Wireless Communications with Nonlinear Energy Harvesting under Hardware Impairment and κ-µ Shadowed Fading

Toi Le-Thanh [1,2,3] and Khuong Ho-Van [1,2,*]

1. Ho Chi Minh City University of Technology (HCMUT), 268 Ly Thuong Kiet Street, District 10, Ho Chi Minh City 700000, Vietnam; lttoi.sdh19@hcmut.edu.vn
2. Vietnam National University Ho Chi Minh City, Linh Trung Ward, Thu Duc District, Ho Chi Minh City 700000, Vietnam
3. Ho Chi Minh City University of Food Industry, 140 Le Trong Tan Street, Tay Thanh Ward, Tan Phu District, Ho Chi Minh City 700000, Vietnam
* Correspondence: hvkhuong@hcmut.edu.vn

Abstract: This paper improves energy efficiency and communications reliability for wireless transmission under κ-µ shadowed fading (i.e., integrating all channel impairments including path loss, shadowing, fading) and hardware impairment by employing a nonlinear energy harvester and multi-antenna power transmitter. To this end, this paper provides explicit formulas for outage probability. Numerous results corroborate these formulas and expose that energy-harvesting nonlinearity, hardware impairment, and channel conditions drastically deteriorate system performance. Notwithstanding, energy-harvesting nonlinearity influences system performance more severely than hardware impairment. In addition, desired system performance is accomplished flexibly and possibly by choosing a cluster of specifications. Remarkably, the proposed communications scheme obtains the optimal performance with the appropriate selection of the time-splitting factor.

Keywords: multiple antennas; energy-harvesting nonlinearity; hardware impairment; κ-µ shadowed fading; performance analysis

1. Introduction

Next-generation wireless networks, e.g., 5G/6G, provide innumerable viable wireless applications for an enormous number of devices, yet induce critical pressures on telecommunications infrastructure, especially in current situations of energy deficiency, in supplying adequate energy for such devices [1–4]. Consequently, countermeasures ameliorating energy efficiency are becoming increasingly needed. One of the emerging countermeasures is to harvest energy available in radio frequency (RF) signals surrounding communications devices. Presently, 5G/6G devices deploy successfully cheap energy harvesting (EH) integrated circuits [5–8]. Nonetheless, a plurality of performance analyses pertinent to RF EH has featured EH to be linear for simplicity [9–12]. Practically, EH circuits are constituted by nonlinear elements, viz. capacitors, inductors, and diodes. As a result, featuring EH needs to account for the nonlinearity (NL) of circuit components. Until now, divergent nonlinear energy-harvesting (NLEH) paradigms have integrated such nonlinearity [13–19].

Wireless communications with energy harvesting (WCwEH), e.g., Figure 1, enables a source S to transmit its information to a destination D. S harvests energy from a power transmitter T to accumulate power for its operation. Here, T can be television/radio broadcasting stations with high and stable transmission power. To further increase the quantity of harvested energy, multiple antennas should be deployed at T, which is the scenario in the current paper.

In practice, hardware impairment (HWi), which may originate from an imperfect design process (viz. in-quadrature-phase imbalances and phase noises) or imperfect hardware elements (viz., amplifier nonlinearities), is present in transceivers [20–22]. HWi plays a role as an interference source and therefore, in degrading system performance significantly [23]. Accordingly, it is mandatory to analyze and evaluate it elaborately in the system configuration process before implementation. Moreover, wireless communications are challenging due to propagation conditions, which include shadowing, fading, and path loss. These conditions happen concurrently and dramatically influence its performance. For WCwEH investigated in our work, propagation conditions also affect the quantity of harvested energy, eventually affecting communications reliability. For performance analysis practically, propagation conditions need to be featured appropriately to match field measurements. The κ-μ shadowed fading paradigm is extensively avouched in characterizing properly simultaneous influences of shadowing, fading, and path loss [24,25]. Remarkably, by varying a parameter group $(\kappa, \mu, \chi, \delta)$ representing such a paradigm, divergent impairment degrees of shadowing, fading, and path loss can be set straightforwardly. The parameter δ indicates channel power including path loss, κ signifies the Rician-K element that represents the Line of Sight effect, χ stands for the shadowing effect, and μ denotes a sum of multi-path sets. As a result, this paradigm features a plurality of general-and-practical propagation conditions, including well-acknowledged fading distributions, viz., Rayleigh, one-sided Gaussian, Rice, Hoyt, Nakagami-m, etc. [25].

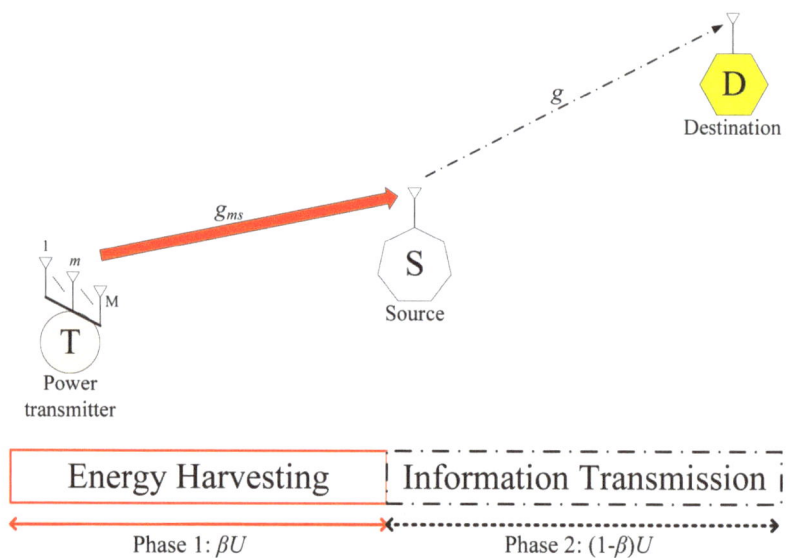

Figure 1. Wireless communications with energy harvesting.

A few works [26–29] have studied wireless communications under concurrent consideration of energy-harvesting nonlinearity (EHNL) and hardware impairment. Nonetheless, these works have revealed several limitations. To be more specific, the authors in ref. [26] considered transmission from S to D under the aid of the reconfigurable intelligent surface operated as a relay. Although the authors in [26] analyzed rate-energy tradeoff, it investigated simple Rayleigh fading channels without accounting for shadowing, and missed the throughput (TP) and the outage probability (OP) analyses. Suffering from the same limitations as [26], the authors in [27] designed beamformers to obtain power efficiency and information security for direct transmission from S to multiple destinations. Nevertheless, ref. [27] studied Rician fading for the channel from T to S. Similar to [26,27] in which no analysis of the TP and the OP was presented, the authors in ref. [28] assessed the TP of the secondary network in the context of cognitive radio over Nakagami-m fading without taking shadowing into account. Recently, the authors in ref. [29] analyzed the TP and the OP of overlay networks over κ-μ shadowed fading which considers fading distributions (Rayleigh [26,27], Nakagami-m [28], Rician [27]) as special cases and accounts for shadowing and path loss. Nevertheless, only [27] considered multiple antennas at T for high energy efficiency and at S for beam-forming implementation, while [26,28,29] investigated the single antenna at T, therefore hardly improving energy-harvesting efficiency. Although multiple antennas at T increase considerably the quantity of harvested energy and therefore enhance the system performance, the performance analysis requires new statistics of harvested energy at S and therefore complicating analytical results. Briefly, the performance analysis for wireless communications through measures of the TP and the OP under simultaneous considerations of EHNL, HWi, multiple antennas at T, flexible-and-general κ-μ fading, shadowing, and path loss has been left open in the literature. This paper will solve this open problem to swiftly evaluate and maximize the reliability of the communication before practical implementation. More specifically, our contributions are presented as follows:

- Our work proposes WCwEH in Figure 1, wherein the power transmitter T employs an arbitrary quantity of antennas for ameliorating energy-harvesting efficiency, ultimately ameliorating communications reliability. To feature properly nonlinear circuit elements in energy harvesters, our work proposes the application of the extensively acknowledged NLEH paradigm in [18].
- To assess the reliability of the communication quickly, our work proposes the TP and the OP analyses for the recommended WCwEH under the consideration of EHNL, multi-antenna power transmitter, HWi, and divergent impairment degrees of shadowing, fading, and path loss in propagation conditions.
- Our work rate maximizes the reliability of communication in diverse realistic contexts. A plurality of results illustrates that EHNL, HWi, and propagation conditions drastically deteriorate system performance. Notwithstanding, EHNL influences the system performance more severely than HWi. In addition, the desired system performance is accomplished flexibly and possibly by choosing a cluster of specifications. Remarkably, the proposed transmission scheme obtains the optimal performance with the appropriate selection of the time-splitting factor.

Our work continues with Section 2, which presents the proposed WCwEH. Then, Section 3 analyzes the TP and the OP for the proposed WCwEH. Subsequently, Section 4 outlines the analyses for four extreme scenarios to facilitate quick performance comparison and highlights the impacts of EHNL, HWi, and multi-antenna deployment. Next, Section 5 provides simulated/analytical results in diverse practical settings. Finally, the paper is concluded in Section 6. Frequently used notations are tabulated in Table 1.

Table 1. Frequently used notations.

Notation	Interpretation
$\mathcal{CN}(0,c)$	complex Gaussian random variable with mean 0 and variance c
$\Gamma(\cdot,\cdot)$	Incomplete upper Gamma function
$C_m^n = \frac{n!}{m!(n-m)!}$	Binomial coefficient
$\bar{F}_N(\cdot)$	complementary cumulative distribution function (CCDF) of N
$\Psi_N(\cdot)$	Moment Generating Function (MGF) of N
$F_N(\cdot)$	cumulative distribution function (CDF) of N
$\mathbb{P}\{\cdot\}$	Probability operator
$\mathbb{E}\{\cdot\}$	Expectation operator
$\Gamma(\cdot)$	Complete Gamma function
$K_i(\cdot)$	Modified Bessel function [30]
$f_N(\cdot)$	probability density function (PDF) of N

2. Wireless Communications with Energy Harvesting

2.1. System Model

Figure 1 demonstrates the basic system model of WCwEH with three devices (T, S, and D). Such a WCwEH may represent direct (uplink/downlink) transmission in mobile communications networks. S is supposed to be power-constrained and therefore it needs to harvest energy from T. T plays a role as a dedicated power transmitter, e.g., television and radio broadcasting stations. In the proposed WCwEH, T supplies energy for S's operations in a time fraction β of a communications frame U, namely Phase 1, and S transmits its information transmission to D in the rest of U, namely Phase 2. To increase the quantity of harvested energy, eventually ameliorating communications reliability, T is assumed to be equipped with M antennas, which is feasible for T to be a high-transmission-power source. Indeed, S can harvest much more energy when T transfers energy through its higher number of antennas. Notwithstanding, since S and D may be mobile devices, the deployment of a single antenna on them is a better assumption. Furthermore, for realistic consideration, all devices are supposed to suffer from HWi.

2.2. Channel Model

We denote g as the channel power gain between S and D, and g_{ms} as the channel power gain between S and the mth transmit antenna of T. We assume slow flat κ-μ shadowed fading channels. More specifically, a parameter set $(\mu, \kappa, \chi, \delta_{tr})$ specifies totally g_{tr} with $g_{tr} \in \{g_{ms}, g\}$ where (κ, μ, χ) were discussed in Section 1 and $\delta_{tr} = \mathbb{E}\{g_{tr}\}$ with $\delta_{tr} \in \{\delta_{ms}, \delta\}$ notates the corresponding channel power. In line with [24], the PDF and the CDF of g_{tr} are respectively expressed to be

$$f_{g_{tr}}(w) = \sum_{n=0}^{N} \frac{Q_n}{Y_{tr}^{\chi_n} \Gamma(\chi_n)} w^{\chi_n - 1} e^{-\frac{w}{Y_{tr}}}, \quad (1)$$

and

$$F_{g_{tr}}(w) = 1 - \sum_{n=0}^{N} \sum_{i=0}^{\chi_n - 1} \frac{Q_n}{Y_{tr}^i i!} w^i e^{-\frac{w}{Y_{tr}}}, \quad (2)$$

where $Y_{tr} = \frac{\delta_{tr}(\chi + \kappa \mu)}{\mu \chi (\kappa + 1)}$, $Q_n = \left(\frac{\kappa \mu}{\chi + \kappa \mu}\right)^{N-n} \left(\frac{\chi}{\chi + \kappa \mu}\right)^n C_n^N$, $\chi_n = \chi - n$, $N = \chi - \mu$ with $\mu \leq \chi$. For compactness, $\mu \leq \chi$ and the same parameter set (κ, μ, χ) for all channels are supposed. In the context that $\mu > \chi$, we analyze similarly to $\mu \leq \chi$ by employing the corresponding symbols in [24] [Table I]. Furthermore, μ and χ are supposed to be integers, which have a slight impact in practicality as comprehended in [24]. Since shadowing and fading are already integrated into the κ-μ shadowed fading paradigm, we only need to embed path loss into this paradigm for wireless channels to be featured by simultaneous influences of fading, shadowing, and path loss. To this end, we model δ_{tr} as $\tau d_{tr}^{-\alpha}$ with α being path loss exponent, d_{tr} being the corresponding transmitter-to-receiver distance, and τ being the fading power at the reference distance of 1 m (m) [13].

Wireless channels between S and the antennas of T are supposed to be independent and identically distributed (i.i.d.). Accordingly, we write the subscript ms pertinent to channel parameters $(g_{ms}, Y_{ms}, \delta_{ms})$ shortly as p in $(g_{ms}, Y_{ms}, \delta_{ms})$ if not causing any confusion, namely $\varpi_{ms} = \varpi_p, \forall m, \varpi = \{g, Y, \delta\}$.

2.3. Signal Model

In Phase 1, T supplies energy for S over the multiple-input single-output channel, dramatically boosting the amount of harvested energy at S. Consequently, S harvests energy as $E = \eta \beta U \bar{P} \sum_{m=1}^{M} g_{ms}$ wherein $0 < \eta < 1$ notates the energy conversion efficiency, $\bar{P}$ is the transmit power of each antenna of T, and $g_{ms} = |h_{ms}|^2$ with h_{ms} being the channel gain between S and the mth transmit antenna of T. Since the duration of Phase 2 is $(1 - \beta)U$, the power for transmission in Phase 2 converted from E is $\frac{E}{(1-\beta)U}$. Conforming to the NLEH paradigm in [18], S transmits information in Phase 2 with the power as

$$P = \begin{cases} \frac{\eta\beta\bar{P}}{1-\beta}\sum_{m=1}^{M}g_{ms} & , \beta\bar{P}\sum_{m=1}^{M}g_{ms} \leq \iota \\ \frac{\eta\beta\iota}{1-\beta} & , \beta\bar{P}\sum_{m=1}^{M}g_{ms} > \iota \end{cases} \quad (3)$$

$$= \begin{cases} GH & , H \leq R \\ J & , H > R \end{cases}$$

where ι is the power saturation threshold, $G = \frac{\eta\beta\bar{P}}{1-\beta}$, $J = \frac{\eta\beta\iota}{1-\beta}$, $R = \frac{\iota}{\beta\bar{P}}$, and $H = \sum_{m=1}^{M}g_{ms}$.

It should be noted that Equation (3) reflects the characteristic of the NLEH. Indeed, the output power of the NLEH is GH, which is proportional linearly to its input power when the input power is below ι; otherwise, the output power of the NLEH is saturated at ι. Furthermore, we note that as ι is high ($\iota \to \infty$), the NLEH reduces to the linear energy harvesting (LEH).

In Phase 2, S transmits its information x with transmit power P to D where $\mathbb{E}\{|x|^2\} = 1$. By accounting for HWi, D receives the signal to be [23]

$$y = h\left(\sqrt{P}x + v\right) + \varsigma, \quad (4)$$

wherein $\varsigma \sim \mathcal{CN}(0,\sigma)$ is additive noise at D, h is the channel gain, $v \sim \mathcal{CN}(0,\rho P)$ is HWi at S and D where ρ is the total HWi at S and D.

The signal-to-interference plus noise ratio (SINR) for D to restore x from Equation (4) is

$$\Lambda = \frac{gP}{\rho gP + \sigma}. \quad (5)$$

where $g = |h|^2$ is the channel power gain.

It is drawn from Equation (5) that HWi plays a role as an interference source generating the quantity of interference as ρgP. This interference causes performance degradation in comparison to hardware perfection.

3. Performance Analysis of WCwEH

The OP of WCwEH is first analyzed in this part. The OP refers to the probability that D decodes unsuccessfully x, i.e., the achieved channel capacity is below the target transmission rate R_0. Subsequently, the OP analysis is extended to achieve the TP analysis. These proposed analyses facilitate the quick OP/TP evaluation without time-consuming simulations.

3.1. Exact Analysis

The communications reliability is represented by the outage probability at D. Therefore, the lower the OP at D, the higher the reliability of the communication. The OP at D is expressed to be

$$\mathcal{O} = \mathbb{P}\{\Lambda < \Lambda_0\}$$
$$= \mathbb{P}\left\{\frac{gP}{\rho gP + \sigma} < \Lambda_0\right\} \quad (6)$$
$$= \begin{cases} \tilde{\mathcal{O}} & , 1 > \Lambda_0\rho \\ 1 & , 1 \leq \Lambda_0\rho \end{cases}$$

where (Since the duration of Phase 2 is $(1-\beta)U$, the channel capacity corresponding to the SINR Λ in Phase 2 is $(1-\beta)\log_2(1+\Lambda)$. The outage event happens if this capacity is below R_0 or the SINR Λ is lower than Λ_0.) $w = gP$, $\Lambda_0 = 2^{R_0/(1-\beta)} - 1$, and

$$\tilde{\mathcal{O}} = \mathbb{P}\left\{w < \frac{\Lambda_0 \sigma}{1 - \Lambda_0 \rho}\right\} \\ = F_w\left(\frac{\Lambda_0 \sigma}{1 - \Lambda_0 \rho}\right). \qquad (7)$$

It is drawn from Equation (6) that since $\Lambda_0 = 2^{R_0/(1-\beta)} - 1$, choosing the target transmission rate R_0, the time-splitting factor β, and the HWi ρ may induce $\Lambda_0 \rho \geq 1$, causing $\mathcal{O}$ to be 1 or leading WCwEH to be in a complete outage. However, this complete outage event can be prevented by selecting properly $\{R_0, \beta, \rho\}$ such that $\Lambda_0 \rho < 1$. This insight can be drawn from the condition $\Lambda_0 \rho < 1$ to avoid the complete outage event as follows. It is seen that $\Lambda_0 \rho < 1$ is equivalent to $R_0 < (1-\beta)\log_2(1+\rho^{-1})$, which means that for WCwEH to prevent the complete outage, the target transmission rate must be upper-bounded properly. The upper bound on R_0 depends on the parameter of energy-harvesting β and the HWi ρ. The higher the β (or ρ), the lower target transmission rate the WCwEH achieves. The higher β means more time for energy harvesting while less time for signal transmission. Similarly, the higher ρ means the HWi is more severe. Furthermore, $\mathcal{O}$ depends on the parameter set $(R_0, \beta, \bar{P}, M, \iota, \eta, \rho)$, which means that S can achieve the desired performance by properly setting this set.

To complete the OP analysis, we must derive the CDF of w, $F_w(z)$, in Equation (7), which is addressed as

$$F_w(z) = \mathbb{P}\{gP \leq z\}. \qquad (8)$$

Invoking P in Equation (3), the CDF of w is further represented to be

$$F_w(z) = \mathbb{P}\left\{g \leq \frac{z}{GH}, H \leq R\right\} + \mathbb{P}\left\{g \leq \frac{z}{J}, H > R\right\} \\ = \int_0^R F_g\left(\frac{z}{Gx}\right) f_H(x) dx + \bar{F}_H(R) F_g\left(\frac{z}{J}\right). \qquad (9)$$

The integral in Equation (9) is approximated exactly by employing the Gaussian–Chebyshev quadrature in [31] as

$$F_w(z) = \frac{\pi R}{2I} \sum_{i=1}^{I} \sqrt{1-\epsilon_i^2} f_H(v_i) F_g\left(\frac{z}{v_i G}\right) + \bar{F}_H(R) F_g\left(\frac{z}{J}\right), \qquad (10)$$

where $\epsilon_i = \cos\left(\frac{2i-1}{2I}\pi\right)$, $v_i = R(\epsilon_i + 1)/2$, and I stands for the accuracy–complexity tradeoff of the Gaussian–Chebyshev quadrature. In Section 5, we demonstrate the results with $I = 200$ that guarantees very high preciseness.

To complete the derivation of Equation (10), we must derive the PDF and the CCDF of $H = \sum_{m=1}^{M} g_{ms}$. To this end, it is noted that the MGF of the sum of M i.i.d random variables is a product of M individual MGFs. Applying this note, one obtains the MGF of H to be

$$\Psi_H(v) = \prod_{m=1}^{M} \Psi_{g_{ms}}(v) \\ = \left[\Psi_{g_p}(v)\right]^M. \qquad (11)$$

In Equation (11), the first equality originated from the statistical independence of M random variables while the second equality is comprehended from their identical distributions. Furthermore, each MGF is represented as

$$\Psi_{g_p}(v) = \int_0^\infty f_{g_p}(y) e^{vy} dy$$

$$= \int_0^\infty \left(\sum_{n=0}^N \frac{Q_n}{\Upsilon_p^{\chi_n} \Gamma(\chi_n)} y^{\chi_n - 1} e^{-\frac{y}{\Upsilon_p}} \right) e^{vy} dy \quad (12)$$

$$= \sum_{n=0}^N \frac{Q_n}{\Upsilon_p^{\chi_n} \Gamma(\chi_n)} \int_0^\infty y^{\chi_n - 1} e^{-\left(\frac{1}{\Upsilon_p} - v\right) y} dy$$

$$= \sum_{n=0}^N Q_n (1 - \Upsilon_p v)^{-\chi_n}.$$

By substituting (12) into (11) and applying the multinomial expansion, one obtains the MGF of H as

$$\Psi_H(v) = \left[\sum_{n=0}^N Q_n (1 - \Upsilon_p v)^{-\chi_n} \right]^M$$

$$= \sum_{\sum_{n=0}^N b_n = M} \frac{M!}{\prod_{n=0}^N b_n!} \prod_{n=0}^N \left[Q_n (1 - \Upsilon_p v)^{-\chi_n} \right]^{b_n} \quad (13)$$

$$= \sum_{\sum_{n=0}^N b_n = M} M! \left\{ \prod_{n=0}^N \frac{(Q_n)^{b_n}}{b_n!} \right\} (1 - \Upsilon_p v)^{-\vartheta},$$

where $\vartheta = \sum_{n=0}^N b_n \chi_n$.

Using the PDF-MGF mapping in [32], one infers the PDF of H to be

$$f_H(z) = \sum_{\sum_{n=0}^N b_n = M} \left(\prod_{n=0}^N \frac{(Q_n)^{b_n}}{b_n!} \right) \frac{M! z^{\vartheta - 1}}{\Upsilon_p^\vartheta \Gamma(\vartheta)} e^{-\frac{z}{\Upsilon_p}}. \quad (14)$$

Subsequently, the CCDF of H is expressed from its definition as $\bar{F}_H(z) = \mathbb{P}\{H \geq z\} = \int_z^\infty f_H(r) dr$. Plugging (14) into $\bar{F}_H(z)$ yields

$$\bar{F}_H(z) = \sum_{\sum_{n=0}^N b_n = M} \frac{M!}{\Upsilon_p^\vartheta \Gamma(\vartheta)} \left(\prod_{n=0}^N \frac{(Q_n)^{b_n}}{b_n!} \right) \int_z^\infty r^{\vartheta - 1} e^{-\frac{r}{\Upsilon_p}} dr. \quad (15)$$

With the aid of ([30], Equation (3.351.2)), one reduces (15) to

$$\bar{F}_H(z) = \sum_{\sum_{n=0}^N b_n = M} \frac{M!}{\Gamma(\vartheta)} \Gamma\left(\vartheta, \frac{z}{\Upsilon_p}\right) \prod_{n=0}^N \frac{(Q_n)^{b_n}}{b_n!}. \quad (16)$$

3.2. Asymptotic Analysis

This subsection analyzes the upper bound in the performance of WCwEH in the range of high transmit power (i.e., $\bar{P} \to \infty$). It should be noted that the energy scavenger becomes saturated completely when $\bar{P} \to \infty$. In other words, $\bar{P} \to J$ when $\bar{P} \to \infty$. Therefore, the CDF of $w = gJ$ reduces to $F_w(z) = F_g\left(\frac{z}{J}\right)$ as $\bar{P} \to \infty$. As a result, the OP of WCwEH is derived as

$$\mathcal{O}^\infty = \begin{cases} F_g\left(\frac{\Lambda_0 \sigma}{[1-\Lambda_0 \rho]J}\right) &, 1 \geq \Lambda_0 \rho \\ 1 &, 1 < \Lambda_0 \rho \end{cases} \quad (17)$$

3.3. Throughput

For WCwEH with delay-limited transmission, the throughput is effortlessly inferred from the OP analysis to be

$$\mathbb{T} = (1 - \beta)R_0(1 - \mathcal{O}). \quad (18)$$

Relied on (18), one observes that the throughput of WCwEH is also jointly affected by the multi-parameter set $(R_0, \beta, \bar{P}, M, \iota, \eta, \rho)$ because this set determines completely $\mathcal{O}$. As a result, a desired throughput is attained by setting these parameters flexibly and properly based on their predetermined value ranges.

4. Extreme Scenarios

This section considers four extreme scenarios for WCwEH: (1) hardware perfection (HWp) and nonlinear energy harvesting; (2) hardware impairment (HWi) and linear energy harvesting; (3) hardware perfection and linear energy harvesting; (4) single-antenna power transmitter $M = 1$. These scenarios facilitate quick comparisons with the analytical results in Section 3 to highlight joint/individual influences of HWi/HWp, NLEH/LEH, and the number of antennas at T. It is worth noting that the performance analyses for these extreme scenarios have not yet been reported, and constitute future contributions of our work.

4.1. Hardware Perfection and Nonlinear Energy Harvesting (HWpNLEH)

Let $\mathcal{O}_1$ denote the OP in the scenario of HWpNLEH (viz., $\rho = 0$ and $\iota < \infty$). By setting $\rho = 0$ in the expressions in Section 3, one obtains $\mathcal{O}_1$ as

$$\mathcal{O}_1 = F_w(\Lambda_0 \sigma). \quad (19)$$

The result in (19) reveals that the multi-parameter set $(R_0, \beta, \bar{P}, M, \iota, \eta)$ completely determines $\mathcal{O}_1$. Moreover, (19) helps study the individual influence of HWi by comparing it with the analytical result from Section 3.

4.2. Hardware Impairment and Linear Energy Harvesting (HWiLEH)

The outage performance of WCwEH in the scenario of HWiLEH (viz., $\rho > 0$ and $\iota \to \infty$) is analyzed in the following for prompt comparison with its nonlinearity counterpart in Section 3, eventually highlighting the individual effect of EHNL on the reliability of the communication. It is noted that the scenario of HWiLEH means $P \to GH$ and therefore, the CDF of w becomes

$$\tilde{F}_w(v) = \Pr\left\{g \leq \frac{v}{GH}\right\}$$
$$= \int_0^\infty F_g\left(\frac{v}{Gz}\right) f_H(z) dz. \quad (20)$$

Plugging $F_g(\cdot)$ in (2) and $f_H(\cdot)$ in (14) into (20) after elaborate manipulations, one simplifies $\tilde{F}_w(v)$ as

$$\tilde{F}_w(v) = 1 - \sum_{n=0}^{N} \sum_{i=0}^{\chi_n-1} \sum_{\substack{\sum_{n=0}^{N} b_n = M}} \frac{Q_n}{i!} \left(\frac{v}{GY}\right)^i \left(\prod_{n=0}^{N} \frac{(Q_n)^{b_n}}{b_n!}\right) \frac{M!}{Y_p^\vartheta \Gamma(\vartheta)} \int_0^\infty z^{\vartheta-i-1} e^{-\frac{v}{YGz} - \frac{z}{Y_p}} dz. \qquad (21)$$

With the aid of [30] [Equation (3.471.9)], the integral in (21) can be solved, and therefore

$$\tilde{F}_w(v) = 1 - \sum_{n=0}^{N} \sum_{i=0}^{\chi_n-1} \sum_{\substack{\sum_{n=0}^{N} b_n = M}} \left(\prod_{n=0}^{N} \frac{(Q_n)^{b_n}}{b_n!}\right) \frac{2Q_n M!}{i!\Gamma(\vartheta)} \left(\frac{v}{YGY_p}\right)^{\frac{\vartheta+i}{2}} K_{\vartheta-i}\left(2\sqrt{\frac{v}{YGY_p}}\right). \qquad (22)$$

Following the procedure in Section 3.1, one derives the outage performance of WCwEH in the scenario of HWiLEH as

$$\mathcal{O}_2 = \begin{cases} \tilde{F}_w\left(\frac{\Lambda_0 \sigma}{1-\Lambda_0 \rho}\right) & , 1 > \Lambda_0 \rho \\ 1 & , 1 \leq \Lambda_0 \rho \end{cases} \qquad (23)$$

The result in (23) is convenient in demonstrating the individual influence of EHNL on the reliability of the communications by comparing it with (6).

4.3. Hardware Perfection and Linear Energy Harvesting (HWpLEH)

Integrating the results in Sections 4.1 and 4.2, one obtains the outage performance of WCwEH in the scenario of HWpLEH as

$$\mathcal{O}_3 = \tilde{F}_w(\Lambda_0 \sigma). \qquad (24)$$

The result in (24) is convenient in exposing the joint influences of EHNL and HWi on the system performance by comparing it with (6).

4.4. Hardware Impairment and Nonlinear Energy Harvesting with $M = 1$ (HWiNLEHw1)

In the scenario of HWiNLEHw1, H becomes g_p and therefore, the CDF of w reduces to

$$\hat{F}_w(v) = \frac{\pi R}{2I} \sum_{i=1}^{I} \sqrt{1-\epsilon_i^2} F_g\left(\frac{v}{\nu_i G}\right) f_{g_p}(\nu_i) + F_g\left(\frac{v}{J}\right) \bar{F}_{g_p}(R). \qquad (25)$$

Consequently, the outage performance of WCwEH in the scenario of HWiNLEHw1 as

$$\mathcal{O}_4 = \begin{cases} \hat{F}_w\left(\frac{\Lambda_0 \sigma}{1-\Lambda_0 \rho}\right) & , 1 > \Lambda_0 \rho \\ 1 & , 1 \leq \Lambda_0 \rho \end{cases} \qquad (26)$$

It should be noted that even though [29] studied the performance of overlay networks over $\kappa - \mu$ shadowed fading under concurrent effects of EHNL and HWi, it did not present the performance analysis for WCwEH with $M = 1$. Therefore, the result in (26) is still novel and convenient in unveiling the individual impact of the number of antennas at T on the system performance by comparing it directly with (6).

5. Illustrative Results

This part presents a plurality of analytical/simulated results to evaluate the OP of WCwEH in numerous parameters. In the following, analytical results (Ana.) are produced by computing the analytical formulas derived in Sections 3 and 4. Moreover, simulated results (Sim.) are generated by Monte Carlo simulations for comparison between analytical and simulated results to corroborate the analytical formulas. Due to linear mapping between the OP and the TP as mentioned in (18), the TP is computed directly from the OP. Consequently, this part focuses solely on the OP. For illustration, devices are located arbitrarily in a two-dimension plane and parameters are selected as in Table 2 unless otherwise addressed. The following figures indicate that (1) the analysis matches exactly the simulation, verifying the preciseness of the derived formulas in Sections 3 and 4; (2) increasing M improves considerably the communications reliability, which is due to the increase in harvested energy, as expected.

Table 2. Selected parameters unless otherwise addressed.

Parameter	Value
Location of T	$(0,0)$ m
Location of S	$(5,5)$ m
Location of D	$(50,0)$ m
Noise power	$\sigma = -90$ dBm
Path loss exponent	$\alpha = 3$
Fading power at the reference distance of 1 m	$\tau = 10^{-2}$
Shadowed fading parameters (κ, χ, μ)	$(3,4,2)$
HWi	$\rho = 0.1$
Power saturation threshold	$\iota = 0$ dBm
Energy conversion efficiency	$\eta = 0.6$
Time-splitting factor	$\beta = 0.4$
Transmit power of each antenna of T	$\bar{P} = 7$ dBW
Target transmission rate	$R_0 = 0.6$ bps/Hz
Quantity of antennas at T	$M = 6$

Figure 2 illustrates the OP against $\bar{P}$, which reveals the considerable reliability improvement (viz., smaller OP) with increasing $\bar{P}$ for five scenarios: (1) hardware impairment and nonlinear energy harvesting (HWiNLEH) with $M = 1$; (2) hardware impairment and nonlinear energy harvesting (HWiNLEH) with $M = 6$; (3) hardware impairment and linear energy harvesting (HWiLEH) with $M = 6$; (4) hardware perfection and nonlinear energy harvesting (HWpNLEH) with $M = 6$; (5) hardware perfection and linear energy harvesting (HWpLEH) with $M = 6$. This is reasonable owing to increasing harvested energy. Additionally, HWi slightly degrades the outage performance (i.e., the OP in the scenario of HWiNLEH/HWiLEH is slightly higher than that in the scenario of HWpNLEH/HWpLEH). Moreover, the effect of EHNL is neglected at low $\bar{P}$ (i.e., the OP in the scenario of HWiNLEH/HWpNLEH is identical to that in the scenario of HWiLEH/HWpLEH for $\bar{P} < 9.5$ dBW). This is because the nonlinear energy harvesting becomes the linear energy harvesting at low $\bar{P}$. However, at high $\bar{P}$ where the nonlinear energy harvester is saturated, EHNL significantly mitigates the outage performance (i.e., the OP in the scenario of HWiNLEH/HWpNLEH is greatly higher than that in the scenario of HWiLEH/HWpLEH for $\bar{P} > 9.5$ dBW) and the performance gap between LEH and NLEH continues to enlarge with increasing $\bar{P}$, showing the detrimental impact of EHNL. Therefore, EHNL impacts the performance of the WCwEH more significantly than HWi.

Figure 3 unveils the influence of the HWi on the outage performance. As per 3GPP LTE [33], the standard range of ρ is from 0.08 to 0.175 where the physical meaning of ρ is error vector magnitude. One notices from Figure 3 that the reliability of the communication is slightly deteriorated by HWi with accreting ρ, which is reasonable because of the increase in the interference caused by HWi. This figure again emphasizes that EHNL impacts the performance of WCwEH more significantly than HWi.

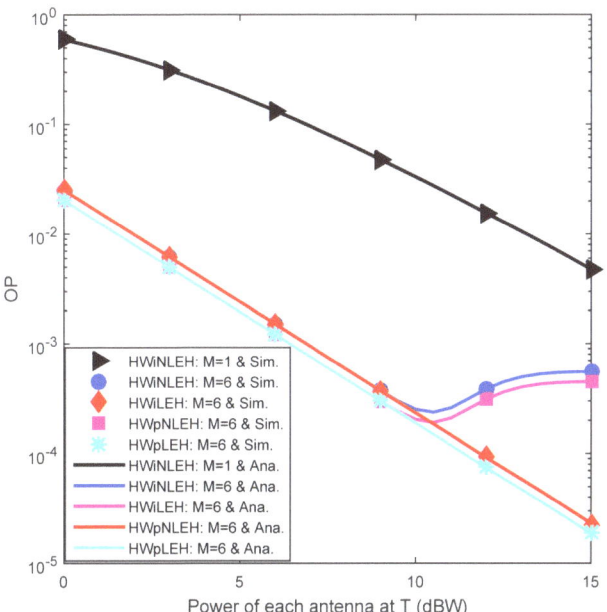

Figure 2. OP versus $\bar{P}$.

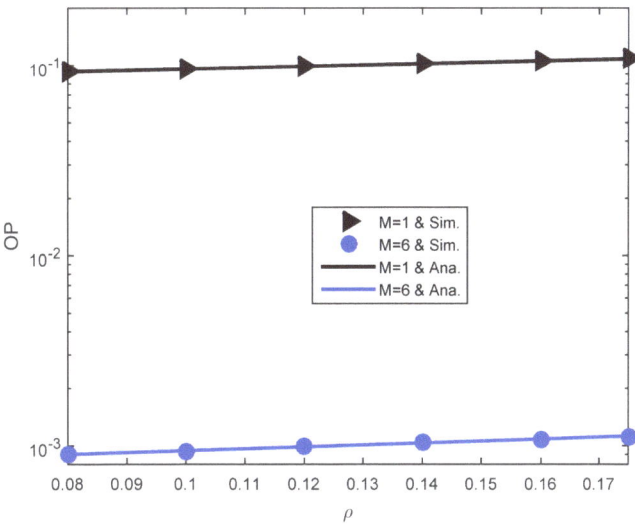

Figure 3. OP versus the HWi degree ρ.

Figure 4 demonstrates the OP against the energy conversion efficiency η (Figure 4a) and the power saturation threshold ι (Figure 4b) at the energy harvester of S. It is expected that increasing η and ι makes S harvest more energy, therefore decreasing the OP. Figure 4 demonstrates accurately this expectation wherein the reliability of the communication is dramatically ameliorated with increasing η and ι. Additionally, S incurs outage performance saturation at large ι at which the NLEH matches the LEH, as expected. Furthermore, S experiences a complete outage for small ι, as predicted.

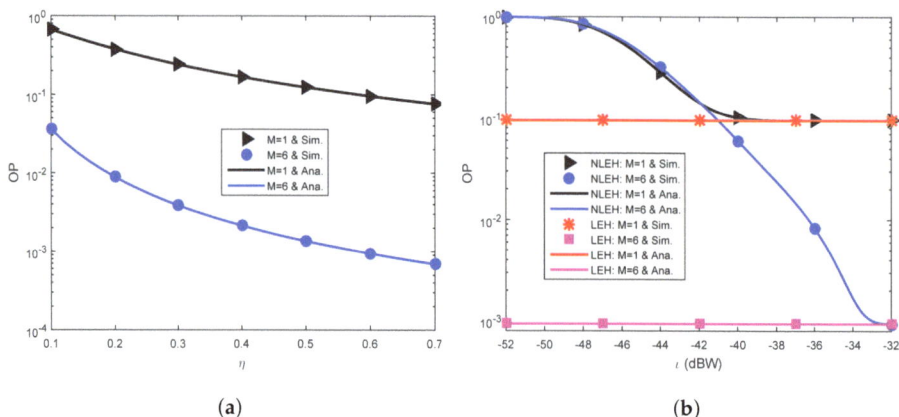

Figure 4. Parameters pertinent to harvested energy. (**a**) OP versus η. (**b**) OP versus ι.

Figure 5 exposes the OP against the time-splitting factor β (Figure 5a) and the target transmission rate R_0 (Figure 5b). Figure 5a shows that β can be optimized to attain optimal communications reliability. The optimal value of β is comprehended to poise the duration for energy harvesting (Phase 1) and the duration for transmission (Phase 2). Moreover, Figure 5a shows that S incurs a complete outage for $\beta \geq 0.827$, which is reasonable as analyzed in Section 3 where the OP is 1 when $\Lambda_0 \rho \geq 1$. Given $R_0 = 0.6$ bps/Hz, $\rho = 0.1$ and $\Lambda_0 = 2^{R_0/(1-\beta)} - 1$, it is obvious that $\Lambda_0 \rho \geq 1$ is equivalent to $\beta \geq 0.827$. Additionally, Figure 5b demonstrates the considerable performance degradation with increasing R_0, as predicted. Furthermore, S also incurs a complete outage for $R_0 \geq 2.076$ bps/Hz, which coincides with the analysis in Section 3 where the OP is 1 when $\Lambda_0 \rho \geq 1$. Given $\beta = 0.4$, $\rho = 0.1$ and $\Lambda_0 = 2^{R_0/(1-\beta)} - 1$, it is obvious that $\Lambda_0 \rho \geq 1$ is equivalent to $R_0 \geq 2.076$ bps/Hz.

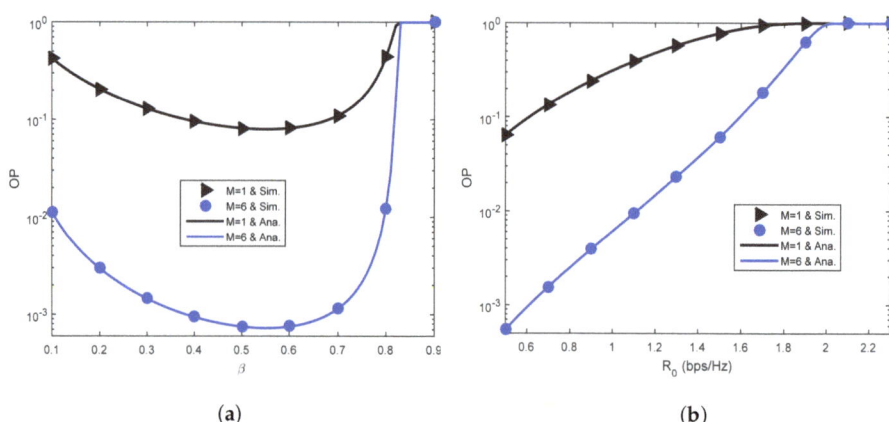

Figure 5. Effects of parameters (β, R_0). (**a**) OP versus β. (**b**) OP versus R_0.

Figure 6 demonstrates the impact of shadowed fading parameters (μ, χ, κ) on the outage performance. It should be noted that this paper considers integer values of (χ, μ) and the scenario of $\chi \geq \mu$. It is observed from Figure 6 that the reliability of the communication is considerably ameliorated with increasing (κ, χ, μ), as anticipated.

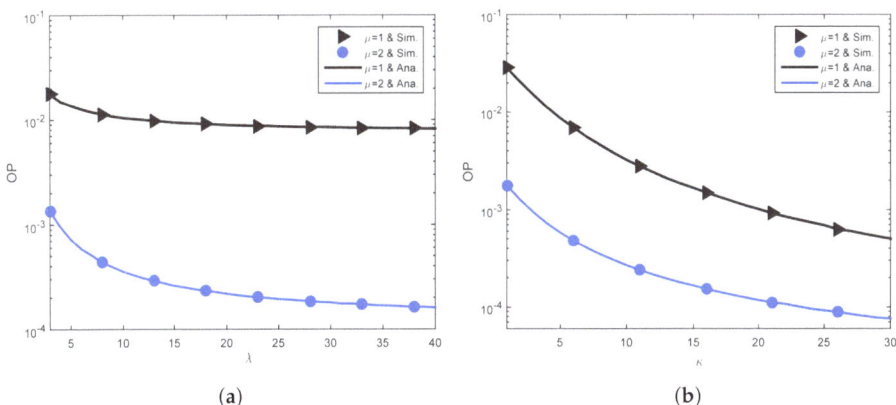

Figure 6. Shadowed fading parameters (μ, χ, κ). (**a**) OP versus χ. (**b**) OP versus κ.

6. Conclusions

Our work presented the outage performance analysis for WCwEH under practical conditions including hardware impairment, multi-antenna configuration, path loss, fading, shadowing, and nonlinear energy harvesting. Various results show that these conditions considerably affect the outage performance. Nonetheless, energy-harvesting nonlinearity influences the reliability of communication more severely than hardware impairment. Moreover, the system performance can be maximized with the appropriate selection of the time-splitting coefficient. Furthermore, the proper selection of the target transmission rate, the hardware impairment level, and the time-splitting factor can prevent the system from a complete outage.

Author Contributions: Conceptualization, K.H.-V.; methodology, T.L.-T.; software, K.H.-V. and T.L.-T.; validation, K.H.-V. and T.L.-T.; formal analysis, K.H.-V. and T.L.-T.; investigation, K.H.-V. and T.L.-T.; resources, K.H.-V. and T.L.-T.; data curation, K.H.-V.; writing—original draft preparation, K.H.-V. and T.L.-T.; writing—review and editing, K.H.-V. and T.L.-T.; supervision, K.H.-V. All authors have read and agreed to the published version of the manuscript.

Funding: This research is funded by Vietnam National University HoChiMinh City (VNU-HCM) under Grant Number B2023-20-08.

Institutional Review Board Statement: Not applicable.

Informed Consent Statement: Not applicable.

Data Availability Statement: Data is contained within the article.

Conflicts of Interest: The authors declare no conflict of interest.

References

1. Mori, S.; Mizutani, K.; Harada, H. Software-Defined Radio-Based 5G Physical Layer Experimental Platform for Highly Mobile Environments. *IEEE Open J. Veh. Technol.* **2023**, *4*, 230–240. [CrossRef]
2. Tarrias, A.; Fortes, S.; Barco, R. Failure Management in 5G RAN: Challenges and Open Research Lines. *IEEE Netw.* **2023**, 1–7. [CrossRef]
3. Kim, W.; Ahn, Y.; Kim, J.; Shim, B. Towards Deep Learning-aided Wireless Channel Estimation and Channel State Information Feedback for 6G. *J. Commun. Netw.* **2023**, *25*, 61–75. [CrossRef]
4. Pei, J.; Li, S.; Yu, Z.; Ho, L.; Liu, W.; Wang, L. Federated Learning Encounters 6G Wireless Communication in the Scenario of Internet of Things. *IEEE Commun. Stand. Mag.* **2023**, *7*, 94–100. [CrossRef]
5. Halimi, M.A.; Khan, T.; Nasimuddin; Kishk, A.A.; Antar, Y.M.M. Rectifier Circuits for RF Energy Harvesting and Wireless Power Transfer Applications: A Comprehensive Review Based on Operating Conditions. *IEEE Microwave Mag.* **2023**, *24*, 46–61. [CrossRef]

6. Derbal, M.C.; Nedil, M. High-Gain Circularly Polarized Antenna Array for Full Incident Angle Coverage in RF Energy Harvesting. *IEEE Access* **2023**, *11*, 28199–28207. [CrossRef]
7. Eltresy, N.A.; Elhamid, A.E.M.A.; Elsheakh, D.M.; Elhennawy, H.M.; Abdallah, E.A. Silver Sandwiched ITO Based Transparent Antenna Array for RF Energy Harvesting in 5G Mid-Range of Frequencies. *IEEE Access* **2021** *9*, 49476–49486. [CrossRef]
8. Fakharian, M.M. RF Energy Harvesting using High Impedance Asymmetric Antenna Array without Impedance Matching Network. *Radio Sci.* **2021**, *56*, 1–10. [CrossRef]
9. Wang, D.; Zhou, F.; Leung, V.C.M. Primary Privacy Preserving With Joint Wireless Power and Information Transfer for Cognitive Radio Networks. *IEEE Trans. Cogn. Commun. Netw.* **2020**, *6*, 683–693. [CrossRef]
10. Ge, L.; Chen, G.; Zhang, Y.; Tang, J.; Wang, J.; Chambers, J.A. Performance Analysis for Multihop Cognitive Radio Networks With Energy Harvesting by Using Stochastic Geometry. *IEEE Internet Things J.* **2020**, *7*, 1154–1163. [CrossRef]
11. Bouabdellah, M.; Bouanani, F.E.; Sofotasios, P.C.; Muhaidat, S.; Costa, D.B.D.; Mezher, K.; Ben-Azza, H.; Karagiannidis, G.K. Cooperative Energy Harvesting Cognitive Radio Networks With Spectrum Sharing and Security Constraints. *IEEE Access* **2019**, *7*, 173329–173343. [CrossRef]
12. Pham-Thi-Dan, N.; Ho-Van, K.; Do-Dac, T.; Vo-Que, S.; Pham-Ngoc, S. Security Analysis for Cognitive Radio Network with Energy Scavenging Capable Relay over Nakagami-m Fading Channels. In Proceedings of the International Symposium on Electrical and Electronics Engineering (ISEE), Ho Chi Minh City, Vietnam, 10–12 October 2019; pp. 68–72.
13. Wang, D.; Rezaei, F.; Tellambura, C. Performance Analysis and Resource Allocations for a WPCN With a New Nonlinear Energy Harvester Model. *IEEE Open J. Commun. Soc.* **2020**, *1*, 1403–1424. [CrossRef]
14. Ni, L.; Da, X.; Hu, H.; Yuan, Y.; Zhu, Z.; Pan, Y. Outage-Constrained Secrecy Energy Efficiency Optimization for CRNs With Non-Linear Energy Harvesting. *IEEE Access* **2019**, *7*, 175213–175221. [CrossRef]
15. Babaei, M.; Aygölü, Ü.; Başaran, M.; Durak-Ata, L. BER Performance of Full-Duplex Cognitive Radio Network With Nonlinear Energy Harvesting. *IEEE Trans. Green Commun. Netw.* **2020**, *4*, 448–460. [CrossRef]
16. Wang, F.; Zhang, X. Secure Resource Allocation for Polarization-Based Non-Linear Energy Harvesting Over 5G Cooperative CRNs. *IEEE Wirel. Commun. Lett.* **2023**. [CrossRef]
17. Zhu, Z.; Wang, N.; Hao, W.; Wang, Z.; Lee, I. Robust Beamforming Designs in Secure MIMO SWIPT IoT Networks With a Nonlinear Channel Model. *IEEE Internet Things J.* **2021**, *8*, 1702–1715. [CrossRef]
18. Solanki, S.; Upadhyay, P.K.; Costa, D.B.D.; Ding, H.; Moualeu, J.M. Performance Analysis of Piece-Wise Linear Model of Energy Harvesting-Based Multiuser Overlay Spectrum Sharing Networks. *IEEE Open J. Comput. Soc.* **2020**, *1*, 1820–1836. [CrossRef]
19. Wang, D.; Men, S. Secure Energy Efficiency for NOMA Based Cognitive Radio Networks With Nonlinear Energy Harvesting. *IEEE Access* **2018**, *6*, 62707–62716. [CrossRef]
20. Aubry, A.; Carotenuto, V.; Maio, A.D.; Farina, A.; Izzo, A.; Moriello, R.S.L. Assessing Power Amplifier Impairments and Digital Pre-Distortion on Radar Waveforms for Spectral Coexistence. *IEEE Trans. Aerosp. Electron. Syst.* **2022**, *58*, 635–650. [CrossRef]
21. Balti, E.; Johnson, B.K. On The Joint Effects of HPA Nonlinearities and IQ Imbalance On Mixed RF/FSO Cooperative Systems. *IEEE Trans. Commun.* **2021**, *69*, 7879–7894. [CrossRef]
22. Papazafeiropoulos, A.; Bjornson, E.; Kourtessis, P.; Chatzinotas, S.; Senior, J.M. Scalable Cell-Free Massive MIMO Systems: Impact of Hardware Impairments. *IEEE Trans. Veh. Technol.* **2021**, *70*, 9701–9715. [CrossRef]
23. Li, X.; Li, J.; Li, L. Performance Analysis of Impaired SWIPT NOMA Relaying Networks Over Imperfect Weibull Channels. *IEEE Syst. J.* **2020**, *14*, 669–672. [CrossRef]
24. Lopez-Martinez, F.J.; Paris, J.F.; Romero-Jerez, J.M. The κ-μ Shadowed Fading Model With Integer Fading Parameters. *IEEE Trans. Veh. Technol.* **2017**, *66*, 7653–7662. [CrossRef]
25. Singh, I.; Singh, N.P. Outage Probability and Ergodic Channel Capacity of Underlay Device-to-Device Communications over κ-μ Shadowed Fading Channels. *Wirel. Netw.* **2020**, *26*, 573–582. [CrossRef]
26. Khalid, W.; Yu, H.; Cho, J.; Kaleem, Z.; Ahmad, S. Rate-Energy Tradeoff Analysis in RIS-SWIPT Systems With Hardware Impairments and Phase-Based Amplitude Response. *IEEE Access* **2022**, *10*, 31821–31835. [CrossRef]
27. Boshkovska, E.; Ng, D.W.K.; Dai, L.; Schober, R. Power-Efficient and Secure WPCNs With Hardware Impairments and Non-Linear EH Circuit. *IEEE Trans. Commun.* **2018**, *66*, 2642–2657. [CrossRef]
28. Shome, A.; Dutta, A.K.; Chakrabarti, S. Throughput Assessment of Non-Linear Energy Harvesting Secondary IoT Network With Hardware Impairments and Randomly Located Licensed Users in Nakagami-m Fading. *IEEE Trans. Veh. Technol.* **2021**, *70*, 7283–7288. [CrossRef]
29. Le-Thanh, T.; Ho-Van, K. Effect of Hardware Imperfections and Energy Scavenging Non-Linearity on Overlay Networks in κ-μ Shadowed Fading. *Arab. J. Sci. Eng.* **2022**, *47*, 14601–14616. [CrossRef]
30. Gradshteyn, I.S.; Ryzhik, I.M. *Table of Integrals, Series and Products*, 6th ed.; Academic: San Diego, CA, USA, 2000.
31. Abramowitz, M.; Stegun, I.A. *Handbook of Mathematical Functions with Formulas, Graphs, and Mathematical Tables*, Tenth Printing ed.; Government Printing Office: Washington, DC, USA, 1972.

32. Simon, M.K.; Alouini, M.S. *Digital Communication over Fading Channels*, 2nd ed.; John Wiley & Sons, Inc.: Hoboken, NJ, USA, 2005.
33. Sesia, S.; Toufik, I.; Baker, M. *LTE—The UMTS Long Term Evolution: From Theory to Practice*, 2nd ed.; Wiley: New York, NY, USA, 2011.

Disclaimer/Publisher's Note: The statements, opinions and data contained in all publications are solely those of the individual author(s) and contributor(s) and not of MDPI and/or the editor(s). MDPI and/or the editor(s) disclaim responsibility for any injury to people or property resulting from any ideas, methods, instructions or products referred to in the content.

Article

Performance Analysis of Wirelessly Powered Cognitive Radio Network with Statistical CSI and Random Mobility

Nadica Kozić [1], Vesna Blagojević [1,*], Aleksandra Cvetković [2] and Predrag Ivaniš [1]

[1] School of Electrical Engineering, University of Belgrade, 11000 Belgrade, Serbia; kn155029p@student.etf.bg.ac.rs (N.K.); predrag.ivanis@etf.rs (P.I.)
[2] Faculty of Mechanical Engineering, University of Nis, 18000 Nis, Serbia; aleksandra.cvetkovic@masfak.ni.ac.rs
* Correspondence: vesna.golubovic@etf.rs

Abstract: The relentless expansion of communications services and applications in 5G networks and their further projected growth bring the challenge of necessary spectrum scarcity, a challenge which might be overcome using the concept of cognitive radio. Furthermore, an extremely high number of low-power devices are introduced by the concept of the Internet of Things (IoT), which also requires efficient energy usage and practically applicable device powering. Motivated by these facts, in this paper, we analyze a wirelessly powered underlay cognitive system based on a realistic case in which statistical channel state information (CSI) is available. In the system considered, the primary and the cognitive networks share the same spectrum band under the constraint of an interference threshold and a maximal tolerable outage permitted by the primary user. To adopt the system model in realistic IoT application scenarios in which network nodes are mobile, we consider the randomly moving cognitive user receiver. For the analyzed system, we derive the closed-form expressions for the outage probability, the outage capacity, and the ergodic capacity. The obtained analytical results are corroborated by an independent simulation method.

Keywords: cognitive radio; ergodic capacity; Internet of Things; outage probability; power beacon; random waypoint mobility model; spectrum sharing; statistical channel state information; underlay; wireless power transfer

Citation: Kozić, N.; Blagojević, V.; Cvetković, A.; Ivaniš, P. Performance Analysis of Wirelessly Powered Cognitive Radio Network with Statistical CSI and Random Mobility. *Sensors* **2023**, *23*, 4518. https://doi.org/10.3390/s23094518

Academic Editors: Onel Luis Alcaraz Lopez and Katsuya Suto

Received: 15 April 2023
Revised: 30 April 2023
Accepted: 1 May 2023
Published: 6 May 2023

Copyright: © 2023 by the authors. Licensee MDPI, Basel, Switzerland. This article is an open access article distributed under the terms and conditions of the Creative Commons Attribution (CC BY) license (https://creativecommons.org/licenses/by/4.0/).

1. Introduction

It is well known that the number of devices and services in the field of wireless communications has been constantly growing, so the introduction of new concepts capable of meeting the set of demands with the resources available has become a necessity [1]. It is expected that 5G and beyond 5G technologies will provide a higher bandwidth, better coverage, reliable connections, energy savings, and an extremely low latency for various classes of users [2]. They also represent an essential component in the development of IoT (Internet of Things) systems, which assume the connection of a high number of devices, sensors, objects, and applications to the Internet. These applications will be able to collect huge amounts of data from various devices and sensors and can be encountered in a wide range of applications, from various commercial, health, and security applications to industrial applications, known as the Industrial Internet of Things [2]. The role of 5G and beyond 5G networks will be to provide high-speed Internet connections for the collection, transmission, control, and processing of data. Due to increasing challenges concerning the provision of 5G services and their applications in various environments, even the extension of conventional telecommunication systems to non-terrestrial ones has been proposed, including satellites and aerial systems [3]. As the integration of satellite–aerial–terrestrial networks enables improved support for diverse IoT applications, the spectrum and energy efficiencies of these systems have also recently attracted considerable attention [4–6].

As the concept of the IoT brings an ever-increasing number of low-power devices, it is obvious that the task of massive data transfer creates the need for additional spectrum resources, more efficient spectrum access, and innovative approaches [7]. According to the current allocation policy, the majority of the spectrum is already allocated, and the usage of novel techniques for efficient spectrum usage is obligatory [8]. Cognitive radio (CR) represents an innovative concept with the potential to solve the problem of spectrum scarcity [9] and has already been recognized in the literature as an important, promising approach for IoT applications [10]. Generally, two types of users can be distinguished in the concept of cognitive radio: primary and secondary users. While the primary users (PUs) are the licensed users of the spectrum, low utilization of the spectrum is prevented by allowing secondary users (SUs) to simultaneously use the resources under predefined conditions [11]. Depending on the spectrum-sharing policy used between two user classes, the following concepts can be distinguished: overlay, interweave, and underlay [12]. In the interweave concept of cognitive radio, secondary users are allowed to transmit when the channel is idle, based on the results of spectrum sensing. In the overlay concept, the secondary user is allowed to transmit under the assumption that it also helps the PU's signal transmission and is based on a cooperation technique. The applications of interweave and overlay approaches in cognitive IoT systems were analyzed in [13,14].

In the underlay concept, the simultaneous communication of SUs with PUs is enabled, provided that harmful interference to the PUs is prevented by using the transmit power adaptation scheme [15]. If the perfect channel state information (CSI) is available, the level of interference occurring at the PU receiver can vary within certain tolerable limits. However, the realistic dynamic environment is characterized by constantly changing conditions, resulting in a practically erroneous or outdated CSI even when using concepts for which the necessary feedback in the system exists [16]. As IoT nodes are usually low-power and computationally limited, the approach based on statistical CSI can be beneficial for energy efficiency as it simplifies the extensive calculations needed for power adaptation in the conventional cognitive underlay concept. A performance analysis for cognitive systems with applied statistical CSI conditions was provided in [16], while a self-sustainable cognitive radio relaying system was analyzed in [17].

Another important issue related to IoT technology is the use of the large number of devices that the system relies on. Powering the devices is also a challenge. On one hand, even in scenarios in which the devices can be equipped with batteries, their lifespans are limited. On the other hand, due to the large number of energy-constrained wireless sensors and the potentially high-risk environment, battery replacement in the devices is not always straightforward. In order to ensure a longer lifetime of these sensors, the focus should be shifted to the use of wireless energy harvesting (EH) techniques, which represent a significant shift in green communications by enabling sensors in a network to harvest energy from various sources. A suitable energy source can be used for harvesting and accumulating energy in sensors, turning it into usable electrical energy. In this way, the lifetime of the batteries can be extended, or their usage can be replaced by a more practical approach. It is known that the EH process can be realized from various energy sources, such as solar, wind, vibration, and radio frequency (RF) signals [18].

An important advantage of using RF signals for EH arises from their capability to carry both energy and information; such an approach is used in the concept of simultaneous wireless information and power transmission (SWIPT) [19,20]. Using this approach, it is possible to enable the powering of sensors that do not have enough energy to transmit and process information. In SWIPT, there are two protocols used to transmit power information wirelessly: time switching (TS) and power splitting (PS) [21,22]. In addition to collecting energy from sources in the environment, energy can also be collected from a source intentionally dedicated to that purpose, called a power beacon (PB) [23,24]. The powering of a cooperative cognitive radio network by a dedicated power beacon was analyzed in [25], while an analysis of a multi-hop cognitive underlay network powered by a PB was presented in [26].

However, in the previously mentioned literature, cognitive radio systems using energy harvesting techniques were analyzed for scenarios in which a cognitive user demonstrated static behavior, which meant that the distances between the network nodes were represented by deterministic constants. This approach is not realistic in the generation of a novel system that is characterized by high user mobility, which significantly affects the performance of the wireless system due to the variability of the reception power [27]. Therefore, proper channel modeling relies on the predicting sensors' movement models, which encompass different approaches [28]. In the literature, the two most represented models are the random direction (RD) and random waypoint (RWP) models. The first model uses a nonuniform spatial distribution [29], while the second one applies uniform spatial distribution [30]. RWP can describe the user's movement through three different patterns, depending on whether the user is moving through 1D, 2D, or 3D systems [31,32].

1.1. Related Work

A performance analysis of a 5G cooperative network with static users and RF energy harvesting was presented in [33], while the performances of a static network whose source was powered by a PB were provided in [34]. The performance of a wirelessly powered network assisted with a dedicated power beacon using a TS protocol was analyzed in [35] for an interference-limited environment. The achievable capacity was calculated for different transmission schemes and power and rate adaptations for the case in which fading undergoes Rayleigh distribution. Another case of analyzing system performances in wirelessly powered communication network with a TS protocol was presented in [36] for the case of Rician fading distribution. The analysis of an SWIPT-based underlay cognitive radio network was presented in [37], while a scenario with multiple power beacons in a cognitive radio network was analyzed in [38].

As previously emphasized, wireless networks with static users were analyzed in most papers. This does not represent the usual behavior of users in the novel generation of wireless communication systems in which users are, in most cases, mobile, e.g., cellular users in cars, trains, and buses and pedestrians on streets. A performance analysis of a wireless system with a mobile receiver was provided for the Nakagami-m fading environment in [32], where the closed-form expressions for the system outage probability (OP) were derived for the applied RWP mobility model. The system performances of a wirelessly powered communication network were provided in [39,40] for the Nakagami-m fading environment, examining the case in which the source is powered by a designated power beacon and the mobility of the receiver node is described using the RWP model.

The impact of mobility on system performance was presented in [31] for a more complex η-μ fading distribution used to model the dynamic behavior of non-homogeneous fading, while the mobile receiver was described using the RWP mobility model. The analytical results for the ergodic channel capacity for generalized fading distribution were presented in [41], and the effects of mobility were evaluated in [42] for the k-μ generalized fading model (including Nakagami-m and Rician fading distribution as special cases); in both papers, the movement of the mobile receiver was described using the RWP model. Another mobile wireless network with Rician fading distribution was discussed in [43], and closed-form expressions for the outage probability in an interference-limited environment with a Rician fading channel were derived. In [44], the impact of mobility on a generalized α-μ fading distribution was analyzed. The authors of [28,31,41,45] analyzed the effects of mobility on the performances of communication systems by engaging the RWP model, while the influence of the RD mobility model was assessed in [46,47]. A comparison of these two models' impact on performance was described in [27,48].

Additionally, [49] expanded the analysis of a system with a mobile receiver in a fading environment to the case in which the transmission of the cognitive node is in accordance with the imposed limitation of a primary network and an available, imperfect CSI [49]. It was assumed that the propagation environment is subject to Rayleigh fading distribution as a special case of Nakagami-m fading distribution.

Our work is motivated by the fact that there is a growing number of connected IoT devices and unsolved issues concerning their supply, as well as the increasing need for efficient solutions for spectrum access. We analyze a system in which the transmitter is wirelessly powered using a PB, while efficient spectrum access and information transfer are enabled by simultaneous spectrum usage with the primary users without jeopardizing the conditions imposed by the primary network. The analysis encompasses a realistic scenario in which the receiver node is mobile, and the obtained performance represents the guidelines for the design of energy- and spectrum-constrained IoT systems with mobile receivers. In Table 1, a summary of related works is shown, demonstrating that there is no available research in the field of wirelessly powered cognitive radio networks with statistical CSI and random mobility, which we provide in this paper, filling the gap in the literature.

Table 1. Summary of the literature.

Reference	Performances	WPT	CR	CSI	Fading Model	Mobility
[24]	Outage probability/average BER/throughput	+	−	−	Nakagami−m	RWP
[26]	Outage probability	+	+	−	Rayleigh	−
[27]	Probability density of the received power	−	−	−	−	RWP/RD
[31]	Outage probability/average BER	−	−	−	Nakagami−m; Nakagami−q (Hoyt), Rayleigh, and one−sided Gaussian distribution	RWP
[32]	Outage probability/average BER	−	−	−	Nakagami−m	RWP
[33]	Rate of IoT−enabled devices	+	−	−	Rayleigh	−
[34]	Outage probability/system throughput	+	−	−	Rayleigh	−
[35]	Outage probability	+	+	Imperfect	Rayleigh	−
[36]	Outage probability	+	+	Instantaneous	Nakagami−m	−
[37]	ergodic channel capacity	−	−	−	Generalized fading	RWP
[38]	Outage probability	−	−	−	$k-\mu$	RWP
[39]	Outage probability/average BER	−	−	−	Rician fading	RWP
[40]	Outage probability/average BER	−	−	−	$\alpha-\mu$ generalized fading	RWP
[41]	average error rate/average channel capacity	−	−	−	$\kappa-\mu$	RWP
[42]	Outage probability	−	−	−	Nakagami−m	RD
[43]	Outage probability	−	−	−	Rayleigh	RD
[44]	Outage probability	−	−	−	Rayleigh	RWP/RD
[45]	Outage probability/average BER	−	+	imperfect	Rayleigh	RWP
[46]	Outage probability, average delay−limited throughput, ergodic capacity, and average delay−tolerant throughput	+	−	−	Nakagami−m	RWP
[47]	Achievable throughput	+	−	+	Nakagami−m	−
[48]	Ergodic capacity	+	−	Absence/partial	Rician	−
This paper	Outage probability/achievable throughput/ergodic throughput	+	+	Statistical	Nakagami−m	RWP

1.2. Contribution and Organization

In this paper we provide a performance analysis of an underlay cognitive radio system that is powered by a dedicated power beacon based on a time-switching protocol. The primary user constraints are defined by an interference threshold and a maximal primary user outage probability, which are satisfied under the assumption that statistical CSI is available to the secondary user of the spectrum. The contributions of this paper are as follows:

- We propose a cognitive radio system model powered by a dedicated power source in which the SU-Rx is mobile and the primary system is protected based on a tolerable interference level. We provide an analysis for the realistic scenario in which statistical CSI is available. Secondary user mobility is modeled using the widely accepted RWP model;

- We derive the analytical expressions for the probability density function of the received SNR and the outage probability of the secondary user;
- Based on the outage probability results, we calculate the outage capacity of the system, which is a relevant performance measure for applications with delay constraints;
- We derive the closed-form ergodic capacity expression relevant for applications with no delay requirements;
- The novel expressions derived in the paper are valid for the general case of the Nakagami-*m* fading environment, which encompasses Rayleigh and approximately Rician fading scenarios as special cases;
- Analytical results are confirmed using an independent simulation method. We analyze the impact of all system and channel parameters to the considered performance metrics.

In the following sections, we present a complete analysis in detail. In Section 2, the system and channel models are presented, in addition to the list of symbols shown in Table 2, which are used for further analysis. The outage performance analysis is provided in Section 3, and the ergodic capacity is analyzed in Section 4. The numerical results and discussion are provided in Section 5, and the conclusion is provided in Section 6.

Table 2. Table of symbols.

Symbol	Explanation
P_{PB}	the power of the signal from the PB
$P_{SU-Tx,max}$	the maximal transmit power of the SU-Tx permitted by the PU
$P_{SU-Tx,EH}$	the maximal transmit power of the SU-Tx enabled by EH
E_H	the energy harvested at the SU-Tx
η	the energy conversion efficiency coefficient
T	the time frame interval
α	the time-splitting factor
$P_{out,PU}$	the maximal tolerable outage probability of the primary network
k_S	the coefficient of adjustment of the SU-Tx power
Q_p	the peak interference threshold of the PU
h_{PB}	the fading envelope in the channel from the PB to the SU-Tx
h_{TR}	the fading envelope in the channel from the SU-Tx to the SU-Rx
g_S	the fading envelope in the channel from the SU-Tx to the PU-Rx
γ_1	the instantaneous channel power gain in the channel from the PB to the SU-Tx
γ_2	the instantaneous channel power gain in the link from the SU-Tx to the SU-Rx
Λ_S	the average channel power gain in the link from the SU-Tx to the PU-Rx
n_{SU-Tx}	the AWGN component at the SU-Tx
x_{PB}	the energy signal sent from the PB
x_{SU-Tx}	the information signal sent from the SU-Tx
y_{SU-Tx}	the received energy signal at the SU-Tx node
y_{SU-Rx}	the received information signal at the SU-Rx node
n_{SU-Rx}	the AWGN component at the SU-Rx
γ	the instantaneous signal-to-noise ratio at the SU-Rx
σ_R^2	the mean power value of the AWGN component at the SU-Rx
B_l, β_l	the coefficients that correspond to topologies of the SU movement in 1D, 2D, or 3D
δ	the path loss exponent
D_{PB}	the distance from the PB to the SU-Tx
r	the distance between the SU-Tx and the SU-Rx
D	the maximal distance between the SU-Tx and the SU-Rx
m_1	Nakagami-*m* fading parameter in the channel from the PB to the SU-Tx
m_2	Nakagami-*m* fading parameter in the channel from the SU-Tx to the SU-Rx
Ω_1	the average channel power gain in the channel from the PB to the SU-Tx
Ω_2	the average channel power gain in the channel from the SU-Tx to the SU-Rx
γ_{th}	the outage threshold

2. System and Channel Model

The wirelessly powered cognitive radio system is shown in Figure 1. The secondary network consists of the energy-constrained secondary transmitter (SU-Tx) and the secondary receiver (SU-Rx), which are allowed to exchange information as long as the interference produced at the primary user receiver is below a predetermined threshold. Further, we assume the SU-Tx does not have its own power supply but harvests energy from the dedicated power beacon. Energy harvesting and information transmission at the SU-Tx are based on the time-switching protocol. According to the applied protocol, all the energy harvested during a scheduled time frame is used for information transmission.

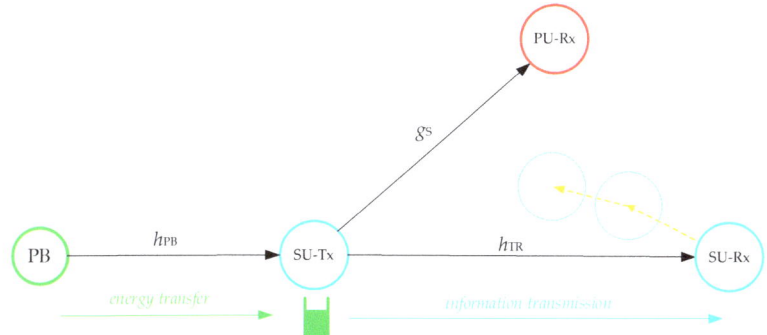

Figure 1. Model of the power-beacon-assisted underlay cognitive network with a random mobility user.

The secondary cognitive transmitter has a fixed position at a distance D_{PB} from the power beacon node. The fading envelope in the channel from the PB to the SU-Tx h_{PB} follows the Nakagami-m distribution, while the channel power gain $\gamma_1 = |h_{PB}|^2$ is distributed in accordance with Gamma distribution. Then, the probability density function (PDF) of a random variable γ_1 is provided by:

$$f_{\gamma_1}(\gamma) = \frac{1}{\Gamma(m_1)}\left(\frac{m_1}{\Omega_1}\right)^{m_1}\gamma^{m_1-1}\exp\left(-\frac{m_1}{\Omega_1}\gamma\right), \qquad (1)$$

where $\Omega_1 = E[|h_{PB}|^2]$ and m_1 denotes the Nakagami-m fading parameter. It is well known that the Nakagami-m model encompasses a wide range of propagation scenarios, including the Rayleigh and Rice models as special cases. When the fading parameter is $m_1 = 1$, Nakagami-m simplifies to the Rayleigh distribution, while in the case in which the propagation has a strong line-of-sight component [50], the parameter m_1 has greater values. The power ratio k between the line-of-sight and the scattered components and the power of the scattered waves σ^2 represent the Rician parameters, and their relation with the Nakagami-m parameters m and Ω can be expressed by the relations $m = \frac{1+k^2}{2k+1}$ and $\sigma^2 = \frac{\Omega}{2}\left(1 - \sqrt{1 - m^{-1}}\right)$.

We analyze the realistic scenario in which the secondary receiver is not at a fixed location and is moving randomly. The movement of the secondary user receiver (SU-Rx) is modeled using the widely accepted RWP mobility model. Depending on the type of movement of the receiver, this model can be used for modeling paths in three different network topologies or dimensions.

As the SU-Rx is mobile, the distance between the fixed secondary transmitter and the receiver is represented by the random variable r. It is assumed that the mobile receiver is within a maximal distance D from the transmitter, i.e., $0 \leq r \leq D$, and the probability density function (PDF) of the distance r between the SU-Tx and SU-Rx is provided by [27]:

$$f_r(r) = \sum_{l=1}^{n}\frac{B_l}{D^{\beta_l+1}}r^{\beta_l}, \qquad (2)$$

where the coefficients B_l and β_l, $l = 1, \ldots, n$, which correspond to various topologies of the SU movement in one, two, or three dimensions, are provided in Table 3 [28].

Table 3. Polynomial coefficients.

Dimension	n	B_l	β_l
1	2	[6, −6]	[1, 2]
2	3	(1/73) [324, −420, 96]	[1, 3, 5]
3	3	(1/72) [735, −1190, 455]	[2, 4, 6]

The secondary network uses the spectrum dedicated to the primary user based on the underlay paradigm, which means that secondary users can perform transmissions concurrent with the primary users of the spectrum as long as the interference generated at the primary receiver does not exceed a permissible threshold. In this paper, we assume a realistic scenario in which only statistical CSI is available to the secondary network, i.e., only the mean channel power gain values (long-term CSI) are known. As the strict interference power constraint cannot be satisfied in this scenario, the PU's outage probability constraint is applied. In this case, the peak interference threshold at the primary receiver can be exceeded, but only with a permitted maximum tolerable outage probability of the primary network. As the cognitive secondary network shares the spectrum with the primary user under the interference limit constraint, the maximal transmit power of the SU is determined by the maximal interference constraints imposed by the primary user of the spectrum.

The transmission block structure is presented in Figure 2 for the case in which energy harvesting and information transmission are performed according to the TS protocol. Similar assumptions were used in [36] and [39]. Within a time frame interval of length T, the interval αT, $\alpha \in \{0, 1\}$ represents the fraction of the block time in which the SU-Tx harvests energy from the PB. The remaining block time, equal to $(1 − \alpha)T$, is used for the transmission of information from the SU-Tx to the SU-Rx.

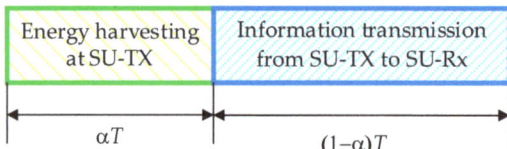

Figure 2. Transmission block structure in the TS scheme.

The received signal at the SU-Tx node is provided by:

$$y_{SU-Tx} = \sqrt{\frac{P_{PB}}{D_{PB}^{\delta}}} h_{PB} x_{PB} + n_{SU-Tx}, \quad (3)$$

where x_{PB} is the energy signal sent from the power beacon, P_{PB} is the power of the signal from the power beacon, D_{PB} is the distance from PB to SU-Tx, and n_{SU-Tx} is an additive white Gaussian noise (AWGN) component. As can be seen from the previous equation, the path-loss effect is included in the analysis, and δ represents the path loss exponent.

Then, the energy harvested at the SU-Tx node can be expressed in the following form [17]:

$$E_H = \eta \alpha T \frac{P_{PB}}{D_{PB}^{\delta}} |h_{PB}|^2, \quad (4)$$

where η ($0 < \eta < 1$) is the energy conversion efficiency coefficient that depends on the characteristics of the energy harvester.

In accordance with the harvested energy and the remaining time used for signal transmission, the maximal available transmit power for the emitting of SU-Tx signal is equal to:

$$P_{SU-Tx,EH} = \frac{E_H}{(1-\alpha)T} = \frac{\eta \alpha P_{PB}}{(1-\alpha)D_{PB}^{\delta}} |h_{PB}|^2. \quad (5)$$

However, the transmit power of the cognitive SU-Tx should not exceed the maximal allowable value:

$$P_{SU-Tx,max} = k_S \frac{Q_p}{\Lambda_S}, \quad (6)$$

which guarantees the fulfillment of the interference outage constraint. The coefficient k_S adjusts the transmit power of the secondary user in accordance with the peak interference threshold Q_p and the average value $\Lambda_S = E\left[|g_S|^2\right]$ of the channel power gain in the link from the SU-Tx to the PU-Rx $|g_S|^2$ such that the following maximal tolerable outage probability of primary network $P_{out,PU}$ is not exceeded

$$\Pr\left\{P_{SU-Tx}|g_S|^2 < Q_p\right\} = 1 - P_{out,PU}. \quad (7)$$

Taking into account both constraints, the transmit power of the secondary user is provided as the minimum of the values $P_{SU-Tx,EH}$ and $P_{SU-Tx,max}$, defined by Expressions (5) and (6), respectively. Therefore, it is defined by the following equation:

$$P_{SU-Tx} = \min(P_{SU-Tx,EH}, P_{SU-Tx,max}) = \begin{cases} \frac{\eta\alpha P_{PB}}{(1-\alpha)D_{PB}^{\delta}}|h_{PB}|^2, & \frac{\eta\alpha P_{PB}}{(1-\alpha)D_{PB}^{\delta}}|h_{PB}|^2 \leq k_S \frac{Q_p}{\Lambda_S}, \\ k_S \frac{Q_p}{\Lambda_S}, & \frac{\eta\alpha P_{PB}}{(1-\alpha)D_{PB}^{\delta}}|h_{PB}|^2 > k_S \frac{Q_p}{\Lambda_S}. \end{cases} \quad (8)$$

The received baseband signal on SU-Rx is equal [33] to

$$y_{SU-Rx} = \sqrt{\frac{P_{SU-Tx}}{r^{\delta}}} h_{TR} x_{SU-Tx} + n_{SU-Rx}, \quad (9)$$

where x_{SU-Tx} is the information signal sent from the SU-Tx, r is the distance from the SU-Tx to the SU-Rx, and n_{SU-Rx} is the AWGN component at the SU-Rx, with a mean power value σ_R^2. The channel gain coefficient from the SU-Tx to the SU-Rx is denoted by h_{TR}.

The instantaneous signal-to-noise ratio (SNR) γ at the SU-Rx is provided by the following equation:

$$\gamma = \frac{P_{SU-Tx}|h_{TR}|^2}{r^{\delta}\sigma_R^2} = \begin{cases} K_1|h_{PB}|^2 \frac{|h_{TR}|^2}{r^{\delta}}, & |h_{PB}|^2 \leq Q, \\ K_2 \frac{|h_{TR}|^2}{r^{\delta}}, & |h_{PB}|^2 > Q, \end{cases} \quad (10)$$

where $K_1 = \frac{\eta\alpha P_{PB}}{\sigma_R^2(1-\alpha)D_{PB}^{\delta}}$, $K_2 = k_S \frac{Q_p}{N_0 B \Lambda_S}$, and $Q = k_S \frac{Q_p}{\Lambda_S K_1}$.

Further, we also assume that the fading envelope in the link between the SU-Tx and the SU-Rx follows Nakagami-m distribution. Taking into account the movement of the SU-Rx and the variable distance r between the SU-Tx and SU-Rx, described by the Equation (2), the final expression for the PDF of the instantaneous channel power gain γ_2 is obtained via ([32], (7)):

$$f_{\gamma_2}(x) = \frac{1}{\Gamma(m_2)}\left(\frac{m_2}{\Omega_2}\right)^{m_2} x^{m_2-1} \sum_{i=1}^{n} \frac{B_i}{m_2\delta + \beta_i + 1} {}_1F_1\left(m_2 + \frac{\beta_i+1}{\delta}, 1+m_2 + \frac{\beta_i+1}{\delta}, -\frac{m_2 x}{\Omega_2}\right), \quad (11)$$

where $\gamma_2 = |h_{TR}|^2 r^{-\delta}$ and $\Omega_2 = E[|h_{TR}|^2 D^{-\delta}]$.

Then, instantaneous SNR expression can be written in the following form:

$$\gamma = \begin{cases} \gamma_{K_1}\gamma_1, & \gamma_1 \leq Q, \\ \gamma_{K_2}, & \gamma_1 > Q, \end{cases} \quad (12)$$

where $\gamma_{K_1} = K_1\gamma_2$ and $\gamma_{K_2} = K_2\gamma_2$. The corresponding probability density functions (PDFs) of the random variables γ_{K_i} and $i = 1, 2$ are provided by $f_{\gamma_{K_i}}(x) = \frac{1}{K_i}f_{\gamma_i}\left(\frac{x}{K_i}\right)$ ([51], (5–7)) and have the form:

$$f_{\gamma_{K_i}}(x) = \frac{1}{K_i^{m_2}}\frac{1}{\Gamma(m_2)}\left(\frac{m_2}{\Omega_2}\right)^{m_2} x^{m_2-1} \sum_{i=1}^{n} \frac{B_i}{m_2\delta+\beta_i+1} \times {}_1F_1\left(m_2 + \frac{\beta_i+1}{\delta}, 1+m_2+\frac{\beta_i+1}{\delta}, -\frac{m_2 x}{K_i\Omega_2}\right). \tag{13}$$

3. Outage Capacity

In the following section, we present the analysis of the outage performances of the wirelessly powered cognitive radio system presented herein. First, we calculate the PDF of the random variable γ, provided by

$$f_\gamma(u) = \int_0^Q f_{u,w}(u,w)dw + \int_Q^\infty f_{\gamma_{K_2},\gamma_1}(u,w)dw, \tag{14}$$

where $f_{u,w}(u,w)$ is the joint PDF of the variables $u = \gamma_{K_1}\gamma_1$ and $w = \gamma_1$, and $f_{\gamma_{K_2},\gamma_1}(u,w)$ is the joint PDF of variables γ_{K_2} and γ_1.

The joint PDF of variables $u = \gamma_{K_1}\gamma_1$ and $w = \gamma_1$ can be obtained using the Jacobian transformation $f_{u,w}(u,w) = |J|f_{\gamma_{K_1}}\left(\frac{u}{w}\right)f_{\gamma_1}(w)$, where $|J| = 1/|w|$. On the other hand, a joint PDF of the independent variables γ_{K_2} and γ_1 can be obtained as $f_{\gamma_{K_2},\gamma_1}(u,w) = f_{\gamma_{K_2}}(u) \cdot f_{\gamma_1}(w)$. The PDF of instantaneous SNR from (14) is:

$$f_\gamma(u) = \int_0^Q |J|f_{\gamma_{K_1}}\left(\frac{u}{w}\right)f_{\gamma_1}(w)dw + \int_Q^\infty f_{\gamma_{K_2}}(u) \cdot f_{\gamma_1}(w)dw. \tag{15}$$

Further, by replacing the PDF expressions (1) and (13) in (15), the following expression is obtained:

$$f_\gamma(u) = \frac{1}{\Gamma(m_1)\Gamma(m_2)}\left(\frac{m_1}{\Omega_1}\right)^{m_1}\left(\frac{m_2}{\Omega_2}\right)^{m_2}\frac{u^{m_2-1}}{K_1^{m_2}}\sum_{i=1}^{n}\frac{B_i}{m_2\delta+\beta_i+1}I_1$$
$$+ \frac{1}{\Gamma(m_1)\Gamma(m_2)}\left(\frac{m_1}{\Omega_1}\right)^{m_1}\left(\frac{m_2}{\Omega_2}\right)^{m_2}\frac{u^{m_2-1}}{K_2^{m_2}}\sum_{i=1}^{n}\frac{B_i}{m_2\delta+\beta_i+1} \tag{16}$$
$$\times {}_1F_1\left(m_2+\frac{\beta_i+1}{\delta}, 1+m_2+\frac{\beta_i+1}{\delta}, -\frac{m_2 u}{K_2\Omega_2}\right)I_2,$$

where corresponding integrals are

$$I_1 = \int_0^Q {}_1F_1\left(m_2+\frac{\beta_i+1}{\delta}, 1+m_2+\frac{\beta_i+1}{\delta}, -\frac{m_2 u}{K_1\Omega_2 w}\right)w^{m_1-m_2-1}\exp\left(-\frac{m_1}{\Omega_1}w\right)dw, \tag{17}$$

and

$$I_2 = \int_Q^\infty w^{m_1-1}\exp\left(-\frac{m_1}{\Omega_1}w\right)dw. \tag{18}$$

The mathematical transformations provided in Appendix A lead to a solution for the integrals I_1 and I_2 and the following final PDF expression for the instantaneous SNR at the SU-Rx:

$$\begin{aligned}
f_\gamma(u) &= C \frac{u^{m_2-1}}{K_1^{m_2}} \sum_{i=1}^{n} \sum_{k=0}^{\infty} \frac{1}{k!} \left(-\frac{m_1}{\Omega_1}\right)^k B_i \frac{\left(m_2 + \frac{\beta_i+1}{\delta}\right)}{m_2 \delta + \beta_i + 1} \\
&\quad \times Q^{m_1 - m_2 + k} G_{3,2}^{1,2} \left(\frac{K_1 \Omega_2 Q}{m_2 u} \middle| \begin{array}{ccc} 1 - m_1 + m_2 - k, & 1, & 1 + m_2 + \frac{\beta_i+1}{\delta} \\ m_2 + \frac{\beta_i+1}{\delta}, & m_2 - m_1 - k & \end{array} \right) \\
&\quad + C \frac{u^{m_2-1}}{K_2^{m_2}} \left(\frac{\Omega_1}{m_1}\right)^{m_1} \Gamma\left(m_1, \frac{m_1}{\Omega_1} Q\right) \sum_{i=1}^{n} \frac{B_i}{m_2 \delta + \beta_i + 1} \\
&\quad \times {}_1F_1\left(m_2 + \frac{\beta_i+1}{\delta}, 1 + m_2 + \frac{\beta_i+1}{\delta}, -\frac{m_2 u}{K_2 \Omega_2}\right),
\end{aligned} \tag{19}$$

where $C = \frac{1}{\Gamma(m_1)} \frac{1}{\Gamma(m_2)} \left(\frac{m_1}{\Omega_1}\right)^{m_1} \left(\frac{m_2}{\Omega_2}\right)^{m_2}$.

Further, the CDF expression for the random variable γ can be obtained via

$$F_\gamma(\gamma_{th}) = \int_0^{\gamma_{th}} f_\gamma(u) du. \tag{20}$$

By replacing (19) in the previous equation, the CDF can be expressed as

$$\begin{aligned}
F_\gamma(\gamma_{th}) &= C \frac{1}{K_1^{m_2}} \sum_{i=1}^{n} \sum_{k=0}^{\infty} \frac{1}{k!} \left(-\frac{m_1}{\Omega_1}\right)^k B_i \frac{\left(m_2 + \frac{\beta_i+1}{\delta}\right)}{m_2 \delta + \beta_i + 1} Q^{m_1 - m_2 + k} I_3 \\
&\quad + C \frac{1}{K_2^{m_2}} \left(\frac{\Omega_1}{m_1}\right)^{m_1} \Gamma\left(m_1, \frac{m_1}{\Omega_1} Q\right) \sum_{i=1}^{n} \frac{B_i}{m_2 \delta + \beta_i + 1} I_4,
\end{aligned} \tag{21}$$

where

$$I_3 = \int_0^{\gamma_{th}} u^{m_2-1} G_{3,2}^{1,2} \left(\frac{K_1 \Omega_2 Q}{m_2 u} \middle| \begin{array}{ccc} 1 - m_1 + m_2 - k, & 1, & 1 + m_2 + \frac{\beta_i+1}{\delta} \\ m_2 + \frac{\beta_i+1}{\delta}, & m_2 - m_1 - k & \end{array} \right) du, \tag{22}$$

and

$$I_4 = \int_0^{\gamma_{th}} u^{m_2-1} {}_1F_1\left(m_2 + \frac{\beta_i+1}{\delta}, 1 + m_2 + \frac{\beta_i+1}{\delta}, -\frac{m_2 u}{K_2 \Omega_2}\right) du. \tag{23}$$

The integrals I_3 and I_4 can be solved in the exact closed form by applying the transformations described in Appendix B. Finally, the expression for the outage probability of the secondary link can be written in the following form:

$$\begin{aligned}
F_\gamma(\gamma_{th}) &= C \frac{\gamma_{th}^{m_2}}{K_1^{m_2}} \sum_{i=1}^{n} \sum_{k=0}^{\infty} \frac{1}{k!} \left(-\frac{m_1}{\Omega_1}\right)^k B_i \frac{\left(m_2 + \frac{\beta_i+1}{\delta}\right)}{m_2 \delta + \beta_i + 1} Q^{m_1 - m_2 + k} \\
&\quad \times G_{3,4}^{2,2} \left(\frac{m_2 \gamma_{th}}{K_1 \Omega_2 Q} \middle| \begin{array}{cccc} 1 - m_2, & 1 - m_2 - \frac{\beta_i+1}{\delta}, & 1 - m_2 + m_1 + k \\ m_1 - m_2 + k, & 0, & -m_2 - \frac{\beta_i+1}{\delta}, & -m_2 \end{array} \right) \\
&\quad + C \frac{\gamma_{th}^{m_2}}{K_2^{m_2}} \left(\frac{\Omega_1}{m_1}\right)^{m_1} \Gamma\left(m_1, \frac{m_1}{\Omega_1} Q\right) \sum_{i=1}^{n} B_i \frac{\left(m_2 + \frac{\beta_i+1}{\delta}\right)}{m_2 \delta + \beta_i + 1} \\
&\quad \times G_{2,3}^{1,2} \left(\frac{m_2 \gamma_{th}}{K_2 \Omega_2} \middle| \begin{array}{ccc} 1 - m_2, & 1 - m_2 - \frac{\beta_i+1}{\delta} \\ 0, & -m_2 - \frac{\beta_i+1}{\delta}, & -m_2 \end{array} \right).
\end{aligned} \tag{24}$$

Another important feature describing the secondary system is the achievable throughput. To obtain the achievable throughput, the outage capacity is calculated using the following equation [52]:

$$C_{OUT} = (1 - P_{OUT}(\gamma_{th})) \log_2(1 + \gamma_{th}), \tag{25}$$

where $P_{OUT}(\gamma_{th}) = F_\gamma(\gamma_{th})$.

The outage capacity is defined as the maximum data rate that can be achieved. Further, the achievable throughput is expressed by [35]:

$$T_{OUT} = (1-\alpha)C_{OUT}. \tag{26}$$

4. Ergodic Capacity

In this section, we analyze the ergodic capacity of the wirelessly powered secondary link, which is one of the most important performance metrics appropriate for applications with no delay requirements. The ergodic capacity represents a maximal long-term rate that can be achieved over a channel with an arbitrary small probability of error [52], and it can be calculated via

$$C_{erg} = \frac{1}{\log_e 2} \int_0^\infty \log_e(1+u) f_\gamma(u) du. \tag{27}$$

Knowing the ergodic capacity, the achievable ergodic throughput can be obtained via

$$T_{erg} = (1-\alpha)C_{erg}. \tag{28}$$

By replacing the PDF represented by Meijer functions using the procedure provided in Appendix C and the integration provided in (27), the expression for the ergodic capacity of a wirelessly powered secondary link is:

$$\begin{aligned}
C_{erg} &= \frac{1}{\log_e 2} \frac{C}{K_1^{m_2}} \sum_{i=1}^{n} \sum_{k=0}^{\infty} \frac{1}{k!} \left(-\frac{m_1}{\Omega_1}\right)^k B_i \frac{\left(m_2 + \frac{\beta_i+1}{\delta}\right)}{m_2\delta+\beta_i+1} Q^{m_1-m_2+k} \\
&\times G_{4,5}^{4,2}\left(\frac{m_2}{K_1\Omega_2 Q} \middle| \begin{array}{cccc} 1-m_2-\frac{\beta_i+1}{\delta}, & -m_2, & 1-m_2, & 1-m_2+m_1+k \\ m_1-m_2+k, & 0, & -m_2, & -m_2, & -m_2-\frac{\beta_i+1}{\delta} \end{array}\right) \\
&+ \frac{1}{\log_e 2} \frac{C}{K_2^{m_2}} \left(\frac{\Omega_1}{m_1}\right)^{m_1} \Gamma\left(m_1, \frac{m_1}{\Omega_1}Q\right) \sum_{i=1}^{n} B_i \frac{\left(m_2+\frac{\beta_i+1}{\delta}\right)}{m_2\delta+\beta_i+1} \\
&\times G_{3,4}^{3,2}\left(\frac{m_2}{K_2\Omega_2} \middle| \begin{array}{ccc} 1-m_2-\frac{\beta_i+1}{\delta}, & -m_2, & 1-m_2 \\ 0, & -m_2, & -m_2, & -m_2-\frac{\beta_i+1}{\delta} \end{array}\right).
\end{aligned} \tag{29}$$

5. Numerical Results

In this section, we present the numerical results of the system performances analyzed in previous chapters. All numerical results were confirmed by an independent Monte Carlo simulation method [53] implemented in MATLAB, based on generated waveform sequences with $L = 10^7$ samples in accordance with [50,54]. The flowchart presented in Figure 3 shows the steps used in the system performance evaluation. Analytical expressions are represented by Equations (24) and (29). Various parameters that the system's performance depended on were changed, and an analysis of the obtained results was carried out.

For the proposed system, the following important secondary system performance metrics were considered: the outage probability, the throughput based on outage capacity, and the throughput based on ergodic capacity. The simulation parameters used for all considered scenarios presented in Figures 4–14 are provided in Table 4. The system and channel parameters presented in Table 4 have significant influence on system performance, and they were varied in the presented analysis in order to provide important conclusions considering the system design.

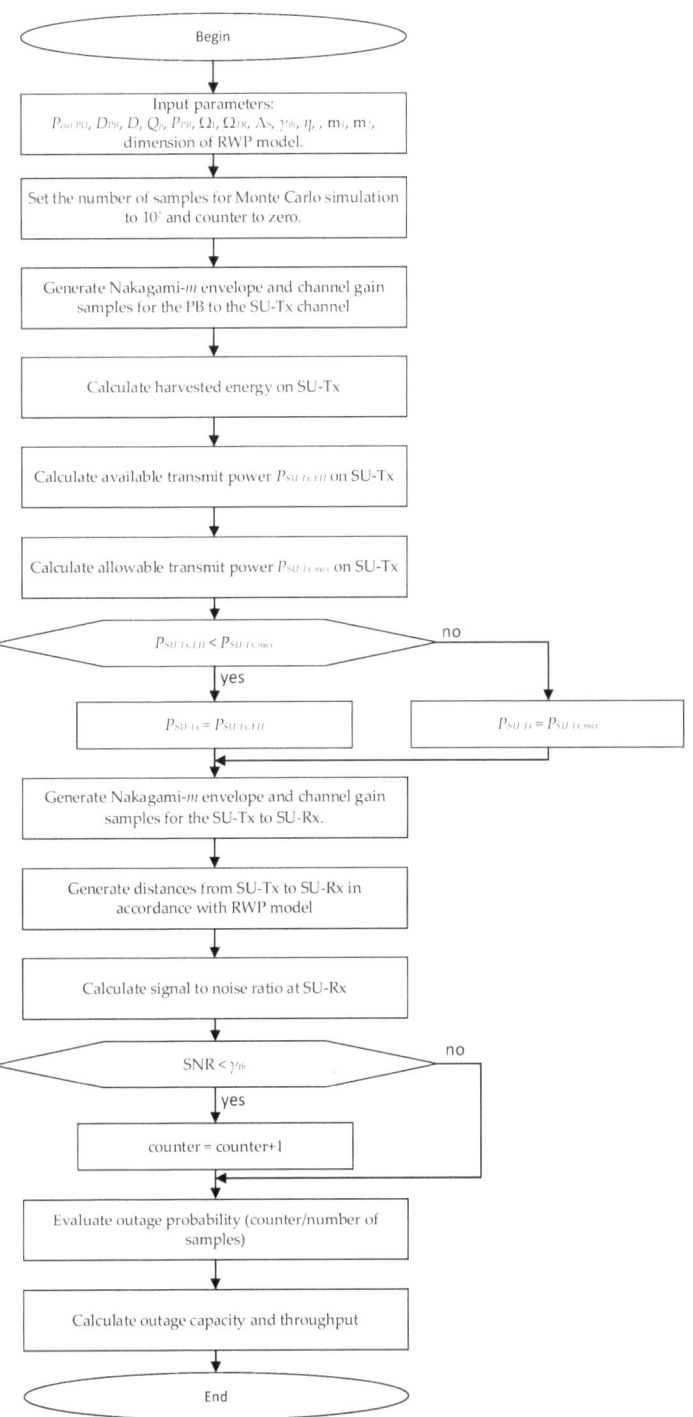

Figure 3. Flowchart of the steps used for the performance evaluation.

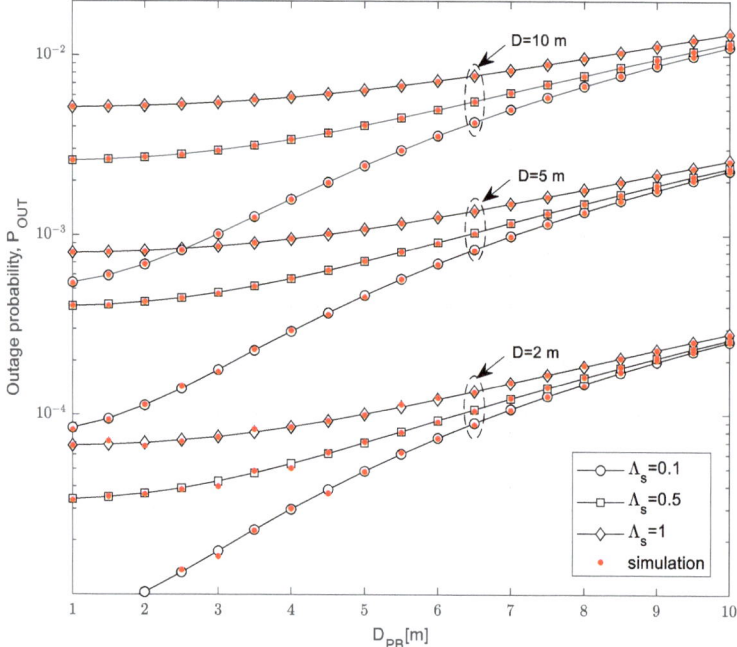

Figure 4. Outage probability vs. distance from the power beacon to the secondary transmitter D_{PB}.

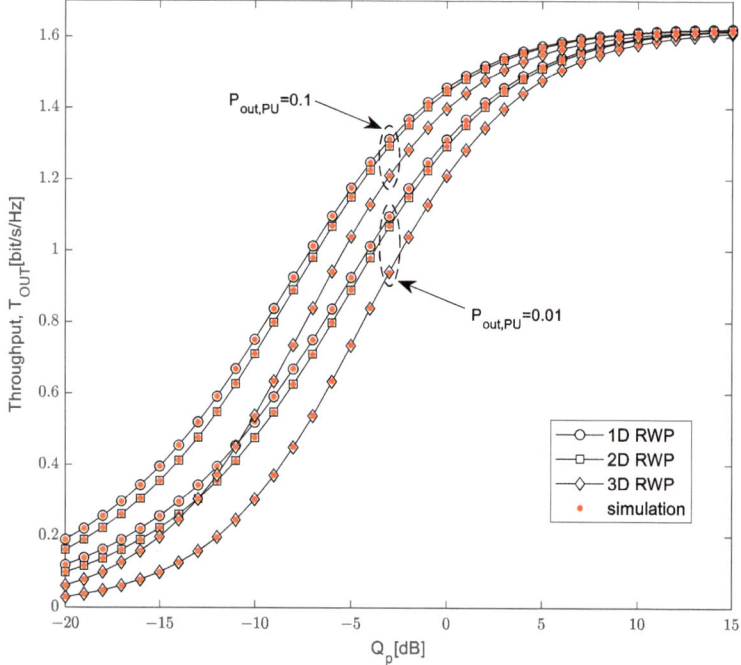

Figure 5. Throughput T_{OUT} vs. interference threshold Q_p for various mobility models.

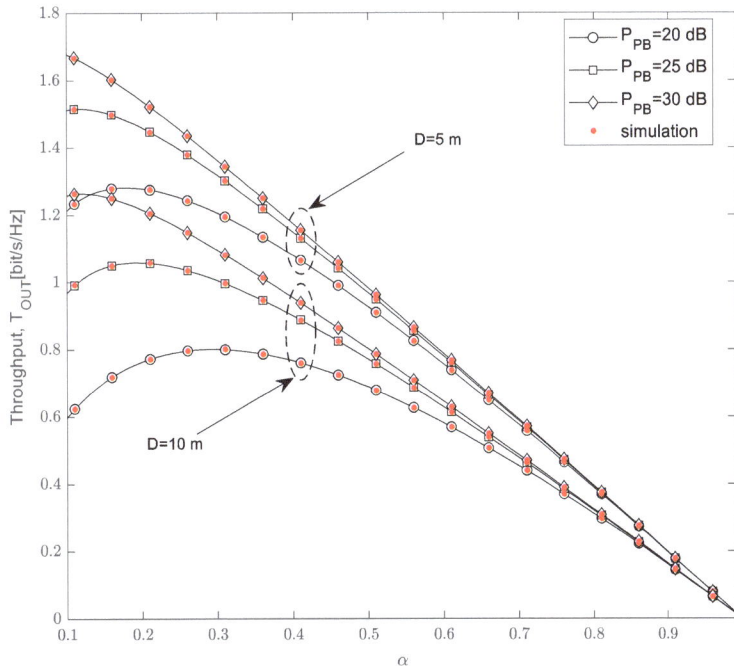

Figure 6. Throughput T_{OUT} vs. time-switching factor α.

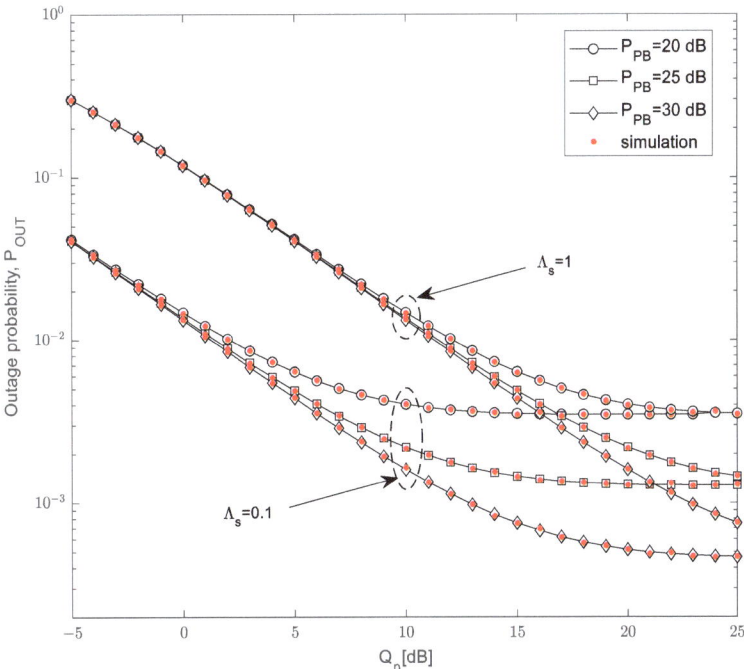

Figure 7. Outage probability vs. interference threshold Q_p for various P_{PB} values.

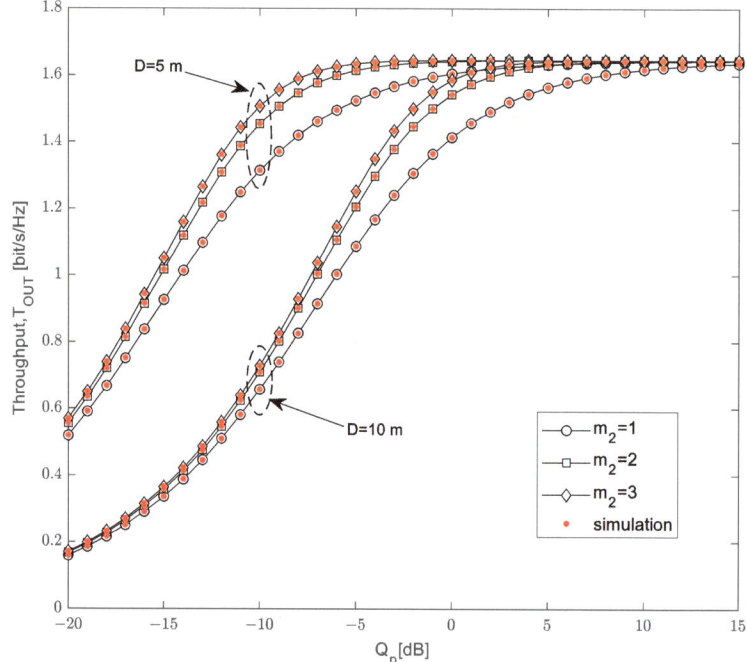

Figure 8. Throughput T_{OUT} vs. interference threshold Q_p for various distances D.

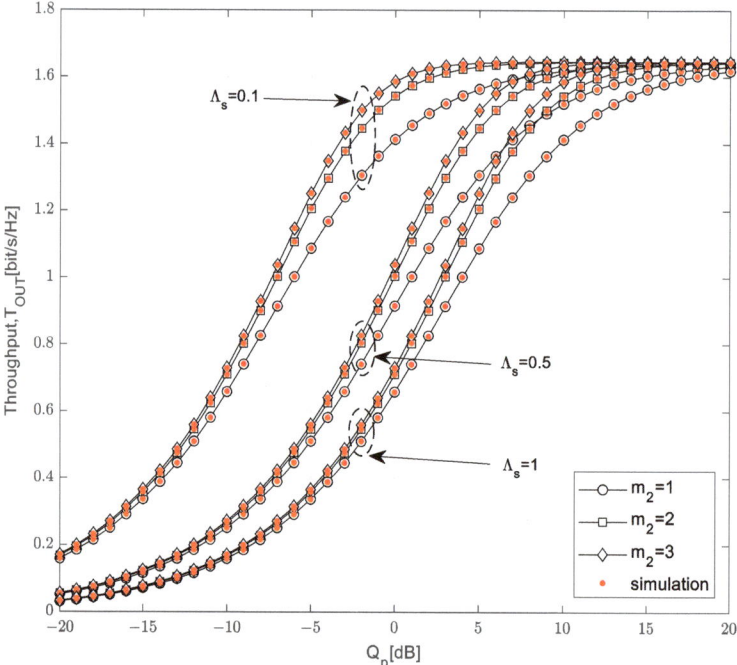

Figure 9. Throughput T_{OUT} vs. interference threshold Q_p for various channel power gains Λ_S.

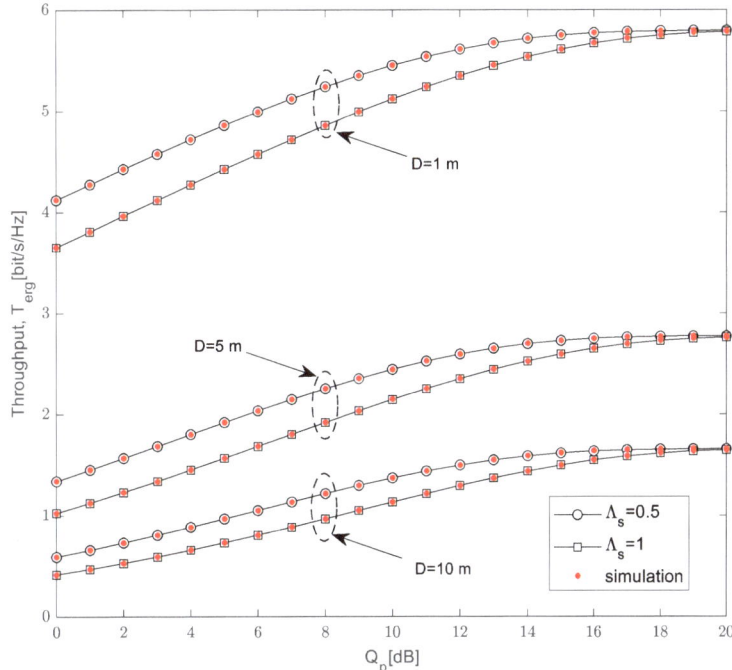

Figure 10. Throughput T_{erg} vs. interference threshold Q_p for various distances D.

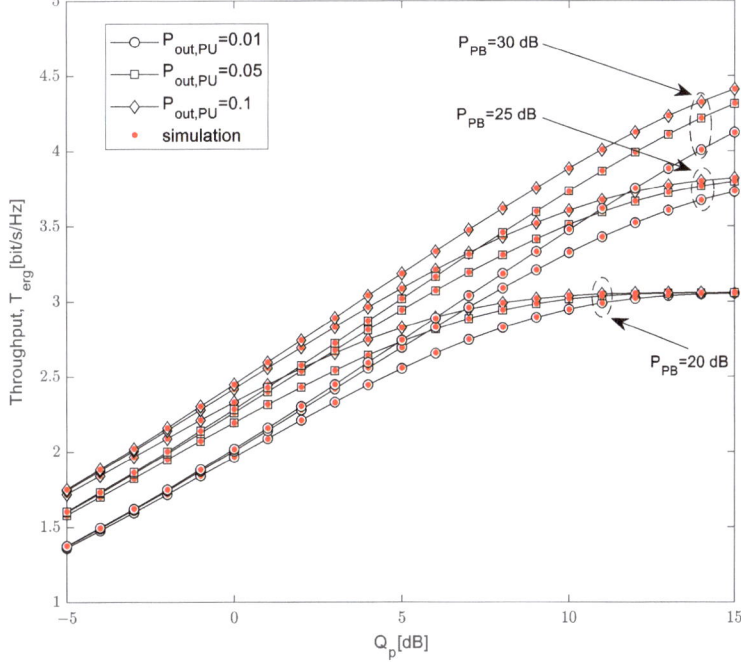

Figure 11. Ergodic throughput vs. interference threshold Q_p for various P_{PB} values.

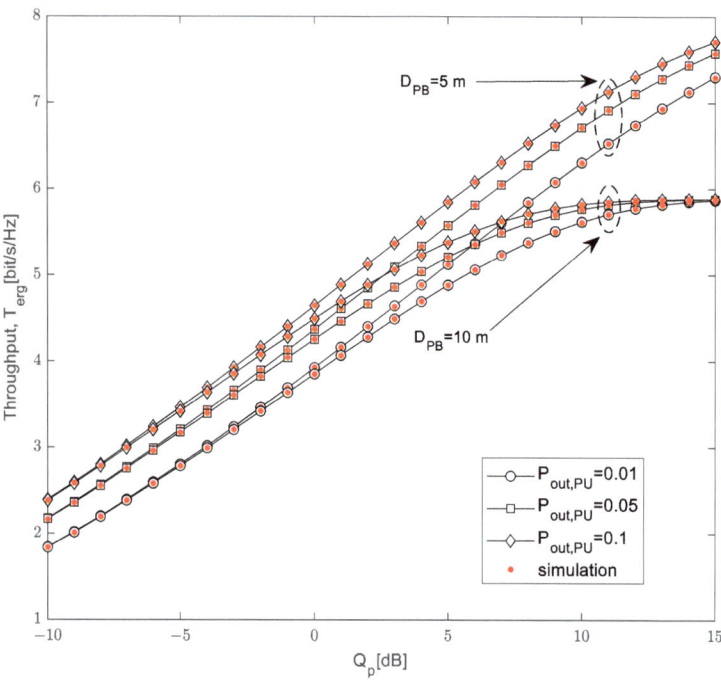

Figure 12. Ergodic throughput vs. interference threshold Q_p for different probability values $P_{out,\,PU}$.

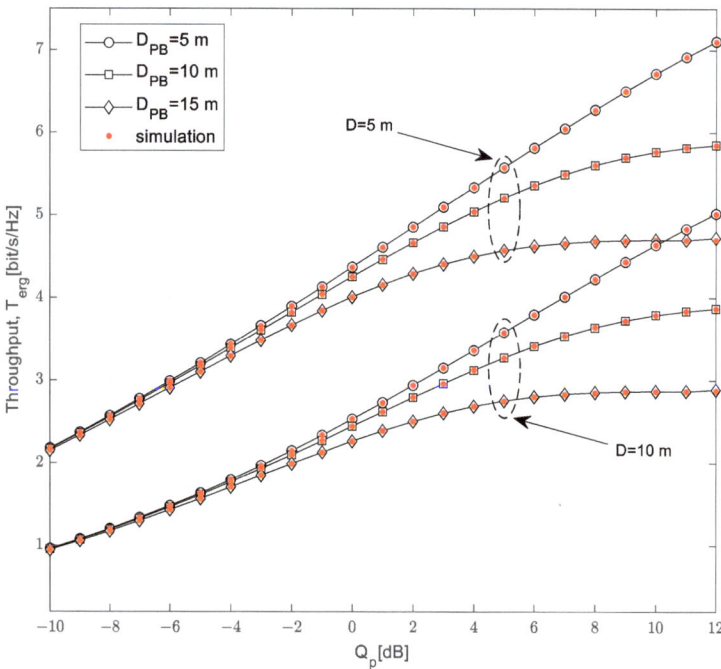

Figure 13. Ergodic throughput vs. interference threshold Q_p for various distances D_{PB} and D.

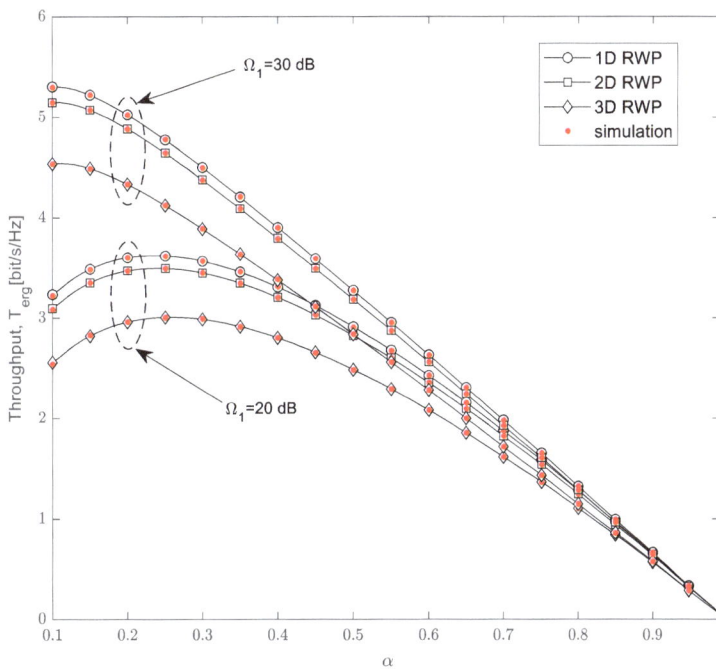

Figure 14. Ergodic throughput vs. time-switching factor α for different mobility models.

Table 4. Simulation parameters.

Parameter	Figure 4	Figure 5	Figure 6	Figure 7	Figure 8	Figure 9	Figure 10	Figure 11	Figure 12	Figure 13	Figure 14
$P_{out,PU}$	0.01	0.01–0.1	0.1	0.1	0.01	0.01	0.05	0.01–0.1	0.01–0.1	0.05	0.1
D_{PB} [m]	1–10	1	10	5	1	1	5	10	5–10	5–15	10
D [m]	2–10	5	5–10	5	5–10	10	1–10	5	5	5–10	5
Q_p [dB]	15	−20 ÷ 20	15	−5 ÷ 25	−20 ÷ 20	−20 ÷ 20	0 ÷ 20	−10 ÷ 15	−20 ÷ 15	−10 ÷ 12	15
P_{PB} [dB]	30	20	20–30	20–30	30	30	20	20–30	30	30	20
Ω_1 [dB]	10	10	20	20	10	10	10	20	10	10	20–30
Ω_{TR} [dB]	30	20	20	20	30	30	20	20	30	30	20
Λ_S	0.1–1	0.1	1	0.1–1	0.1	0.1–1	0.5–1	0.1	0.5	0.5	1
γ_{th} [dB]	−5	5	5	−5	5	5	-	-	-	-	-
η	0.9	0.9	0.9	0.9	0.9	0.9	0.9	0.9	0.9	0.9	0.9
α	0.5	0.2	0.1–0.99	0.5	0.2	0.2	0.5	0.5	0.2	0.2	0.1–0.99
m_1	1	1	1	1	1	1	1	1	1	1	1
m_2	1	1	1	1	1–3	1–3	1	1	1	1	1
RWP	1D	1D–3D	1D	2D	1D	1D	2D	2D	2D	2D	1D–3D

Figure 4 shows the dependence of the outage probability of the distance D_{PB} from the PB to the SU-Tx. The curves on the graph are presented for different values of the maximum distance from the SU-Tx to the SU-Rx and the mean channel power gain Λ_S in the channel from the SU-Tx to the PU-Rx for the outage threshold equal $\gamma_{th} = -5$ dB. The time-switching factor was $\alpha = 0.5$, which means that equal time intervals were dedicated for the wireless power transfer and the information transfer in the secondary system. In the considered scenario, the 1D path model was assumed, meaning that the SU-Rx moves according to the RWP model along a straight line. The limitations imposed by the primary networks were: the interference threshold $Q_p = 15$ dB and the outage probability of the primary network $P_{\text{out,PU}} = 0.01$.

From the obtained results, it can be noted that the increase in the distance between the PB and the SU-Tx also increases the outage probability values. The reason for this effect is the fact that the increase in the distance from the PB to the SU-Tx decreases the received signal power at the SU-Tx and the harvested energy on the SU-Tx, leading to its lower transmission power. In addition, greater values of the maximum distance from the SU-Tx to the SU-Rx have a degrading impact on the outage probability performances as the received power at the SU-Rx is smaller and the SNR at the SU-Rx is lower. The propagation characteristics in the channel from the SU-Tx to the PU-Rx also significantly impact outage performances. With a decrease in the mean channel power gain value in the channel from the SU-Tx to the PU-Rx, the outage probability of the SU system decreases because the limitation of the transmitted power at the SU-Tx is weaker. By increasing the distance from the PB to the SU-Tx, the value of the mean channel power gain in the channel from the secondary transmitter to the primary receiver has a less significant impact on the outage probability. In this case, a smaller amount of energy is harvested, resulting in a smaller transmitted power of the SU when the interference limitation of the transmitted power does not have a dominant impact. On the other hand, small values of the distance D_{PB} result in high amounts of harvested energy when the interference constraint imposed by the PU is dominant. Therefore, in this case, when the channel power gain is high, variations in the values of D_{PB} do not dominantly influence outage performances, and they have similar values in the range of up to 3 m.

The dependence of the throughput T_{OUT} on the interference threshold Q_p is analyzed in Figure 5 for two values of the permitted primary network outage probability, $P_{out,PU}$, equal to 0.1 and 0.01, respectively. In this figure, we also consider scenarios with different SU-Rx mobility models.

In accordance with the expectations, better system performance was achieved when the permitted outage probability of the PU network and the required interference threshold had greater values. With a more lenient condition for the harmful interference on the primary receiver, which was expressed through higher values of the interference threshold and the outage probability of the primary network, a higher value of transmission power was allowed on the SU-Tx. For interference threshold values higher than 10 dB, the throughput values entered the saturation range, and they were not dominantly dependent on the outage probability of the primary network. In this scenario, the transmission power of the SU solely depended on the harvested energy and not on limitations caused by harmful interference on the primary user. Finally, Figure 5 shows the throughput for different mobility models, demonstrating performances in the cases of SU-Rx movement along the line path, circle surface, or inside the sphere. It can be noted that the best results are obtained when movement of SU-Rx can be described using the 1D model path.

The dependence of throughput on the time-switching factor α is demonstrated in Figure 6. Additionally, the impact of the transmission power of the PB and the maximum distance from the SU-Tx to the SU-Rx are analyzed for the case in which the SU-Rx moves along 1D path model.

The throughput of the secondary network increases by increasing the time-switching factor α up to a certain value; after this value, it begins to decrease as although a higher α value results in a higher amount of harvested energy at the SU-Tx, the dominant impact has a smaller amount of time dedicated to the transmission of information. For the same value of time-switching factor α, higher throughput values are achieved if the transmission power of the PB is higher because more energy is harvested at the SU-Tx in this case, resulting in a higher transmit power of the SU. The influence of the PB's power values is greater in the case in which $D = 10$ m than in the case in which $D = 5$ m. In addition, the smaller value of the maximum distance from the SU-Tx to the SU-Rx leads to greater throughput values as the receiver power at the SU-Rx is higher, resulting in a higher received SNR at the SU-Rx.

The dependence of outage probability on the interference threshold Q_p for different values of the transmission power of the PB and different values of the mean channel power gain from the SU-Tx to the PU-Rx is shown in Figure 7, demonstrating the case in which the primary network outage probability is $P_{out,PU} = 0.1$ and the outage threshold is $\gamma_{th} = -5$ dB, while the secondary receiver moves over the 2D area with a maximum distance from the SU-Tx of $D = 5$ m.

The outage probability of the SU decreases with the increase in the interference threshold Q_p, as the SU-Tx can transmit with higher permitted power. With high values of the interference threshold Q_p, the outage probability enters the saturation region. For smaller values of the PB transmit power, the dependence enters the saturation region for a smaller Q_p, while with higher values of the PB power, the saturation occurs for a higher interference threshold Q_p. For high values of the interference threshold, the transmit power of the SU-Tx depends completely on the harvested energy, so in this scenario only the PB power has an influence on the collected energy. Furthermore, more energy results in a higher transmit power of the SU-Tx, which leads to a smaller outage probability. For small values of the interference threshold Q_p, the outage probability is not dependent on the power of the PB because the limitation caused by harmful interference on the PU-RX has the dominant impact. It can be noted that better system performances are obtained for lower values of mean channel power gain in the channel from the SU-Tx to the PU-Rx, which is the effect also shown in Figure 4. Additionally, for a higher interference threshold and the same power value of the PB, the outage probability in the saturation region becomes independent of the mean channel power gain in the channel from the SU-Tx to the PU-Rx because the transmission power of the SU-Tx depends only on the harvested energy from the PB.

The throughput T_{OUT} is presented in Figure 8 as a function of the interference threshold Q_p. The influence of the maximum distance between the SU-Tx and the SU-Rx is analyzed, as is the impact of the Nakagami-m fading parameter m_2 in the SU channel (from the SU-Tx to the SU-Rx).

The throughput T_{OUT} of the secondary network increases with the increase in the interference threshold, while saturation occurs for larger values of the interference threshold. In the region of a smaller Q_p with no saturation effect, the throughput depends on all analyzed parameters. In accordance with the expectations, a smaller throughput is achieved if the maximum distance from the SU-Tx to the SU-Rx is higher: in that case, the mean value of the received power at the SU-Rx is smaller. A higher value of the Nakagami-m fading parameter m_2 in the channel from the SU-Tx to the SU-Rx leads to a higher value of the achievable throughput. It can be noted that the throughput increases with the increase in the interference threshold to a certain limit, when the saturation region begins. In the case in which the maximum distance from the SU-Tx to the SU-Rx is $D = 5$ m, the saturation occurs at a lower Q_p, and a further increase in the interference threshold does not change the throughput values. At high values of the interference threshold, there is no significant difference between the throughput values for the maximum distance $D = 5$ m and $D = 10$ m and for the analyzed values of the Nakagami-m fading parameter m_2 in the SU link. By increasing the interference threshold, the limitation conditioned by the harmful interference on the primary user becomes weaker, so the secondary transmitter can use all the harvested energy for transmission.

The dependence of the throughput T_{OUT} on the interference threshold Q_p is shown in Figure 9. In this scenario, the values of the mean channel power gain in the channel from the SU-Tx to the PU-Rx, as well as the Nakagami-m fading parameter m_2 in the SU link, are varied. The mean channel power gain in Nakagami-m channel from the PB to the SU-Tx is $\Omega_1 = 10$ dB, the distance from the PB to the SU-Tx is $D_{PB} = 1$ m, and the transmission power of the PB is $P_{PB} = 30$ dB. It is assumed that the SU-Rx movement can be described with a 1D model. The maximum distance from the SU-Tx to the SU-Rx is $D = 5$ m, the outage probability of the primary network is $P_{out,PU} = 0.01$, the time-switching factor is $\alpha = 0.2$, and the outage threshold is $\gamma_{th} = 5$ dB.

For higher values of the interference threshold Q_p, the throughput reaches the saturation value, the interference threshold value at which the saturation region begins depends on the mean channel power gain value of the channel from the SU-Tx to the PU-Rx. Both factors, the interference threshold Q_p and the mean channel power gain in the channel from the SU-Tx to the PU-Rx, have influences on the limiting values conditioned by harmful interference on the primary user. A greater value of the Nakagami-m fading parameter m_2 in the secondary user channel leads to a higher achievable throughput. Observing the behavior of curves for very small values of the interference threshold Q_p, it can be concluded that for the same parameter of the mean channel power gain, Λ_S, approximately the same values of throughput are achieved for different values of the fading parameter m_2, as the allowed transmitted power from the secondary transmitter is very small in all cases.

The curves in Figure 10 show the dependence of the ergodic throughput on the interference threshold Q_p for different values of the maximum distance from SU-Tx to SU-Rx and the mean channel power gain in the channel from the SU-Tx to the PU-Rx. The time-switching factor is $\alpha = 0.5$, the mean channel power gain in the Nakagami-m channel from the PB to the SU-Tx is $\Omega_1 = 10$ dB, the distance from the PB to the SU-Tx is $D_{PB} = 5$ m, and the transmission power of the PB is $P_{PB} = 20$ dB. It is assumed that the SU-Rx moves along a 2D path model, while the mean channel power gain in the Nakagami-m channel from the SU-Tx to the SU-Rx is $\Omega_{TR} = 20$ dB, and the outage probability of the primary network is $P_{\text{out,PU}} = 0.05$.

By increasing the interference threshold, the ergodic throughput of the secondary network increases for an interference threshold smaller than 15 dB, while for values greater than 15 dB, the saturation effect occurs. A higher value of the interference threshold Q_p leads to a greater value of the ergodic throughput because the SU-Tx is allowed to transmit with more power, so the received SNR at SU-Rx is higher. For values greater than 15 dB, the ergodic throughput depends solely on the harvested energy. By further increasing the interference threshold, the ergodic throughput becomes independent of the mean channel power gain in the channel from the SU-Tx to the PU-Rx. A smaller ergodic throughput is achieved if the maximum distance from the SU-Tx to the SU-Rx is higher because in that case, the received power at the SU-Rx is smaller. If the mean channel power gain in channel from the SU-Tx to the PU-Rx is higher, a smaller ergodic throughput is achieved because of the interference limitation at the primary user, leading to the fact that the smaller transmit power of the SU-Tx is allowed.

Figure 11 shows the dependence of the ergodic throughput on the interference threshold Q_p. System performances are analyzed for different values of the transmission power of the PB and the outage probability of primary network $P_{\text{out,PU}}$.

With a high value of the interference threshold, saturation occurs for smaller values the of transmission power of the PB, while for larger values of transmission power of the PB, the ergodic throughput enters the saturation region.

For a higher interference threshold, the ergodic throughput begins to depend on the energy harvested, which is smaller for lower values of the PB's transmission power. On the other hand, the ergodic throughput is independent of the PB's transmission power for lower values of the interference threshold. In that case, the transmission power on the SU-Tx depends on the limitations caused by harmful interference on the PU, so only the outage probability of the primary network has an influence on the results. A higher value of the ergodic throughput is achieved if the outage probability of the primary network $P_{\text{out,PU}}$ is higher. This is because with a higher value of the interference threshold and a higher value of outage probability of the primary network, a higher transmission power is allowed for the secondary user, with a lenient condition of harmful interference on the primary receiver.

The dependence of the ergodic throughput of the interference threshold Q_p is presented in Figure 12. The analysis was performed based on the varying values of the distance from the PB to the SU-Tx and the outage probability of the primary network $P_{\text{out,PU}}$.

For the interference threshold lower than -5 dB, it can be noted that the ergodic throughput is independent of the distance from the PB to the SU-Tx. In that case, the transmission power of the SU-Tx is limited by the condition of harmful interference on the PU-Rx and depends on the interference threshold and outage probability of the primary network. When increasing the interference threshold to above -5 dB, less ergodic throughput is achieved if the distance from the PB to the SU-Tx is higher because in that case, the harvested energy at the SU-Tx, is smaller and less transmission power on the SU-Tx is available. With a higher value of the interference threshold and a higher value of the outage probability of the primary network, a larger transmission power is allowed on the SU, so higher values of the ergodic throughput are achieved. In the case in which the distance from the power beacon to the secondary transmitter is 10 m, saturation occurs when the interference threshold crosses a threshold approximately equal to 10 dB. In this case, the transmission power of the SU-Tx depends on the harvested energy, which is smaller when there is an increasing distance between the PB and the SU-Tx.

In Figure 13, the dependence of the ergodic throughput on the distance from the PB to the SU-Tx, D_{PB}, the maximum distance from the secondary transmitter to the secondary receiver D, and the interference threshold Q_p are presented.

A smaller ergodic throughput is achieved if the maximum distance from the SU-Tx to the SU-Rx is higher because in this case, the received power at the SU-Rx is smaller. As in Figure 12, the same behavior of the ergodic throughput dependence on the distance from the PB to the SU-Tx D_{PB} can be noted. For an interference threshold lower than -5 dB, it can be concluded that the ergodic throughput is independent of the distance from the PB to the SU-Tx, while above that interference threshold value, the ergodic throughput decreases with an increasing distance from the PB to the SU-Tx until saturation occurs in cases in which that distance increases to 15 m and the interference threshold is above 5 dB.

The dependence of the ergodic throughput on the time-switching factor is shown in Figure 14. System performances are analyzed for different values of the mean channel power gain in the Nakagami-m channel from the PB to the SU-Tx, a fixed value of the power of the PB, $P_{PB} = 20$ dB, and different mobility models used for the description of the SU-Rx movement. The outage probability of the primary network is $P_{out,PU} = 0.1$, the mean channel power gain in the channel from the SU-Tx to the PU is $\Lambda_S = 1$, the distance from the PB to the SU-Tx is $D_{PB} = 10$ m, and the maximum distance from the SU-Tx to the SU-Rx is $D = 5$ m.

The ergodic throughput of the secondary network grows to some value of the time-switching factor; after this point, it decreases. It is possible to determine the peak of the ergodic throughput and the optimal value of the time-switching factor. If the mean channel power gain in the Nakagami-m channel from the PB to the SU-Rx is 10 dB, the peak of the ergodic throughput is achieved when the value of the time-switching factor is around 0.25 for each of the mobility models applied to the SU-Rx. The highest peak is reached by the 1D mobility model, which achieves better results than the other models. In the case in which the mean channel power gain in the Nakagami-m channel from the PB to the SU-Tx has a higher value, the peak of the ergodic throughput is reached for smaller values of the time-switching factor. The ergodic throughput is larger if the mean channel power gain in the Nakagami-m channel from the PB to the SU-Tx is higher, but increasing the time-switching factor leads to ergodic throughput independence on the mean channel power gain in the Nakagami-m channel from the PB to the SU-Tx.

6. Conclusions

In this paper, a power-beacon-assisted cognitive system is analyzed based on spectrum sharing and available statistical CSI for the scenario in which the SU-Tx has a stationary position and the SU-Rx node is mobile. System performances are analyzed, and novel closed-form expressions are derived for the outage probability, the outage throughput, and the ergodic capacity. The theoretical expressions are verified using an independent simulation method.

Based on the obtained results, the influence of the interference limitation imposed by the primary network is discussed. The performance of the observed cognitive system depends on both PU system limitations, as well as the transmit power of the PB, the distances between the network nodes, and the channel power gain values. The best system performances are obtained when the permitted outage probability of the PU network and the required interference threshold have higher values. It has been observed that for very small values of the interference threshold Q_p, system performances are independent of the PB's power and the distance from the PB to the SU-Tx, as the limitation caused by the permitted interference on the PU-Rx is dominant with respect to the harvested energy. Furthermore, in the upper limit case, in which the interference threshold Q_p has a great value, the saturation effect occurs as the system performance depends dominantly on the energy harvested from the PB. The provided analysis of the considered spectrally efficient system provides important guidelines for the design of future networks and IoT scenarios in which a higher presence of receiver mobility is expected. This analysis also represents the basis for the optimization of system parameters for various scenarios. Finally, as network nodes are computationally and energy constrained, future work will investigate issues related to the physical layer security of the analyzed system.

Author Contributions: Conceptualization, N.K. and V.B.; methodology, N.K., V.B. and P.I.; software, N.K.; validation, N.K., V.B. and P.I.; formal analysis, N.K., V.B. and A.C.; investigation, N.K.; resources, V.B.; data curation, N.K.; writing—original draft preparation, N.K.; writing—review and editing, V.B., A.C. and P.I.; visualization, N.K., A.C. and P.I.; supervision, V.B. and P.I.; All authors have read and agreed to the published version of the manuscript.

Funding: This research received no external funding.

Acknowledgments: Vesna Blagojević and Predrag Ivaniš acknowledge the support of the Science Fund of the Republic of Serbia, grant No 7750284 (Hybrid Integrated Satellite and Terrestrial Access Network—hi-STAR).

Conflicts of Interest: The authors declare no conflict of interest.

Appendix A

In order to obtain the solution for the integral I_1 in the closed form, we apply following transformations. By using ([55], (1.211-1)) and with the help of ([56], (07.20.26.0006.01)), the integral I_1 can be written as

$$I_1 = \frac{\Gamma\left(1+m_2+\frac{\beta_i+1}{\delta}\right)}{\Gamma\left(m_2+\frac{\beta_i+1}{\delta}\right)} \sum_{k=0}^{\infty} \frac{1}{k!}\left(-\frac{m_1}{\Omega_1}\right)^k \\ \times \int_0^Q G_{1,2}^{1,1}\left(\frac{m_2 u}{wK_1\Omega_2}\bigg|\begin{array}{c}1-m_2-\frac{\beta_i+1}{\delta}\\0,\;-m_2-\frac{\beta_i+1}{\delta}\end{array}\right) w^{m_1-m_2+k-1} dw. \quad (A1)$$

By applying transformation ([56], (07.34.17.0012.01)), we further achieve

$$G_{1,2}^{1,1}\left(\frac{m_2 u}{wK_1\Omega_2}\bigg|\begin{array}{c}1-m_2-\frac{\beta_i+1}{\delta}\\0,\;-m_2-\frac{\beta_i+1}{\delta}\end{array}\right) = G_{2,1}^{1,1}\left(\frac{wK_1\Omega_2}{m_2 u}\bigg|\begin{array}{c}1,\;1+m_2+\frac{\beta_i+1}{\delta}\\m_2+\frac{\beta_i+1}{\delta}\end{array}\right), \quad (A2)$$

thus, the integral I_1 can be written in following form:

$$I_1 = \frac{\Gamma\left(1+m_2+\frac{\beta_i+1}{\delta}\right)}{\Gamma\left(m_2+\frac{\beta_i+1}{\delta}\right)} \sum_{k=0}^{\infty} \frac{1}{k!}\left(-\frac{m_1}{\Omega_1}\right)^k \\ \times \int_0^Q G_{2,1}^{1,1}\left(\frac{wK_1\Omega_2}{m_2 u}\bigg|\begin{array}{c}1,\;1+m_2+\frac{\beta_i+1}{\delta}\\m_2+\frac{\beta_i+1}{\delta}\end{array}\right) w^{m_1-m_2+k-1} dw, \quad (A3)$$

which can finally be solved by using transformation ([56], (07.34.21.0085.01))

$$I_1 = \left(m_2+\frac{\beta_i+1}{\delta}\right)\sum_{k=0}^{\infty}\frac{1}{k!}\left(-\frac{m_1}{\Omega_1}\right)^k \\ \times Q^{m_1-m_2+k}G_{3,2}^{1,2}\left(\frac{K_1\Omega_2 Q}{m_2 u}\bigg|\begin{array}{c}1-m_1+m_2-k,\;1,\;1+m_2+\frac{\beta_i+1}{\delta}\\m_2+\frac{\beta_i+1}{\delta},\;m_2-m_1-k\end{array}\right). \quad (A4)$$

The solution of the integral I_2 can be obtained by transformation $x = \frac{m_2}{\Omega_2}w$ and applying ([55], (8.350.2)), when we obtain

$$I_2 = \left(\frac{\Omega_1}{m_1}\right)^{m_1}\Gamma\left(m_1,\frac{m_1}{\Omega_1}Q\right). \quad (A5)$$

Appendix B

In order to obtain the solution for the integral I_3, we apply the transformation of the MeijerG-function provided by ([56], (07.34.17.0012.01)):

$$I_3 = \int_0^{\gamma_{th}} u^{m_2-1} G_{2,3}^{2,1}\left(\frac{m_2 u}{K_1 \Omega_2 Q} \bigg| \begin{matrix} 1-m_2-\frac{\beta_i+1}{\delta}, & 1-m_2+m_1+k \\ m_1-m_2+k, & 0, & -m_2-\frac{\beta_i+1}{\delta} \end{matrix}\right) du. \quad (A6)$$

Then, by using ([56], (07.34.21.0084.01)), the integral I_3 is solved in the exact closed form:

$$I_3 = \gamma_{th}^{m_2} G_{3,4}^{2,2}\left(\frac{m_2 \gamma_{th}}{K_1 \Omega_2 Q} \bigg| \begin{matrix} 1-m_2, & 1-m_2-\frac{\beta_i+1}{\delta}, & 1-m_2+m_1+k \\ m_1-m_2+k, & 0, & -m_2-\frac{\beta_i+1}{\delta}, & -m_2 \end{matrix}\right). \quad (A7)$$

We are solving integral I_4 by transforming the hypergeometric function into a Meijer G-function using equation ([56], (07.20.26.0006.01))

$$I_4 = \frac{\Gamma\left(1+m_2+\frac{\beta_i+1}{\delta}\right)}{\Gamma\left(m_2+\frac{\beta_i+1}{\delta}\right)} \int_0^{\gamma_{th}} u^{m_2-1} G_{1,2}^{1,1}\left(\frac{m_2 u}{K_2 \Omega_2} \bigg| \begin{matrix} 1-m_2-\frac{\beta_i+1}{\delta} \\ 0, & -m_2-\frac{\beta_i+1}{\delta} \end{matrix}\right) du, \quad (A8)$$

while the final expression is obtained by using identity ([56], (07.34.21.0084.01))

$$I_4 = \left(m_2 + \frac{\beta_i+1}{\delta}\right) \gamma_{th}^{m_2} G_{2,3}^{1,2}\left(\frac{m_2 \gamma_{th}}{K_2 \Omega_2} \bigg| \begin{matrix} 1-m_2, & 1-m_2-\frac{\beta_i+1}{\delta} \\ 0, & -m_2-\frac{\beta_i+1}{\delta}, & -m_2 \end{matrix}\right). \quad (A9)$$

Appendix C

In (27), we express the logarithm function as a Meijer G-function, using ([56], (01.04.26.0003.01)) as

$$\ln(1+u) = G_{2,2}^{1,2}\left(u \bigg| \begin{matrix} 1,1 \\ 1,0 \end{matrix}\right). \quad (A10)$$

Based on the previously derived closed-form Equation (19) for the PDF of the SNR at the secondary receiver, we transform the argument of the Meijer G-function in the PDF expression using ([56], (07.34.16.0002.01)), and we express the hypergeometric function in the PDF expression using the Meijer G-function, as in ([56], (07.20.26.0006.01)). Thus, the obtained PDF expression can be written in the following form:

$$\begin{aligned}
f_\gamma(u) &= C \frac{u^{m_2-1}}{K_1^{m_2}} \sum_{i=1}^n \sum_{k=0}^\infty \frac{1}{k!} \left(-\frac{m_1}{\Omega_1}\right)^k B_i \frac{\left(m_2+\frac{\beta_i+1}{\delta}\right)}{m_2 \delta + \beta_i + 1} \\
&\quad \times Q^{m_1-m_2+k} G_{2,3}^{2,1}\left(\frac{m_2 u}{K_1 \Omega_2 Q} \bigg| \begin{matrix} 1-m_2-\frac{\beta_i+1}{\delta}, & 1-m_2+m_1+k \\ m_1-m_2+k, & 0, & -m_2-\frac{\beta_i+1}{\delta} \end{matrix}\right) \\
&\quad + C \frac{u^{m_2-1}}{K_2^{m_2}} \left(\frac{\Omega_1}{m_1}\right)^{m_1} \Gamma\left(m_1, \frac{m_1}{\Omega_1} Q\right) \\
&\quad \times \sum_{i=1}^n \frac{B_i}{m_2\delta+\beta_i+1} \left(m_2+\frac{\beta_i+1}{\delta}\right) G_{1,2}^{1,1}\left(\frac{m_2 u}{K_2 \Omega_2} \bigg| \begin{matrix} 1-m_2-\frac{\beta_i+1}{\delta} \\ 0, & -m_2-\frac{\beta_i+1}{\delta} \end{matrix}\right).
\end{aligned} \quad (A11)$$

By replacing previous Equations (A10) and (A11) in (27), we obtain

$$
\begin{aligned}
C_{erg} &= \frac{1}{\ln 2} \frac{C}{K_1^{m_2}} \sum_{i=1}^{n} \sum_{k=0}^{\infty} \frac{1}{k!} \left(-\frac{m_1}{\Omega_1}\right)^k B_i \frac{\left(m_2 + \frac{\beta_i+1}{\delta}\right)}{m_2\delta + \beta_i + 1} Q^{m_1-m_2+k} \\
&\times \int_0^{\infty} u^{m_2-1} G_{2,2}^{1,2}\left(u \left|\begin{array}{c} 1,1 \\ 1,0 \end{array}\right.\right) G_{2,3}^{2,1}\left(\frac{m_2 u}{K_1 \Omega_2 Q} \left|\begin{array}{ccc} 1-m_2-\frac{\beta_i+1}{\delta}, & 1-m_2+m_1+k \\ m_1-m_2+k, & 0, & -m_2-\frac{\beta_i+1}{\delta} \end{array}\right.\right) du \\
&+ \frac{1}{\ln 2} \frac{C}{K_2^{m_2}} \left(\frac{\Omega_1}{m_1}\right)^{m_1} \Gamma\left(m_1, \frac{m_1}{\Omega_1} Q\right) \sum_{i=1}^{n} B_i \frac{\left(m_2 + \frac{\beta_i+1}{\delta}\right)}{m_2\delta + \beta_i + 1} \\
&\times \int_0^{\infty} u^{m_2-1} G_{2,2}^{1,2}\left(u \left|\begin{array}{c} 1,1 \\ 1,0 \end{array}\right.\right) G_{1,2}^{1,1}\left(\frac{m_2 u}{K_2 \Omega_2} \left|\begin{array}{cc} 1-m_2-\frac{\beta_i+1}{\delta} \\ 0, & -m_2-\frac{\beta_i+1}{\delta} \end{array}\right.\right) du.
\end{aligned}
\tag{A12}
$$

Finally, using ([56], (07.34.21.0011.01)) on both integrals, we obtain the closed-form expression provided by (29).

References

1. Jameel, F.; Hamid, Z.; Jabeen, F.; Zeadally, S.; Javed, M.A. A Survey of Device-to-Device Communications: Research Issues and Challenges. *IEEE Commun. Surv. Tutor.* 2018, 20, 2133–2168. [CrossRef]
2. Dangi, R.; Lalwani, P.; Choudhary, G.; You, I.; Pau, G. Study and Investigation on 5G Technology: A Systematic Review. *Sensors* 2022, 22, 26. [CrossRef] [PubMed]
3. Lin, Z.; Lin, M.; de Cola, T.; Wang, J.-B.; Zhu, W.-P.; Cheng, J. Supporting IoT with Rate-Splitting Multiple Access in Satellite and Aerial-Integrated Networks. *IEEE Internet Things J.* 2021, 8, 11123–11134. [CrossRef]
4. Lin, Z.; Lin, M.; Champagne, B.; Zhu, W.-P.; Al-Dhahir, N. Secrecy-Energy Efficient Hybrid Beamforming for Satellite-Terrestrial Integrated Networks. *IEEE Trans. Commun.* 2021, 69, 6345–6360. [CrossRef]
5. Lin, Z.; Lin, M.; Wang, J.-B.; de Cola, T.; Wang, J. Joint Beamforming and Power Allocation for Satellite-Terrestrial Integrated Networks with Non-Orthogonal Multiple Access. *IEEE J. Sel. Top. Signal Process.* 2019, 13, 657–670. [CrossRef]
6. Lin, Z.; Niu, H.; An, K.; Wang, Y.; Zheng, G.; Chatzinotas, S.; Hu, Y. Refracting RIS-Aided Hybrid Satellite-Terrestrial Relay Networks: Joint Beamforming Design and Optimization. *IEEE Trans. Aerosp. Electron. Syst.* 2022, 58, 3717–3724. [CrossRef]
7. Khan, S.; Alvi, A.N.; Javed, M.A.; Roh, B.-H.; Ali, J. An Efficient Superframe Structure with Optimal Bandwidth Utilization and Reduced Delay for Internet of Things Based Wireless Sensor Networks. *Sensors* 2020, 20, 1971. [CrossRef]
8. Hu, F.; Chen, B.; Zhu, K. Full spectrum sharing in cognitive radio networks toward 5G: A survey. *IEEE Access* 2018, 6, 15754–15776. [CrossRef]
9. Yau, K.-L.A.; Qadir, J.; Wu, C.; Imran, M.A.; Ling, M.H. Cognition-Inspired 5G Cellular Networks: A Review and the Road Ahead. *IEEE Access* 2018, 6, 35072–35090. [CrossRef]
10. Awin, F.A.; Alginahi, Y.M.; Abdel-Raheem, E.; Tepe, K. Technical Issues on Cognitive Radio-Based Internet of Things Systems: A Survey. *IEEE Access* 2019, 7, 97887–97908. [CrossRef]
11. Liang, Y.-C.; Chen, K.-C.; Li, G.Y.; Mahonen, P. Cognitive radio networking and communications: An overview. *IEEE Trans. Veh. Technol.* 2011, 60, 3386–3407. [CrossRef]
12. Goldsmith, A.; Jafar, S.A.; Maric, I.; Srinivasa, S. Breaking spectrum gridlock with cognitive radios: An information theoretic perspective. *Proc. IEEE* 2009, 97, 894–914. [CrossRef]
13. Liu, X.; Xueyan, Z. NOMA-based resource allocation for cluster-based cognitive industrial internet of things. *IEEE Trans. Ind. Inform.* 2019, 16, 5379–5388. [CrossRef]
14. Ansere, J.A.; Han, G.; Wang, H.; Choi, C.; Wu, C. A Reliable Energy Efficient Dynamic Spectrum Sensing for Cognitive Radio IoT Networks. *IEEE Internet Things J.* 2019, 6, 6748–6759. [CrossRef]
15. Kumar, B.; Dhurandher, S.K.; Woungang, I. A survey of overlay and underlay paradigms in cognitive radio networks. *Int. J. Commun. Syst.* 2017, 31, 2. [CrossRef]
16. Jarrouj, J.; Blagojevic, V.; Ivanis, P. Outage Probability and Ergodic Capacity of Spectrum-Sharing Systems with MRC Diversity. *Frequenz* 2016, 70, 157–171. [CrossRef]
17. Kozić, N.; Blagojević, V.; Ivaniš, P. Performance Analysis of Underlay Cognitive Radio System with Self-Sustainable Relay and Statistical CSI. *Sensors* 2021, 21, 3727. [CrossRef] [PubMed]
18. Zeadally, S.; Shaikh, F.K.; Talpur, A.; Sheng, Q.Z. Design architectures for energy harvesting in the Internet of Things. *Renew. Sustain. Energy Rev.* 2020, 128, 109901. [CrossRef]
19. Özyurt, S.; Coşkun, A.F.; Büyükçorak, S.; Karabulut Kurt, G.; Kucur, O. A Survey on Multiuser SWIPT Communications for 5G+. *IEEE Access* 2022, 10, 109814–109849. [CrossRef]
20. Ashraf, N.; Sheikh, S.A.; Khan, S.A.; Shayea, I.; Jalal, M. Simultaneous Wireless Information and Power Transfer with Cooperative Relaying for Next-Generation Wireless Networks: A Review. *IEEE Access* 2021, 9, 71482–71504. [CrossRef]
21. Nasir, A.A.; Zhou, X.; Durrani, S.; Kennedy, R.A. Relaying protocols for wireless energy harvesting and information processing. *IEEE Trans. Wirel. Commun.* 2013, 12, 3622–3636. [CrossRef]
22. Blagojevic, V.M.; Cvetkovic, A.M.; Ivanis, P. Performance analysis of energy harvesting DF relay system in generalized-K fading environment. *Phys. Commun.* 2018, 28, 190–200. [CrossRef]

23. Liu, L.; Zhang, R.; Chua, K.-C. Wireless information transfer with opportunistic energy harvesting. In Proceedings of the 2012 IEEE International Symposium on Information Theory, Cambridge, MA, USA, 1–6 July 2012; pp. 950–954. [CrossRef]
24. Phan, V.-D.; Nguyen, T.N.; Tran, M.; Trang, T.T.; Voznak, M.; Ha, D.-H.; Nguyen, T.-L. Power Beacon-Assisted Energy Harvesting in a Half-Duplex Communication Network under Co-Channel Interference over a Rayleigh Fading Environment: Energy Efficiency and Outage Probability Analysis. *Energies* **2019**, *12*, 2579. [CrossRef]
25. Nobar, S.K.; Mehr, K.A.; Niya, J.M.; Tazehkand, B.M. Cognitive radio sensor network with green power beacon. *IEEE Sens. J.* **2017**, *17*, 1549–1561. [CrossRef]
26. Xu, C.; Zheng, M.; Liang, W.; Yu, H.; Liang, Y.-C. Outage Performance of Underlay Multihop Cognitive Relay Networks with Energy Harvesting. *IEEE Commun. Lett.* **2016**, *20*, 1148–1151. [CrossRef]
27. Govindan, K.; Zeng, K.; Mohapatra, P. Probability density of the received power in mobile networks. *IEEE Trans. Wirel. Commun.* **2011**, *10*, 3613–3619. [CrossRef]
28. Camp, T.; Boleng, J.; Davies, V. A survey of mobility models for ad hoc network research. *Wirel. Commun. Mob. Comput.* **2002**, *2*, 483–502. [CrossRef]
29. Bettstetter, C.; Resta, G.; Santi, P. The node distribution of the random waypoint mobility model for wireless ad hoc networks. *IEEE Trans. Mob. Comput.* **2003**, *2*, 256–269. [CrossRef]
30. Nain, P.; Towsley, D.; Lui, B.; Liu, Z. Properties of random direction models. *Proc. IEEE INFOCOM* **2005**, *3*, 1897–1907. [CrossRef]
31. Meesa-Ard, E.; Pattaramalai, S. Evaluating the Mobility Impact on the Performance of Heterogeneous Wireless Networks Over η–μ Fading Channels. *IEEE Access* **2021**, *9*, 65017–65032. [CrossRef]
32. Aalo, V.A.; Mukasa, C.; Efthymoglou, G.P. Effect of Mobility on the Outage and BER Performances of Digital Transmissions over Nakagami-m Fading Channels. *IEEE Trans. Veh. Technol.* **2016**, *65*, 42715–42721. [CrossRef]
33. Amjad, M.; Chughtai, O.; Naeem, M.; Ejaz, W. SWIPT-Assisted Energy Efficiency Optimization in 5G/B5G Cooperative IoT Network. *Energies* **2021**, *14*, 2515. [CrossRef]
34. Tin, P.T.; Dinh, B.H.; Nguyen, T.N.; Ha, D.H.; Trang, T.T. Power Beacon-Assisted Energy Harvesting Wireless Physical Layer Cooperative Relaying Networks: Performance Analysis. *Symmetry* **2020**, *12*, 106. [CrossRef]
35. Badarneh, O.S. A comprehensive analysis of the achievable throughput in interference-limited wireless-powered networks with nonlinear energy harvester. *Trans. Emerg. Telecommun. Technol.* **2020**, *31*, e4141. [CrossRef]
36. Zhao, F.; Lin, H.; Zhong, C.; Hadzi-Velkov, Z.; Karagiannidis, G.K.; Zhang, Z. On the Capacity of Wireless Powered Communication Systems Over Rician Fading Channels. *IEEE Trans. Commun.* **2018**, *66*, 404–417. [CrossRef]
37. Tin, P.T.; Phan, V.D.; Nguyen, T.N.; Tu, L.T.; Minh, B.V.; Voznak, M.; Fazio, P. Outage Analysis of the Power Splitting Based Underlay Cooperative Cognitive Radio Networks. *Sensors* **2021**, *21*, 7653. [CrossRef] [PubMed]
38. Le, N.P. Outage Probability Analysis in Power-Beacon Assisted Energy Harvesting Cognitive Relay Wireless Networks. *Wirel. Commun. Mob. Comput.* **2017**, *2017*, 2019404. [CrossRef]
39. Badarneh, O.S.; Da Costa, D.B.; Nardelli, P.H.J. Wireless-Powered Communication Networks with Random Mobility. *IEEE Access* **2019**, *7*, 166476–166492. [CrossRef]
40. Badarneh, O.S.; Benevides da Costa, D.; Nardelli, P.H.J. Transmit Antenna Selection in Wireless-Powered Communication Networks. In Proceedings of the IEEE 30th Annual International Symposium on Personal, Indoor and Mobile Radio Communications (PIMRC), Istanbul, Turkey, 8–11 September 2019; pp. 1–6. [CrossRef]
41. Aalo, V.A.; Bithas, P.S.; Efthymoglou, G.P. Ergodic Capacity of Generalized Fading Channels with Mobility. *IEEE Open J. Veh. Technol.* **2022**, *3*, 15–25. [CrossRef]
42. Meesa-Ard, E.; Pattaramalai, S.; Madapatha, M.D.C. Evaluating the Impact of Mobility over k–μ Generalized Fading Channels in Digital Communication. In Proceedings of the 2018 8th International Conference on Electronics Information and Emergency Communication (ICEIEC), Beijing, China, 15–17 June 2018; pp. 35–39. [CrossRef]
43. Li, C.; Yao, J.; Wang, H.; Ahmed, U.; Du, S. Effect of Mobile Wireless on Outage and BER Performances Over Rician Fading Channel. *IEEE Access* **2020**, *8*, 91799–91806. [CrossRef]
44. Meesa-Ard, E.; Pattaramalai, S. Analyzing the Impact of Mobility over α-μ Generalized Fading Channels in Wireless Communication. In Proceedings of the 3rd International Conference on Computer and Communication Systems (ICCCS), Nagoya, Japan, 27–30 April 2018; pp. 318–322. [CrossRef]
45. Aalo, V.A.; Bithas, P.S.; Efthymoglou, G.P. On the Impact of User Mobility on the Performance of Wireless Receivers. *IEEE Access* **2020**, *8*, 197300–197311. [CrossRef]
46. Yao, J.; Li, C.; Du, S.; Wu, W.; Gao, R. Outage Probability over Nakagami−m Fading Channel in the Random Direction Mobile Model. In Proceedings of the 2020 International Conference on Information and Communication Technology Convergence (ICTC), Jeju, Republic of Korea, 21–23 October 2020; pp. 427–430. [CrossRef]
47. Das, M.; Sahu, B. Effect of MRC Diversity on Outage Probability in Mobile Networks. In Proceedings of the 2019 Global Conference for Advancement in Technology (GCAT), Bangalore, India, 18–20 October 2019; pp. 1–4. [CrossRef]
48. Ju, P.; Song, W.; Jin, A.-L. Exact Outage Probability for a Wireless Diversity Network with Spatially Random Mobile Relays. *IEEE Commun. Lett.* **2014**, *18*, 1641–1644. [CrossRef]
49. Odeyemi, K.O.; Owolawi, P.A.; Olakanmi, O.O. On the performance of underlay cognitive radio system with random mobility under imperfect channel state information. *Int. J. Commun. Syst.* **2020**, *33*, e4561. [CrossRef]

50. Yacoub, M.D.; Bautista, J.E.V.; Guerra de Rezende Guedes, L. On higher order statistics of the Nakagami-m distribution. *IEEE Trans. Veh. Technol.* **1999**, *48*, 790–794. [CrossRef]
51. Papoulis, A. *Probability, Random Variables, and Stochastic Processes*; McGraw-Hill: New York, NY, USA, 1991.
52. Goldsmith, A. *Wireless Communications*; Cambridge University Press: Cambridge, UK, 2005.
53. Jeruchim, M.C.; Balaban, P.; Shanmugan, K.S. *Simulation of Communication Systems: Modeling, Methodology and Techniques*; Springer Science & Business Media: Berlin, Germany, 2006.
54. Zheng, Y.R.; Xiao, C. Simulation models with correct statistical properties for Rayleigh fading channels. *IEEE Trans. Commun.* **2003**, *51*, 920–928. [CrossRef]
55. Gradshteyn, I.S.; Ryzhik, I.M. *Table of Integrals, Series and Products*, 5th ed.; Academic Press Inc.: San Diego, CA, USA, 1994.
56. Available online: https://functions.wolfram.com/ (accessed on 20 December 2022).

Disclaimer/Publisher's Note: The statements, opinions and data contained in all publications are solely those of the individual author(s) and contributor(s) and not of MDPI and/or the editor(s). MDPI and/or the editor(s) disclaim responsibility for any injury to people or property resulting from any ideas, methods, instructions or products referred to in the content.

Article

Resource Allocation for Secure MIMO-SWIPT Systems in the Presence of Multi-Antenna Eavesdropper in Vehicular Networks

Vieeralingaam Ganapathy [1], Ramanathan Ramachandran [1] and Tomoaki Ohtsuki [2,*]

[1] Department of Electronics and Communication Engineering, Amrita School of Engineering, Amrita Vishwa Vidyapeetham, Coimbatore 641112, India; g_vieeralingaam@cb.students.amrita.edu (V.G.); r_ramanathan@cb.amrita.edu (R.R.)
[2] Department of Information and Computer Science, Keio University, Tokyo 108-8345, Japan
* Correspondence: ohtsuki@keio.jp

Abstract: In this paper, we optimize the secrecy capacity of the legitimate user under resource allocation and security constraints for a multi-antenna environment for the simultaneous transmission of wireless information and power in a dynamic downlink scenario. We study the relationship between secrecy capacity and harvested energy in a power-splitting configuration for a nonlinear energy-harvesting model under co-located conditions. The capacity maximization problem is formulated for the vehicle-to-vehicle communication scenario. The formulated problem is non-convex NP-hard, so we reformulate it into a convex form using a divide-and-conquer approach. We obtain the optimal transmit power matrix and power-splitting ratio values that guarantee positive values of the secrecy capacity. We analyze different vehicle-to-vehicle communication settings to validate the differentiation of the proposed algorithm in maintaining both reliability and security. We also substantiate the effectiveness of the proposed approach by analyzing the trade-offs between secrecy capacity and harvested energy.

Keywords: vehicular networks; SWIPT; MIMO; secrecy capacity; convex optimization

Citation: Ganapathy, V.; Ramachandran, R.; Ohtsuki, T. Resource Allocation for Secure MIMO-SWIPT Systems in the Presence of Multi-Antenna Eavesdropper in Vehicular Networks. *Sensors* **2023**, *23*, 8069. https://doi.org/10.3390/s23198069

Academic Editor: Davy P. Gaillot

Received: 28 August 2023
Revised: 14 September 2023
Accepted: 22 September 2023
Published: 25 September 2023

Copyright: © 2023 by the authors. Licensee MDPI, Basel, Switzerland. This article is an open access article distributed under the terms and conditions of the Creative Commons Attribution (CC BY) license (https://creativecommons.org/licenses/by/4.0/).

1. Introduction

Simultaneous wireless information and power transfer (SWIPT) is an optimistic and robust method for sustainable power supply to wireless networks [1]. The lifetime of a typical node in wireless networks can be extended by energy harvesting. Solar and RF energy are popular sources of energy in the environment. RF energy sources are used for energy harvesting to make a system independent of weather conditions. Also, the RF signals can carry information and energy simultaneously [1], hence the name SWIPT.

The challenges of controlling the data rate and desired quality of service necessitate a cooperative multiple-input multiple-output (CMIMO) system [2]. CMIMO systems can improve the performance of wireless networks and wireless sensor networks. A comprehensive overview of various supporting technologies, such as compressive sensing and SWIPT, can be augmented to CMIMO to improve the throughput and energy efficiency performance. The challenges involved in the augmented CMIMO systems, energy-saving techniques, and overview of protocol layers are analyzed in [3]. To increase the capacity and coverage of wireless networks, heterogeneous networks with macrocells and femtocells have become a popular architecture. However, the regulation of interference between these layers still poses a serious difficulty and requires efficient mitigation techniques. In [4], a joint optimization problem of maximizing the sum rates to inter-tier interference under resource allocation constraints is proposed to reduce the interference to macrocell users.

The use of different SWIPT system settings in a variety of applications, including Internet of Things (IoT) and biosensors, has prompted researchers to investigate the reliability and sustainability of the SWIPT system by analyzing the rate–energy trade-off. In the MIMO environment, the transmit power affects the performance of the SWIPT system.

The performance analysis can be conducted by calculating the mean square error (MSE) of the detected/received symbol [5] and also by analyzing the relationship between the transmit power and the interference regimes present at the receiver end [6]. Since the transmit power is directly related to the information rate of a legitimate receiver, the security aspect of wireless communication systems must be analyzed. For example, the usability of MIMO systems without secure protocols will compromise the system; therefore, secure transmission algorithms in the presence of multiple eavesdroppers are studied in [7]. The keyless physical layer security (PLS) represents a subset of theoretical paradigms in wireless communications that optimize the inherent properties of the physical layer to ensure secrecy/reliability (without the need for traditional cryptographic keys). A comprehensive survey based on keyless PLS for ensuring secrecy in the wireless network environment is studied in [8], where the security challenges in IoT and V2X networks are discussed. Also in [9], a survey is conducted of the impact of security on the sixth-generation (6G) wireless technologies, network architecture and potential applications. A detailed analysis related to the security requirements, distributed machine learning (DML) is also studied.

A multi-agent deep reinforcement learning (DRL)-based approach is studied in [10] for the vehicular edge-computing network, where a secure resource allocation strategy is proposed by optimizing the transmission power, spectrum and computation resource allocation components. The DRL approach proposed in [10] significantly reduces the delay while maintaining the confidentiality probability. In the context of IoT networks, work in [11] explores the cooperative potential of non-orthogonal multiple access (NOMA) for SWIPT, addressing the challenges of imperfect successive interference cancellation (SIC) and using deep learning to optimize the throughput performance. In [12], a novel semi-supervised intrusion detection method based on federated learning is proposed to improve the quality of the predicted outputs (thereby avoiding incorrect predictions and achieving lower communication overhead). The recent work proposed in [13] explores the potential of deep learning applications in device-to-device (D2D) unmanned aerial vehicle (UAV) communication, leveraging the use of both SWIPT and multi-agent deep Q-networks (MADQN) to improve energy efficiency based on the design of reward functions. In the context of the rapidly evolving and highly dynamic vehicular ad hoc networks (VANETs) landscape, as studied in [14], the deep neural network framework for anomaly detection contributes to jointly enhance the security and reliability of these networked systems.

Dynamic settings and the mobility of devices in wireless network environments (such as unmanned aerial vehicle (UAV) and ad hoc networks) present additional difficulties and opportunities for ensuring effective and reliable data transmission and resource management. Vehicle clustering provides a solution for dynamic wireless networks, enabling effective resource management and solving mobility issues. For ad hoc networks, the vehicle-clustering algorithm plays a key role in establishing effective communication and resource sharing among vehicles. In [15], a clustering algorithm is proposed to minimize the total energy consumption of vehicles based on their direction and entropy using a fuzzy C-means algorithm. As discussed in [16], three-dimensional trajectory planning in UAV networks is crucial to enable seamless cooperation and data exchange/collection among UAVs, optimize their flight paths, and efficiently utilize scarce resources.

The transmit power, along with the number of antennas, plays a significant role in determining the secrecy rate of the system. Due to the characteristics of RF signals in SWIPT systems, separate and co-located energy-harvesting/information-decoding (EH/ID) receivers with corresponding precoder designs are being developed [17]. The characterization of different capacity scaling techniques (in general, the capacity values vary depending on the base station (BS) antennas, signal-to-noise ratio (SNR), beamforming architecture, and coherence block size) are investigated for the massive MIMO systems under strong spatial correlation regimes as shown in [18]. Based on the optimization of transmit beamforming vectors and power-splitting ratios, the battery depletion phenomenon is mitigated by formulating an optimal resource allocation strategy based on the transmit power optimization algorithm [19]. The channel characteristics affect the received energy in a wireless network, thus affecting the energy-harvesting capability of the intended receiver. The amount of energy harvested for passively powered sensor networks is investigated in [20], where the circuit is designed and relationships between the DC power and received power are analyzed.

Vehicular networks are becoming increasingly interconnected and autonomous, leading to the need for reliable, secure, and sustainable communication mechanisms. Recent research has shown that integrating SWIPT into vehicular networks offers promising solutions to energy sustainability concerns [21,22]. As relays increasingly become "hubs"—collecting, sending, and receiving large amounts of information—the energy and security requirements of in-vehicle communication systems are increased dramatically [23]. Integrating vehicle networks with SWIPT promises a dual benefit: reliable communications, and reduced dependence on external power sources. This synergy of security and energy harvesting can significantly increase the potential autonomy of vehicular communication systems. The impact of distance and control over the power-splitting factor on physical layer security is studied in [24]. The authors also investigated jamming techniques to improve security (as the distance between jammer and eavesdropper changes).

1.1. Motivation

The presence of a sophisticated eavesdropper equipped with multiple antennas adds another layer of complexity. Multi-antenna eavesdroppers can intercept signals from different spatial paths, posing a significant threat to MIMO communications and SWIPT systems. A nuanced approach to resource allocation is essential to ensure both efficient energy harvesting and secure data transmission in such scenarios. Our work aims to address the overlapping challenges of energy and PLS management in vehicular networks, laying the foundation for a more resilient and sustainable vehicular communication system. With the significant modernization and increased applicability of MIMO-SWIPT vehicular networks in industrial applications, ensuring secure communication in the presence of potential eavesdroppers becomes a paramount concern. The potential vulnerabilities posed by multi-antenna eavesdroppers in such vehicular networks necessitate the development of robust resource allocation strategies to secure wireless power and information transmission. By optimizing the transmit power allocation, our work aims to enhance the secrecy capacity and harvested energy in MIMO-SWIPT systems, thus ensuring reliable and secure communication for vehicular applications. Our research aims to contribute to the advancement of secure resource allocation techniques, enabling sustainable and secure wireless communication integration in vehicular networks while addressing security challenges in dynamic communication environments.

1.2. Contributions

In this paper, we present a resource allocation strategy for a MIMO SWIPT system in vehicular networks to optimize the information capacity of multi-antenna Bob under resource allocation constraints in the presence of a multi-antenna passive eavesdropper. The information capacity maximization problem with secrecy rate and quality-of-service constraints is formulated for different positions of Bob, and transmit power/power-splitting (PS) ratio values are achieved using the divide-and-conquer strategy. The following dynamic scenario is considered: Bob is moving from position-A (Pos-A) to position-B (Pos-B). Also, Pos-B is further away from Alice (compared to Pos-A) because Bob traverses from Pos-A to Pos-B away from Alice in the presence of a multi-antenna eavesdropper (which is positioned relatively close to Alice). The contributions of this paper are as follows:

- We present two transmission scenarios in SWIPT-enabled vehicular networks in the presence of a multi-antenna eavesdropper: Bob is at Pos-A, and Bob is at Pos-B.
- We formulate corresponding optimization problems for two scenarios: when Bob is at Pos-A and when Bob is at Pos-B under resource allocation constraints.
- We maximize the information capacity under the secrecy capacity, harvested energy requirements, and other resource allocation constraints.

- We propose a divide-and-conquer strategy by splitting the optimization problem into two sub-problems (alternating the choice of optimization variables). By optimizing the power-splitting ratio, our proposed algorithm can allocate the harvested energy to the EH/ID receiver and maintain sufficient secrecy capacity.
- We provide simulation results, highlighting the differences between the transmission scenarios considered. We also validate the need for the proposed scheme by analyzing the performance under the two considered scenarios.

1.3. Related Work

The transmit power parameter is one of the key variables in determining the information rate for wireless communication systems. The transmit power can be optimized by using beamforming resource allocation strategies. Since we incorporated SWIPT in the MIMO system, there is also a need to optimize the value of the power-splitting ratio. The system is modeled without the inclusion of SWIPT, and is studied in the works [6,18], where the beamforming problem for cognitive radio and MIMO environments is developed. In [6], the transmit power for secondary users is minimized, while the power for the primary user is increased. The problem is reformulated as a relaxed semidefinite problem, and transmit power/SNR analysis is performed for primary and secondary users.

In [18], the achievable rate in a MIMO environment is analyzed. The rate-scaling characteristics are studied for different values of antennas. The analytical structure of the information rate depends on the selected channel characteristics. An investigation of the use of multiple transmit/receive antennas for a single-user communication model is studied for fading and non-fading channels [25]. The capacities and error exponents of both fading and non-fading channels are formulated. The result shows that the use of multiple antennas significantly increases the information capacity of a single user.

By incorporating a SWIPT perspective into the wireless communication model, the rate–energy (R-E) characteristics are analyzed as in [1,3], and the corresponding beamforming vector optimization strategy is used to improve the efficiency (by preventing depletion and ensuring the latency/harvesting required amount of energy) as in [19]. The characteristics of EH/ID receivers are studied for various practical designs. The co-located and separated EH/ID receivers and their relationship with information rates are analyzed. The reliability of the practical parameters present in the receiver circuit must be introspected and incorporated in the study of the rate–energy trade-off. The effect of the circuit specification on the R-E region is analyzed in [3]. A multi-antenna SWIPT framework is studied in [19], where both (i) the semi-definite relaxation (SDR) approach combined with fractional programming (FP), and (ii) successive convex approximation (SCA) are proposed to achieve a robust and energy-efficient resource allocation strategy to mitigate battery depletion.

The hardware imperfections and distortions occur due to imbalance and nonlinearity in the high-power amplifier (HPA) and phase noise, which can degrade the quality of the EH/ID receivers. In [26], the estimation of distortions present in a SWIPT power-splitting system is investigated. The nonlinearity and total harmonic distortion present in the output of the HPA are analyzed using the recursive least squares technique. An analysis of the received SNR and harvested energy is performed for different harmonic distortion characteristics. In [27], a PS-based SWIPT system is studied under hardware impairments and in-phase and quadrature imbalances. An optimization problem is proposed to maximize the harvested energy under a SNR constraint in the presence of hardware impairments. An optimal value of harvested energy is achieved using bio-inspired algorithms in a Rayleigh fading environment.

The energy-harvesting capability can also cause interference in the communication environment, so the selection of optimal users based on a resource allocation strategy is developed in [28]. The use and design of appropriate variables in the resource allocation algorithms guarantee the selection of sensors and users depending on the defined optimization problem under the energy-harvesting wireless network environment [28].

The number of deployable nodes and transmission strategies is increasing in a wireless network environment, so several relay selection strategies need to be explored. In [29], a novel and efficient relay node selection scheme based on a power-splitting ratio for decoding and forward cooperative relay networks under the SWIPT framework is proposed. A power-splitting scheme is formulated to express the rate–energy exchange for the energy used for information decoding. The expression for the probability of failure is also derived, and a performance metric is analyzed in terms of energy conversion efficiency and SNR at different thresholds of the data rates.

The secrecy rate of the system is analyzed as a min-max problem for different ranges of SNR [30] in the presence of an active eavesdropper attack. The effect of multiple antennas between the sender, receiver, and eavesdropper on the secrecy rate is also analyzed. A trade-off between the average harvested energy and secrecy rate is analyzed for secure SWIPT in cell-free MIMO systems with multiple access points (APs) transmitting information to users. In [31], the authors extensively analyze the phenomenon of active pilot attack (by eavesdroppers) to compromise the base station channel and propose a methodology to achieve a secure link in intelligent reflecting surface (IRS)-assisted MIMO systems. The authors also successfully suppress the active eavesdropper using beamforming optimization (which is also validated in the trade-off curve between the secrecy capacity and eavesdropper position).

To achieve a secure wireless communication environment, a cognitive radio transmission system for decode and forward UAV with energy harvesting at the source and relay nodes is proposed [32]. The distribution of the probability of the non-zero secrecy rate of the system under the time-shared protocol is studied. A performance analysis based on the optimal secrecy rate selection strategy and the optimal antenna selection strategy at the destination node of the secondary network is investigated. The resource allocation algorithm of the SWIPT system can be extended to the security aspect of the algorithm by incorporating an analysis involving two significant characteristics: secrecy throughput, and transmission schemes, such as delay-constrained transmission and delay-tolerant transmission [33].

A resource allocation strategy is proposed to achieve a secure information rate and green power transmission for mobile receivers with distributed antennas connected to a central processor [34]. The total network transmission power is minimized under quality-of-service constraints. An optimal iterative algorithm based on generalized Bender's decomposition is proposed.

Most practical applications require the optimization of rate and energy variables, so such variants of optimization problems are biconvex. An extensive survey of the theory of biconvex sets is given in [35], and biconvex optimization problems are solved by exploiting the properties of biconvex sets. The mathematical approach of convex optimization always aims to find optimal or suboptimal values of the variables under consideration. The penalty convex–concave approach is used to perform first-order convex approximation to solve the bilinear difference of convex problem [5]. The SDR approach for the co-located receiver is developed to achieve optimal parameter values [17].

As the technology evolves, the simulation of MIMO wireless environments requires an accurate, practical model for arriving at an optimal solution, so the nonlinear EH models are studied in [3]. The various aspects of resource allocation algorithms are developed to analyze the characteristics of MIMO system models. Resource allocation problems based on the system models provide an optimal solution to several key variables. This also helps to understand the relationship between these variables. As the deployability of SWIPT systems increases, there is a need to characterize the trade-off between the information and energy-harvesting parameters, which provides a detailed analysis on the SWIPT, WPT-enabled systems [1,3].

In summary, vehicular networks require a high rate of information exchange for synchronous and concurrent operations. Wireless networks are also growing in almost all practical aspects, so MIMO-based capabilities are being used to enhance various physical layer aspects of these systems. This use of MIMO technologies and the increasing robust use of RF signals have led to the development of SWIPT-enabled MIMO systems. From the SWIPT perspective of the wireless system, the receiver is required to maintain an optimal amount of energy-harvesting and information-decoding capability, as this feature determines the lifetime and reliability of the entire transmission.

To the best of the authors' knowledge, the study of optimization problems under different transmission scenarios for the MIMO-SWIPT system in the presence of a multi-antenna eavesdropper has not been studied. To elaborate the specific novelty of our work, we compared it with the previous works. The main differences are as follows:

- In [3], the main objective is to maximize the harvested energy for time-switching and power-splitting scenarios for the MIMO broadcasting system. The authors analyzed the trade-off between information capacity and energy for separated and co-located SWIPT receivers. In our proposed work, we included the maximization of the information capacity of Bob with the co-located SWIPT system in the presence of a multi-antenna eavesdropper. Our work focuses on the secrecy capacity–harvested energy trade-off under dynamic vehicular settings.
- In [30], the authors consider active eavesdropping scenarios and solve the maximized ergodic secrecy capacity problem using semidefinite relaxation programming. We considered dynamic MIMO-SWIPT settings in the presence of multi-antenna eavesdroppers under a worst-case scenario when Bob traverses from Pos-A to Pos-B (hence different from [30]).
- In [36], the authors propose a joint channel estimation and transmit power allocation strategy to maximize the average signal-to-error-plus-noise ratio (SENR) under resource allocation constraints. The authors use a Karush–Khun–Tucker (KKT)-based solution to arrive at the solution. Unlike [36], we proposed a "divide-and-conquer strategy" by reformulating the non-convex problem into convex sub-problems based on maximizing the information capacity in the presence of the secrecy capacity and other resource allocation constraints under dynamic vehicular environment.

The rest of the paper is organized as follows. Section 2 presents the problem formulation, where the system model for the MIMO-SWIPT system in the presence of a multi-antenna eavesdropper and the problem definition for both transmission scenarios are discussed. Section 3 presents the proposed solution for both transmission scenarios and a description of the algorithm. Section 4 describes the result obtained for both transmission scenarios. The complexity analysis is also presented in this section. Section 5 concludes the paper.

2. Problem Formulation

This section presents the system model and problem definition for two of the transmission modes, formulating the maximization of the information capacity under the secrecy capacity, harvested energy requirements, and other resource allocation constraints.

2.1. System Model

The system model of a dynamic SWIPT-based MIMO for a co-located EH/ID receiver for a power-splitting architecture in the presence of a multi-antenna eavesdropper is considered as shown in Figure 1. We consider a multi-antenna transmitter (Alice) with multiple antennas and a co-located energy-harvesting/information-decoding (EH/ID) receiver with multiple antennas in the presence of a multi-antenna passive eavesdropper (Eve) with the same number of antennas as the receiver. The number of antennas at the transmitter is N_T, and the numbers of antennas at the receiver and eavesdropper are N_R and N_E, respectively. Let $H_1 \in \mathbb{C}^{N_R \times N_T}$ be the channel matrix between Alice and Bob at position-A. When Bob moves from position-A (Pos-A) to position B (Pos-B), the channel fading state changes from H_1 to H_2. Let $X \in \mathbb{C}^{N_T \times N_R}$ be the transmitted signal from Alice. The covariance matrix of X is given by $E[XX^H] = \text{Tr}(Q)$, where H_E^1 and H_E^2 ($\in \mathbb{C}^{N_E \times N_T}$) denote the channel matrices between Alice and the eavesdropper (Eve), while Bob is at Pos-A and Pos-B, respectively. It is assumed that Eve is close to Alice.

We consider a large-scale fading model where D_i is given by $(\frac{d_i}{d_0})^{-\alpha}$. Here, d_i is the distance between Alice and a corresponding receiver (Bob or Eve). We consider the reference distance d_0 to be 10 m. The variable α denotes the path loss exponent with α (we take it to be three). So, we have $D_A^s = \left(\frac{d_A^s}{d_0}\right)^{-\alpha}$ for Alice-[Bob at Pos-A]; $D_B^s = \left(\frac{d_B^s}{d_0}\right)^{-\alpha}$ for Alice-[Bob at Pos-B]; and $D_E = \left(\frac{d_E}{d_0}\right)^{-\alpha}$ for Alice-Eve. Using D_A^s (for H_1), D_B^s (for H_2), and D_E (for H_E^1 & H_E^2), small-scale fading is incorporated using both line-of-sight and non-line-of-sight (LOS and NLOS) components. Since Bob traverses from Pos-A to Pos-B, and Pos-B is further away from Alice (than Pos-A), the value of d_B^s is always greater than d_A^s.

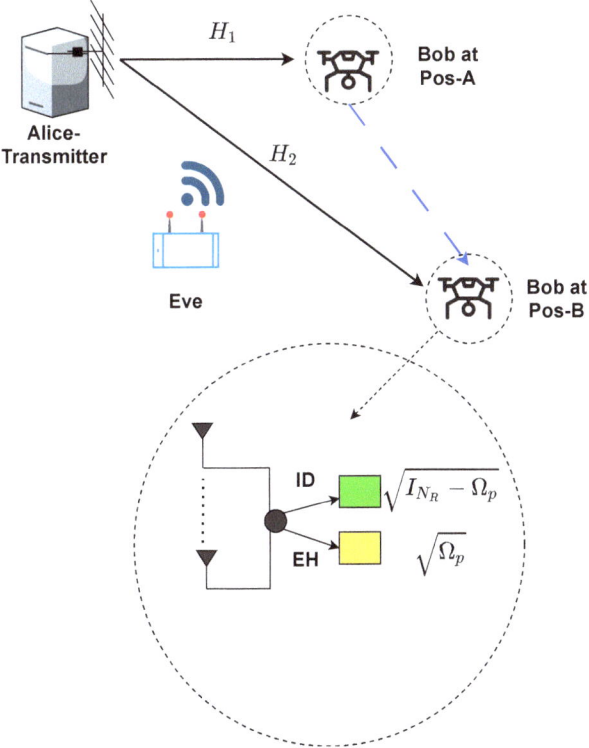

Figure 1. System model: In the downlink transmission mode, Alice transmits information and energy signal to multi-antenna co-located EH/ID legitimate receiver (called Bob). Multi-antenna eavesdropper attempts to tap information from the Alice as the Bob moves from position A to B.

2.2. Problem Definition

We aim to optimize the transmit power matrix contained in the information rate maximization problem for Bob at Pos-A in the presence of an eavesdropper. We derive a closed-form expression for obtaining the energy-harvesting values based on the nonlinear EH models and the channel fading state H_1. Theoretically, the information rate for a SWIPT system depends on the parameters: transmit power and power sharing ratio [3]. We formulate an information rate maximization problem under transmit power and secrecy capacity constraints. The received signal is modeled as a function of the power-splitting ratio $(\sqrt{\Omega_p})$ and thus determines the value of the transmit power matrix, which is used to arrive at a harvested energy value. The term "$(\sqrt{I_{N_R} - \Omega_p})$" in (2) indicates the received signal allocated for information decoding (since we considered the co-located SWIPT system). Theoretically, I_{N_R} is the upper bound for the power-splitting ratio, forming an identity matrix of dimension $N_R \times N_R$. If $\sqrt{\Omega_p}$ is allotted for energy harvesting, then the remaining $\left(\sqrt{I_{N_R} - \Omega_p}\right)$ is naturally allotted for information decoding. The received signals at the EH Y_E and ID Y_I receivers can be given by

$$Y_E^1 = \left(\sqrt{\Omega_p}\right) H_1 X, \tag{1}$$

$$Y_I^1 = \left(\sqrt{I_{N_R} - \Omega_p}\right) H_1 X + N_I^1, \tag{2}$$

where $N_I^1 \in \mathbb{C}^{N_R \times N_R} \sim \mathcal{CN}(0, \tilde{\sigma}_i^2 I_{N_R})$ is an additive complex Gaussian noise received at the ID portion of Bob at Pos-A. Let ρ_i denote the power-splitting ratio for the i-th receiver antenna, defining (Ω_p) as $\mathrm{diag}(\rho_1, \rho_2, \ldots, \rho_{N_R})$, $0 \leq \rho_i \leq 1$, where i varies from 1 to N_R. The received signal at the eavesdropper side Y_{eve} is given by

$$Y_{eve}^1 = H_E^1 X + N_{eve}^1, \tag{3}$$

where $N_{eve}^1 \in \mathbb{C}^{N_E \times N_E} \sim \mathcal{CN}(0, \tilde{\sigma}^2 I_{N_E})$ is also an additive complex Gaussian noise at Eve. Similarly, the received signals at Bob (when he is at Pos-B) are given by

$$Y_E^2 = \left(\sqrt{\Omega_p}\right) H_2 X, \tag{4}$$

$$Y_I^2 = \left(\sqrt{I_{N_R} - \Omega_p}\right) H_2 X + N_I^2. \tag{5}$$

where $N_I^2 \in \mathbb{C}^{N_R \times N_R} \sim \mathcal{CN}(0, \tilde{\sigma}_i^2 I_{N_R})$ is the additive complex Gaussian noise received at the ID portion of Bob at Pos-A. Here, the noise between Alice and the corresponding EH/ID node is varied as the position is changed. The received signal at Eve's side Y_{eve}^2 is given by

$$Y_{eve}^2 = H_E^2 X + N_{eve}^2. \tag{6}$$

where $N_{eve}^2 \in \mathbb{C}^{N_E \times N_E} \sim \mathcal{CN}(0, \tilde{\sigma}^2 I_{N_E})$ is also an additive complex Gaussian noise at Eve. The optimization problem is formulated to maximize the information rate under the constraints of transmit power and secrecy capacity for Bob at Pos-A. The $P1 - A$ problem is given by

$$P1 - A : \max_{Q, \Omega_p} \log_2 \left\| I_{N_R} + \frac{(I_{N_R} - \Omega_p)^{\frac{1}{2}} H_1 Q H_1^H (I_{N_R} - \Omega_p)^{\frac{1}{2}}}{\hat{\sigma}_i^2} \right\| \tag{7}$$

$$C1 : \mathrm{Tr}(Q) \leq P,$$

$$C2 : 0 \preccurlyeq \Omega_p \preccurlyeq I_{N_R},$$

$$C3 : Q \succcurlyeq 0,$$

$$C4 : \log_2 \left\| I_{N_R} + \frac{(I_{N_R} - \Omega_p)^{\frac{1}{2}} H_1 Q H_1^H (I_{N_R} - \Omega_p)^{\frac{1}{2}}}{\hat{\sigma}_i^2} \right\|$$

$$- \log_2 \left\| I_{N_E} + \frac{H_E^1 Q H_E^{1^H}}{\hat{\sigma}^2} \right\| \geq R,$$

$$C5 : \mathrm{Tr}\left(\Omega_p H_1 Q H_1^H\right) \geq E.$$

Constraint C1 guarantees an upper bound on the value of the transmit power available in the environment. The power-splitting ratio is also bounded between zero and one to determine the distribution of the received signal for energy harvesting or information decoding. Constraint C3 is included to guarantee a positive semidefinite property for the transmit power matrix Q. Constraint C4 is the secrecy capacity constraint that guarantees the secure transmission of information. Constraint C5 is included to satisfy the harvested energy requirement E. When Bob is at Pos-B (which is further away from Alice compared to Pos-A), the optimization problem is modified to include the channel fading H_2, so we have a problem $P2 - A$, where H_1 is replaced by H_2 and H_E^1 is replaced by H_E^2. The noise between Alice–Eve and Alice–Bob at Pos-B also changes. The optimization problem for Bob at Pos-B is given by

$$P1 - B : \max_{Q,\Omega_p} \log_2 \left\| I_{N_R} + \frac{(I_{N_R} - \Omega_p)^{\frac{1}{2}} H_2 Q H_2^H (I_{N_R} - \Omega_p)^{\frac{1}{2}}}{\tilde{\sigma}_i^2} \right\| \tag{8}$$

$$C1 : \mathrm{Tr}(Q) \leq P,$$

$$C2 : 0 \preccurlyeq \Omega_p \preccurlyeq I_{N_R},$$

$$C3 : Q \succcurlyeq 0,$$

$$C4 : \log_2 \left\| I_{N_R} + \frac{(I_{N_R} - \Omega_p)^{\frac{1}{2}} H_2 Q H_2^H (I_{N_R} - \Omega_p)^{\frac{1}{2}}}{\tilde{\sigma}_i^2} \right\|$$

$$- \log_2 \left\| I_{N_E} + \frac{H_E^2 Q H_E^{2H}}{\tilde{\sigma}^2} \right\| \geq R,$$

$$C5 : \mathrm{Tr}\left(\Omega_p H_2 Q H_2^H\right) \geq E.$$

Although the denotation of the transmit power matrix Q remains the same, we may obtain different values of Q due to the change in channel fading states (and also due to the need to maintain positive secrecy).

3. Proposed Solution

In this section, we derive the solution to the formulated optimization problems and obtain the optimal values of transmit power matrices and power-splitting ratios for the corresponding transmission modes.

3.1. Transmission Mode Pos-A

The problem $P1 - A$ (7) is a non-convex NP-hard problem due to its dependence on the transmit power matrix and the power splitting ratio. Also, the problem $P1$ is in the form of a product of variables, which requires a reformulation strategy with appropriate constraints to obtain an optimal solution. The problem $P1$ has two variables Ω_p and Q, by transforming the problem into two sub-problems [35]: one sub-problem $P2 - A$ to solve Q for a given value of Ω_p and another sub-problem $P3 - A$ to solve Ω_p for a given value of Q. We guarantee optimal values of the transmit power matrix Q and Ω_p.

3.1.1. Sub-Problem P2

Keeping the transmit power matrix Q as the only variable, the subproblem $\hat{P}2 - A$ is given by

$$\hat{P}2 - A : \max_Q \log_2 \left\| I_{N_R} + \frac{(I_{N_R} - \Omega_p)^{\frac{1}{2}} H_1 Q H_1^H (I_{N_R} - \Omega_p)^{\frac{1}{2}}}{\hat{\sigma}_i^2} \right\| \tag{9}$$

$$C1 : \mathrm{Tr}(Q) \leq P,$$

$$C2 : Q \succcurlyeq 0,$$

$$C3 : \log_2 \left\| I_{N_R} + \frac{(I_{N_R} - \Omega_p)^{\frac{1}{2}} H_1 Q H_1^H (I_{N_R} - \Omega_p)^{\frac{1}{2}}}{\hat{\sigma}_i^2} \right\|$$

$$- \log_2 \left\| I_{N_E} + \frac{H_E^1 Q H_E^{1H}}{\hat{\sigma}^2} \right\| \geq R,$$

$$C4 : \mathrm{Tr}\left(\Omega_p H_1 Q H_1^H\right) \geq E.$$

Here, the constraint C3 is in the form of a difference of concave functions and must be reformulated. Let

$$I_{N_R} + \frac{(I_{N_R} - \Omega_p)^{\frac{1}{2}} H_1 Q H_1^H (I_{N_R} - \Omega_p)^{\frac{1}{2}}}{\hat{\sigma}_i^2} \triangleq R_\Omega, \tag{10}$$

$$I_{N_E} + \frac{H_E^1 Q {H_E^1}^H}{\hat{\sigma}^2} \triangleq R_E. \tag{11}$$

From (10), we obtain

$$H_1 Q H_1^H = \left((I_{N_R} - \Omega_p)^{\frac{1}{2}}\right)^{-1} \left(\hat{\sigma}_i^2 (R_\Omega - I_{N_R})\right) \left((I_{N_R} - \Omega_p)^{\frac{1}{2}}\right)^{-1}, \tag{12}$$

Let T_{EH} be the variable denoting harvested energy. After incorporating (12), we have

$$T_{EH} = \mathrm{Tr}\left(\Omega_p \left((I_{N_R} - \Omega_p)^{\frac{1}{2}}\right)^{-1} \left(\hat{\sigma}_i^2 (R_\Omega - I_{N_R})\right) \left((I_{N_R} - \Omega_p)^{\frac{1}{2}}\right)^{-1}\right). \tag{13}$$

Also, (11) can be rewritten as

$$Q = (H_E^1)^\dagger (\hat{\sigma}^2 (R_E - I_{N_E})) \left({H_E^1}^H\right)^\dagger. \tag{14}$$

The inverse of a non-square matrix requires the use of a pseudoinverse or Moore–Penrose inverse. Thus, the inverse of H_E^1 is represented by $(H_E^1)^\dagger$ (we can only use $(H_E^1)^{-1}$ if H_E^1 is a square matrix). The Moore–Penrose inverse of $(H_E^1)^\dagger$ is given by $(H_E^1)^H \left(H_E^1 (H_E^1)^H\right)^{-1}$. Using (10) and (11), the constraint C3 from the subproblem $\tilde{P}2 - A$ can be rewritten as

$$\log_2 \|R_\Omega\| - \log_2 \|R_E\| \geq R. \tag{15}$$

The first-order Taylor approximation is performed on the C3 [37], and C3 is modified into

$$\log_2 \|R_\Omega\| - \log_2 \|R_E\| \approx \log_2 \|R_\Omega^*\|$$
$$+ \mathrm{Tr}((R_\Omega^*)^{-1}(R_\Omega - R_\Omega^*))$$
$$- \log_2 \|R_E^*\| - \mathrm{Tr}((R_E^*)^{-1}(R_E - R_E^*)). \tag{16}$$

Since $\log \|R_\Omega\|$ is concave on $R_\Omega \geq 0$, the approximation symbol $\approx$ can be replaced by $\leq$. Using the same simplifications, the subproblem can be rewritten as $\hat{P}2 - A$ as follows:

$$\hat{P}2 - A : \max_{R_\Omega, R_E} \log_2 \|R_\Omega\| \tag{17}$$

$$C1 : \mathrm{Tr}((H_E^1)^\dagger (\sigma^2(R_E - I_{N_E}))({H_E^1}^H)^\dagger) \leq P,$$

$$C2 : R_E \succeq I_{N_E},$$

$$C3 : \log_2 \|R_\Omega^*\| + \mathrm{Tr}((R_\Omega^*)^{-1}(R_\Omega - R_\Omega^*))$$
$$- \log_2 \|R_E^*\| - \mathrm{Tr}((R_E^*)^{-1}(R_E - R_E^*)) \geq R,$$

$$C4 : T_{EH} \geq E.$$

The use of the solution variable R_E from the problem $P2 - A$ (17) is substituted in (14) to obtain the intermediate transmit power matrix $Q^\#$. The transmit power matrix $Q^\#$ (computed from (14)) is used in the following subproblems to obtain the final solution.

3.1.2. Sub-Problem P3

Keeping the power-splitting ratio Ω_p as the only variable, the subproblem $\hat{P}3 - A$ is given by,

$$\hat{P}3 - A : \max_{\Omega_p} \log_2 \left\| I_{N_R} + \frac{(I_{N_R} - \Omega_p)^{\frac{1}{2}} H_1 Q^{\#} H_1^H (I_{N_R} - \Omega_p)^{\frac{1}{2}}}{\hat{\sigma}_i^2} \right\| \tag{18}$$

$$C1 : 0 \preccurlyeq \Omega_p \preccurlyeq I_{N_R},$$

$$C2 : \log_2 \left\| I_{N_R} + \frac{(I_{N_R} - \Omega_p)^{\frac{1}{2}} H_1 Q^{\#} H_1^H (I_{N_R} - \Omega_p)^{\frac{1}{2}}}{\hat{\sigma}_i^2} \right\|$$

$$- \log_2 \left\| I_{N_E} + \frac{H_E^1 Q^{\#} H_E^{1H}}{\sigma^2} \right\| \geq R,$$

$$C3 : \mathrm{Tr}\left(\Omega_p H_1 Q^{\#} H_1^H\right) \geq E.$$

where $Q^{\#}$ is a constant, and Ω_p is a variable. The objective function of the subproblem $\hat{P}3 - A$ can be expressed as

$$W_{\Omega} \triangleq I_{N_R} + \frac{(I_{N_R} - \Omega_p)^{\frac{1}{2}} H_1 Q^{\#} H_1^H (I_{N_R} - \Omega_p)^{\frac{1}{2}}}{\hat{\sigma}_i^2}. \tag{19}$$

Here, W_{Ω} is a function that depends on the variable Ω_p. Similarly, in the case of Eve, we can express the information rate between Alice and Eve as

$$W_E \triangleq I_{N_E} + \frac{H_E^1 Q^{\#} H_E^{1H}}{\sigma^2}. \tag{20}$$

The variables W_{Ω} and W_E are positive semidefinite matrices, and the subproblem $\hat{P}3 - A$ (18) can be recast into

$$P3 - A : \max_{W_{\Omega}, W_E, \Omega_p} \log_2 \|W_{\Omega}\| \tag{21}$$

$$C1 : W_{\Omega} \succcurlyeq 0,$$
$$C2 : W_E \succcurlyeq 0,$$
$$C3 : \|W_{\Omega}\| \succcurlyeq \|W_E\|,$$
$$C4 : W_{\Omega} \preccurlyeq I_{N_R} + \frac{H_1 Q^{\#} H_1^H}{\hat{\sigma}_i^2},$$
$$C5 : W_E \preccurlyeq I_{N_E} + \frac{H_E^1 Q^{\#} H_E^{1H}}{\sigma^2},$$
$$C6 : \mathrm{Tr}\left(\Omega_p H_1 Q^{\#} H_1^H\right) \geq E.$$

After obtaining the value of W_{Ω}, the value of Ω_p is calculated by rearranging the terms in (19). Thus, Ω_p is given by

$$\Omega_p = diag(I_{N_R} - (W_{\Omega} - I_{N_R})D_i^{-1}). \tag{22}$$

where $D_i = \frac{H_1 Q^{\#} H_1^H}{\hat{\sigma}_i^2}$, and the term $diag$ denotes the diagonal elements of the matrix. Here, the constraint C3 guarantees a positive value of the secrecy capacity, and the constraints C4 and C5 denote boundary conditions for the variables W_{Ω} and W_E (since we are maximizing the objective, it is necessary to bound the values of the capacities, as they can become overbound, resulting in an unbounded problem). The values of the parameters Q and Ω_p are obtained by solving $P2 - A$ (17), $P3 - A$ (21) and (22).

3.2. Transmission Mode Pos-B

The problem $P1 - B$ (8) is a non-convex maximization problem; similar to the procedure shown in Sections 3.1.1 and 3.1.2, we divide the problem into separate problems and solve for the transmit power matrix and the power-splitting ratio. The final reformulated problems for solving the maximization problem in (8) include the following procedure:

$$P2 - B : \max_{r_\Omega, r_E} \log_2 \|r_\Omega\| \tag{23}$$

$$C1 : \text{Tr}((H_E^2)^\dagger (\tilde{\sigma}^2 (r_E - I_{N_E}))(H_E^{2H})^\dagger) \leq P,$$
$$C2 : r_E \succcurlyeq I_{N_E},$$
$$C3 : \log_2 \|r_\Omega^*\| + \text{Tr}((r_\Omega^*)^{-1}(r_\Omega - r_\Omega^*))$$
$$- \log_2 \|r_E^*\| - \text{Tr}((r_E^*)^{-1}(r_E - r_E^*)) \geq R,$$
$$C4 : T_{EH}' \geq E.$$

Here, r_Ω is given by

$$I_{N_R} + \frac{(I_{N_R} - \Omega_p)^{\frac{1}{2}} H_2 Q H_2^H (I_{N_R} - \Omega_p)^{\frac{1}{2}}}{\tilde{\sigma}_i^2} \triangleq r_\Omega. \tag{24}$$

The variable r_E is given by

$$I_{N_E} + \frac{H_E^2 Q H_E^{2H}}{\tilde{\sigma}^2} \triangleq r_E. \tag{25}$$

Similar to (13), we have a corresponding expression for denoting the harvested energy constraint:

$$T_{EH}' = \text{Tr}\left(\Omega_p \left((I_{N_R} - \Omega_p)^{\frac{1}{2}}\right)^{-1} \left(\tilde{\sigma}_i^2 (r_\Omega - I_{N_R})\right)\left((I_{N_R} - \Omega_p)^{\frac{1}{2}}\right)^{-1}\right). \tag{26}$$

By solving (23) and (25), we obtain the intermediate transmit power matrix $Q^\#$ from the solution variable r_E. Using the transmit power matrix $Q^\#$, we solve the upcoming problem $P3 - B$ to obtain the power-splitting ratio:

$$P3 - B : \max_{w_\Omega, w_E, \Omega_p} \log_2 \|w_\Omega\| \tag{27}$$

$$C1 : w_\Omega \succcurlyeq 0,$$
$$C2 : w_E \succcurlyeq 0,$$
$$C3 : \|w_\Omega\| \succcurlyeq \|w_E\|,$$
$$C4 : w_\Omega \preccurlyeq I_{N_R} + \frac{H_2 Q^\# H_2^H}{\tilde{\sigma}_i^2},$$
$$C5 : w_E \preccurlyeq I_{N_E} + \frac{H_E^2 Q^\# H_E^{2H}}{\tilde{\sigma}^2},$$
$$C6 : \text{Tr}\left(\Omega_p H_2 Q^\# H_2^H\right) \geq E.$$

Similarly, the expressions for w_Ω and w_E are as follows:

$$w_\Omega \triangleq I_{N_R} + \frac{(I_{N_R} - \Omega_p)^{\frac{1}{2}} H_2 Q^\# H_2^H (I_{N_R} - \Omega_p)^{\frac{1}{2}}}{\tilde{\sigma}_i^2}, \tag{28}$$

And,

$$w_E \triangleq I_{N_E} + \frac{H_E^2 Q^\# H_E^{2H}}{\tilde{\sigma}^2}. \tag{29}$$

Thus, from (28), the value for Ω_p also changes, and it is given by

$$\Omega_p = diag(I_{N_R} - (w_\Omega - I_{N_R})d_i^{-1}). \tag{30}$$

Here, d_i can be written as $d_i = \frac{H_2 Q^\# H_2^H}{\sigma_i^2}$. After solving $P2 - B$ (23), $P3 - B$ (27) and (30), we obtain the values of the transmit power matrix Q and the power-splitting ratio. By solving the optimization problems of the two transmission modes, we can calculate the value of the harvested energy. The closed-form expression for the harvested energy at Bob for is calculated using

$$\text{Tr}(\Omega_p^\# H Q^\# H^H). \tag{31}$$

Using (31), the amount of harvested energy available is then calculated.

3.3. Algorithm Description

The secure resource allocation algorithm for transmission modes (when Bob is at Pos-A and Pos-B) based on the "divide-and-conquer" approach is summarized in Algorithm 1. The intermediate counter value $I_c(i)$ is defined as the collection of optimization variables $Q^\#$ and $\Omega_p^\#$ at iteration-i. At the beginning of the algorithm, we initialize the maximum number of iterations $L_{\max}$, the initial value of the intermediate counter $I_c(0)$, the transmit power budget P, and the desired secrecy capacity R. We also initialize a threshold (T_{I_c}) for the intermediate counter (I_c). The values of the optimization variables $(\Omega_p^\#$ and $Q^\#)$ are computed based on the solutions of the problems $P2 - A$ and $P3 - A$ for Bob at Pos-A (the problems $P2 - B$ and $P3 - B$ are solved for Bob at Pos-B). The values of the optimization variables are stored in $I_c(i)$ (when the algorithm is in the i-th iteration). The difference between the current and previous values of I_c is calculated to determine the threshold. Theoretically, we can conclude that the algorithm converges if the difference is less than or equal to the defined threshold T_{I_c}.

The expressions $Q^\#$ and $\Omega_p^\#$ are obtained by solving different problems $[P2 - A; P3 - A]$ when Bob is at Pos-A and $[P2 - B; P3 - B]$ when Bob is at Pos-B. Since we made the difference explicit in Algorithm 1, we excluded the use of multiple subscripts for the sake of conciseness and brevity.

Algorithm 1 Secure resource allocation for the transmission modes Pos-A and Pos-B

1: **Initialization**: Choose L_{max} along with P, R values and initialize the values of $i, I_c(0)$ and T_{I_c};
2: **repeat**
   - Mode Pos-A: Calculate the value of Q using the Problem $P2 - A$ for a given value of Ω_p and set Q as $Q^\#$;
   - Mode Pos-B: Calculate the value of Q using the Problem $P2 - B$ for a given value of Ω_p and set Q as $Q^\#$;
   - Mode Pos-A: For the value of $Q^\#$, solve for the value of Ω_p using the Problem $P3 - A$, and set Ω_p as $\Omega_p^\#$;
   - Mode Pos-B: For the value of $Q^\#$, solve for the value of Ω_p using the Problem $P3 - B$, and set Ω_p as $\Omega_p^\#$;
3: Now calculate the value of harvested energy by (31) using the values of $Q^\#$ and $\Omega_p^\#$ for the corresponding modes and store in $I_c(i+1)$.;
4: Set $i = i + 1$;
5: **until** i value reaches L_{max} or $I_c(i) - I_c(i-1) \leq T_{I_c}$
6: Return $Q^\#$ and $\Omega_p^\#$ as the final optimal solution.

---

4. Results and Discussion

In this section, we present the simulation results to illustrate the trade-off between harvested energy and secrecy capacity for a single-user multi-antenna SWIPT system in the presence of a multi-antenna eavesdropper. The Matlab and CVX tools are used to obtain the following results. The simulation results are performed to evaluate the proposed algorithm when Bob is present at Pos-A and Pos-B. The channel fading states H_1 and H_2 are generated over 1000 channel realizations. The differences between H_1 and H_2 are based on the distance between Alice and Bob. We generated the channel state conditions by varying the distance of Bob as it moves away from Alice. For Pos-A, the distance between Alice and Bob is 30 m, while for Pos-B, the distance between Alice and Bob is 80 m. The eavesdropper is stationary, and its distance from Alice is assumed to be 20 m. Table 1 shows the essential sets of simulation parameters used to generate the results. The transmit power available at Alice is varied in the range of 5 to 40 dBm. The expressions for the harvested energy in the nonlinear EH models [3] are as follows:

$$E_{nonlinear} = \frac{\frac{M}{1+\exp(-a(\text{Tr}(\Omega_p^\# H Q^\# H^H)-b))} - \frac{M}{1+\exp(ab)}}{1 - \frac{1}{1+\exp(ab)}}. \tag{32}$$

Here, M denotes the maximum energy harvested at the receiver when the EH circuit is saturated, and (a, b) denotes the circuit parameters for the nonlinear EH model considered. For the nonlinear EH model, the circuit parameters $[a = 6400, b = 0.003]$ and $M = 0.024$ are chosen as given in [3] and the harvested energy is then calculated, where the circuit parameters a, b and the maximum harvested energy M simulate the effects caused by the constraints such as current leakage and hardware sensitivity [3,20]. The noise at Bob is -30 dBm (when Bob is at Pos-A). A worst-case scenario is considered by assuming that the noise at Bob increases to 1 dBm (while at Pos-B). The noise at the eavesdropper is assumed to be -10 dBm.

Table 1. Simulation parameters.

Parameters	Values
Number of transmit antennas at Alice	2, 4, 8, 16
Number of eavesdropper antennas	2, 4, 8
Noise Variance at Bob in Pos-A and Bob in Pos-B	−30 dBm and 1 dBm
Circuit Parameter value a	6400
Circuit Parameter value b	0.003
Transmit power values	5 to 40 dBm
Maximum harvested energy requirement	20 dBm
Path loss exponent	4
Rician factor	3 dB
Reference distance	10 m

4.1. Performance Analysis Based on Trade-Off between Harvested Energy and Secrecy Capacity

4.1.1. Bob at Pos-A

In Figure 2, we present the trade-off between the secrecy capacity and harvested energy for varying transmit power and transmit antenna values at Alice. The simulations are performed when the channel fading state is H_1 and Bob is at Pos-A. The co-located SWIPT setting for Bob implies that the received signal is split between energy harvesting and information decoding. As the formulated problem is information capacity maximization, the solution to the variables Ω_p and Q are "fine tuned" so that the overall information capacity (thereby, the secrecy capacity) is maximized. This means that due to the maximization of the information capacity, there will be a decrease in the value of Ω_p, which further decreases the value of the harvested energy (as Ω_p is directly proportional to harvested energy).

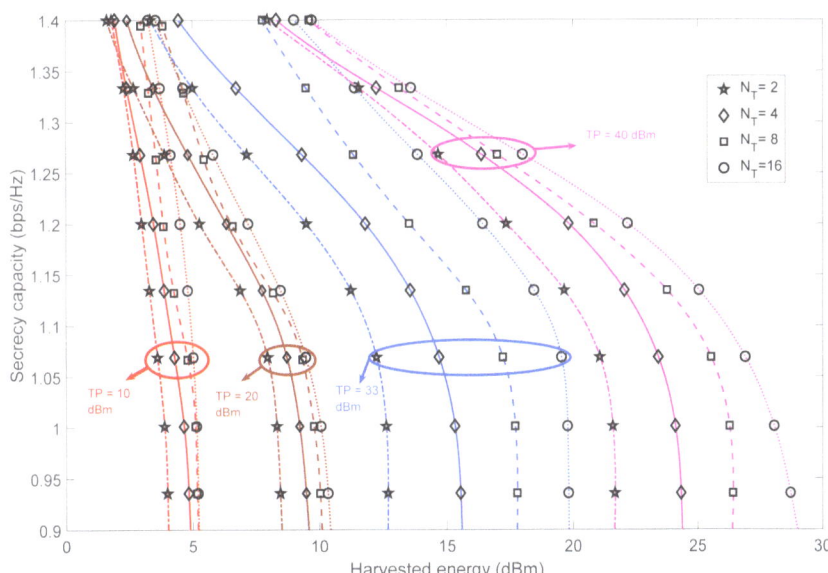

Figure 2. The trade-off between secrecy capacity and harvested energy for varying values of transmit antennas at Alice and transmit power when Bob is in Pos-A. The number of receiver antennas at Bob and eavesdropper are fixed at two.

The achievable harvested energy values and corresponding range are small for the lower transmit power regimes due to the directly proportional relationship between the transmit power matrix, harvested energy and the transmit power constraint $C1$ in $P1 - A$ (7). However, the harvested energy range becomes wider as the transmit power value increases. This increase in range is due to the inherent properties of the transmit power matrix Q. Increasing the number of transmit antennas at Alice increases the matrix dimension of Q, which further amplifies the harvested energy (which can be verified from the harvested energy expression in (31)). Hence, we can infer that the increase in the number of transmit antennas is the least noticeable (due to the narrow range of harvested energy values) when the transmit power is low. Comparing within the higher transmit power regime, the increase in the harvested energy decreases the secrecy capacity (for all the combinations of transmit antennas at Alice). Overall, the trend of the secrecy capacity is steep. This is due to the more stringent constraint imposed by the formulated problem.

4.1.2. Bob at Pos-B

Figure 3 illustrates the relationship between the secrecy capacity and harvested energy for different transmit power values and antennas. The analysis focuses on the channel fading state H_2, where Bob is moving away from Alice. Theoretically, this results in an overall decrease in the upper bound of both the secrecy capacity and the harvested energy. Also, increasing the harvested energy decreases the secrecy capacity values. We can conclude that the difference in harvested energy is significantly high for the high transmit power regime. In summary, increasing the transmit power increases the harvested energy values. We can also observe that increasing the number of transmit antennas at Alice has a smaller effect on the change in the harvested energy values. This difference in the consequences of varying the values of transmit antennas and transmit power can be attributed to the influence of the transmit power matrix on the proposed algorithm. The steepness of the trend of the trade-off is significantly high for the lower transmit power regime. From this observation, we can infer that the secrecy capacity decreases significantly as the harvested energy value increases. This is due to the lack of transmit power availability to maintain the required harvested energy due to the co-located nature of the SWIPT system.

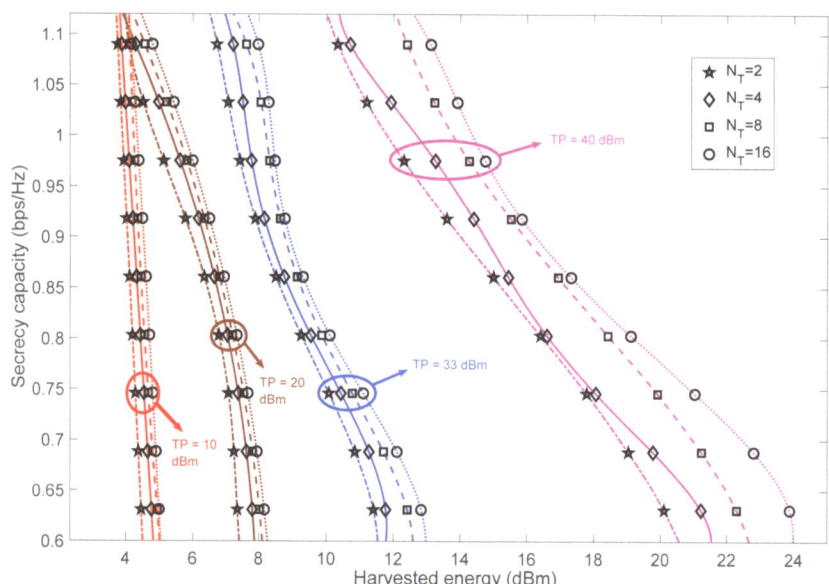

Figure 3. The trade-off between secrecy capacity and harvested energy for varying values of transmit antennas at Alice and transmit power when Bob is in Pos-B. The number of receiver antennas at Bob and eavesdropper are fixed at two.

4.1.3. Comparison with Benchmarks

Figure 4 illustrates the secrecy capacity performance when the transmit power and transmit antennas (of Alice) are fixed for SDP, KKT, and the proposed algorithms. The harvested energy is computed by the solution variables Q and the power-splitting ratio Ω_p (which can be verified from (31)). We observe a steep decrease in the secrecy capacity values as the harvested energy increases. This is due to the co-located nature of Bob's EH/ID. However, when comparing Pos-A and Pos-B, significant differences in the (achievable) harvested energy can be observed. This is due to the distance variation and channel fading state of Pos-A and Pos-B of Bob. We can also infer that to achieve positive secrecy capacity values between Pos-A and Pos-B, there is a significant drop in the harvested energy values between Pos-A and Pos-B. This phenomenon can be attributed to the information capacity maximization (of Bob), where our goal is to optimize the information capacity while maintaining the harvested energy and secrecy capacity requirements. Our algorithm successfully maintained the required secrecy capacity values by minimizing the harvested energy value (via the optimization of the Ω_p variable). That is, a decrease in Ω_p reduces the harvested energy, thereby maintaining the required secrecy capacity.

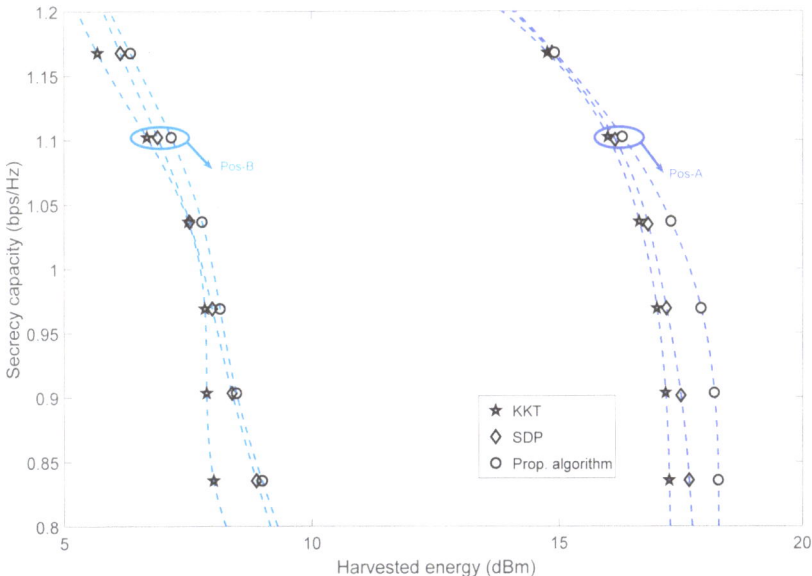

Figure 4. The trade-off between secrecy capacity and harvested energy for fixed value of transmit antennas (taken as 8) at Alice for Pos-A and Pos-B, analyzed for different algorithms—SDP [30] and KKT [36]. Here, transmit power is fixed at 33 dBm.

4.2. Analysis of Secrecy Capacity for Varying Number of Eavesdropper Antennas

4.2.1. Bob at Pos-A

Figure 5 shows the secrecy capacity for different numbers of eavesdropper antennas. As the number of eavesdropper antennas increases, the secrecy capacity value decreases. This is because the information capacity value of the eavesdropper increases as the number of eavesdropper antennas increases (thus decreasing the secrecy capacity value). However, when the number of eavesdropper antennas is eight, the secrecy capacity barely exceeds the secrecy capacity of 0.5 bps/Hz. This means that simultaneously increasing the number of eavesdropper antennas and decreasing the transmit power has a negative effect on the secrecy capacity values. However, the increase in transmit power has a positive effect on the secrecy capacity values, resulting in an overall increase in the secrecy capacity (seen throughout the N_E regime). The secrecy capacity difference for all the algorithms is almost the same for higher transmit power values. However, for lower values of transmit power, the differences in performance based on secrecy capacity can be observed. For all the simulated configurations of N_E, the proposed algorithm can achieve higher values of secrecy capacity, with a maximum difference in performance observed when $N_E = 4$ and the transmit power is at 8 dBm.

4.2.2. Bob at Pos-B

Figure 6 shows the effect of the number of eavesdropper antennas on the secrecy capacity for varying transmit power values when Bob is at Pos-B. The cause of the overall decrease in secrecy can be attributed to the increase in the resulting distance between Alice and Bob. This increase in distance reduces the value of the secrecy capacity (compared to the value of the secrecy capacity in Figure 5). Because of the change in Bob's trajectory (as he moves further away from Alice), the beamforming technique is ineffective. This is evident from the fact that the secrecy capacity value does not exceed 1.5 bps/Hz, even at a transmit power of 40 dBm. The decrease in secrecy capacity means that the overall information capacity of the eavesdropper increases as the number of antennas on the eavesdropper side increases. Because of the directly proportional relationship between the transmit power and the information capacity of Bob, increasing the transmit power increases the value of the secrecy capacity. This is true for all algorithms. The significant performance difference between different algorithms is observed when the transmit power is 8 dBm. This means that our proposed algorithm can maintain a relatively high secrecy capacity value, even when the transmit power is reduced.

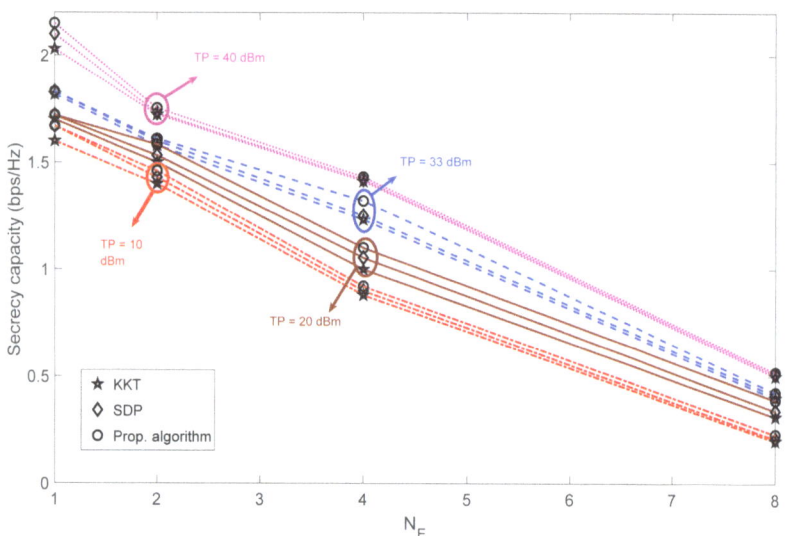

Figure 5. The trade-off between secrecy capacity for varying values of transmit power and eavesdropper antennas when Bob is in Pos-A. The number of antennas at the eavesdropper side is assumed to be varying $N_E = 1, 2, 4, 8$. Algorithms compared are SDP [30] and KKT [36].

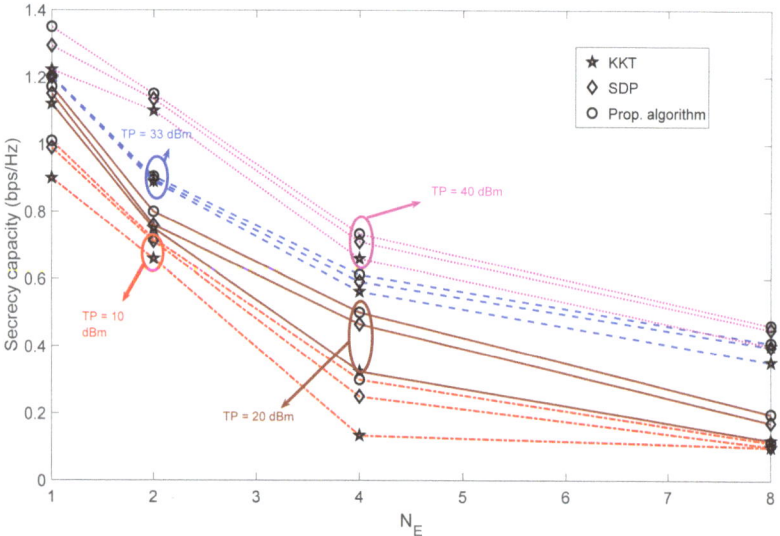

Figure 6. The trade-off between secrecy capacity for varying values of transmit power and eavesdropper antennas when Bob is at Pos-B. The number of antennas at eavesdropper side is assumed to be varying $N_E = 1, 2, 4, 8$. Algorithms compared are SDP [30] and KKT[36].

4.3. Non-Zero Secrecy Capacity Probability Analysis

4.3.1. Bob at Pos-A

Figure 7 shows the relationship between the non-zero secrecy capacity probability and the transmit power when Bob is at Pos-A. The non-zero probability of secrecy can be defined as the probability of achieving positive values of secrecy for the simulated set of iterations. The secrecy capacity requirement is relaxed [refer Appendix A], while the harvested energy requirement is increased to a limit (for a given transmit power). This analysis provides the probability of achieving positive secrecy capacity values for the proposed algorithm while maintaining a significant harvested energy requirement. The security aspect of the proposed algorithm is tested by maintaining a stricter harvested energy requirement, thereby accessing the number of iterations (we used 1000 iterations) where the secrecy capacity values fall below zero. We can conclude that the probability of obtaining a positive secrecy capacity increases as the transmit power increases. This is because the transmit power plays a significant role in achieving the more stringent harvested energy requirement and maintaining positive secrecy capacity values. The number of transmit antennas in Alice helps increase the probability of achieving positive secrecy capacity as shown in Figure 7. Therefore, to maintain non-zero secrecy capacity for a lower power regime, we must guarantee more transmit antennas for Alice. However, the difference in probability is not significant at a high transmit power.

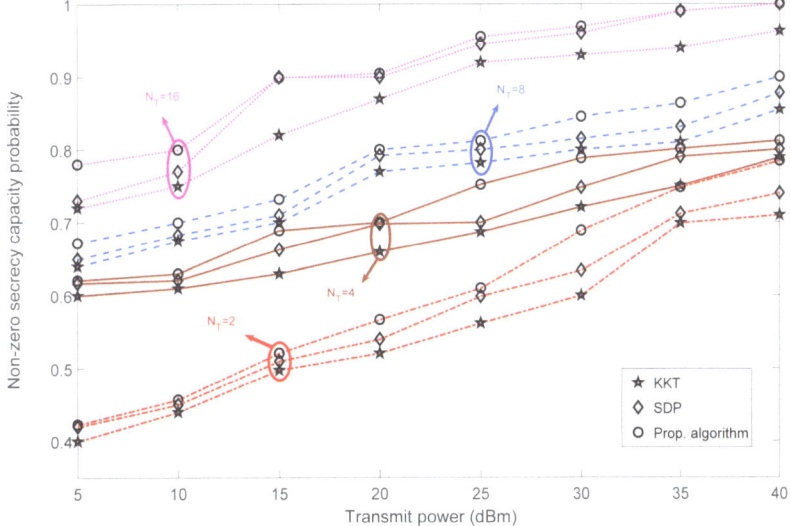

Figure 7. Analysis of non-zero secrecy capacity probability versus transmit power for varying transmit antennas at Alice when Bob is at Pos-A for different algorithms—SDP [30] and KKT [36].

Comparing different algorithms, our proposed algorithm maintains a higher probability value for the entire transmit power regime. However, for the higher transmit power regime, the probability of achieving positive secrecy capacity is almost the same for $N_T = 16$. Overall, we can conclude that under a high transmit power regime, our proposed algorithm has a high probability of generating positive secrecy capacity values compared to other algorithms.

4.3.2. Bob at Pos-B

Figure 8 illustrates the effect of transmit power on the achievable secrecy capacity for different values of transmit antennas. For a lower transmit power regime, the probability of achieving secrecy capacity is less than 0.7 for all combinations of transmit antennas.

This is due to two reasons: one is the channel fading state H_2, which negatively affects Bob's information capacity value, and the other is the inherent nature of low transmit power (which also reduces the value of secrecy capacity). As the transmit power increases, the probability of achieving a positive secrecy capacity increases for all transmit antenna configurations. We can observe that the difference between the probabilities of $N_T = 16$ and other N_T configurations is large. This suggests that to achieve positive secrecy capacity, it is only appropriate to deploy a large number of transmit antennas for Alice for a worst-case scenario (when the eavesdropper is in a better state than Bob). Unlike Figure 7 for Bob at Pos-A, the probabilistic analysis shows a clear difference in the simulated values for all the transmit power combinations considered when Bob is at Pos-B. This means that when Bob is farther away from Alice, our proposed algorithm can maintain a higher probability of achieving positive secrecy for the entire transmit power regime. For all N_T combinations, the probability value differences between SDP and the proposed algorithm are small for higher transmit power values. However, for the lower transmit power regime, the difference in probability values between the different algorithms is relatively high. In summary, for higher values of N_T, the probability of achieving positive secrecy capacity increases.

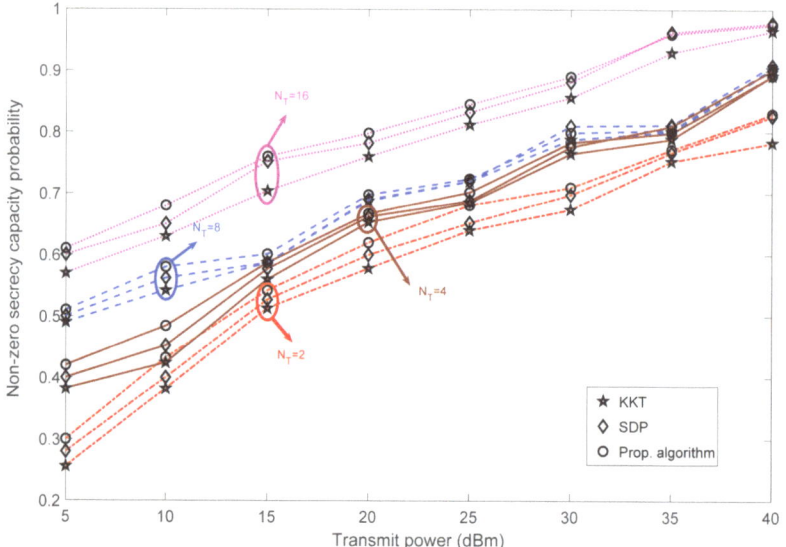

Figure 8. Analysis of non-zero secrecy capacity probability versus transmit power for varying transmit antennas at Alice when Bob is at Pos-B for different algorithms—SDP [30] and KKT [36].

4.4. Computational Complexity

The complexity analysis is performed for the proposed subproblems P2-A (17) and P3-A (21) (the computational complexity for Pos-A and Pos-B is identical; therefore, it is analyzed only for Pos-A). The increased computational complexity of the proposed algorithm is due to the presence of multiple matrices in the formulated problems. Table 2 shows the computational complexity of subproblems P2-A and P3-A, where N_T, N_E, and N_R denote the number of antennas present at Alice, Eve, and Bob, respectively. The accuracy bounds for the proposed KKT and SDP algorithms are denoted as ϵ_q, ϵ_o (for individual subproblems), $\epsilon_q \epsilon_o$, and ϵ, respectively. Table 2 gives an approximate upper bound on the complexity per iteration for the subproblems. Also, the computational complexity remains the same for both scenarios (when Bob is at Pos-A and when Bob is at Pos-B). There are special algorithms that can have lower complexity than the KKT algorithm. However, global optimality is not guaranteed when using such low-complexity algorithms. The interior-point approximation strategy can achieve low complexity, but this resource allocation strategy primarily leads to locally optimal solutions. Our proposed "divide-and-conquer approach" is a combination of sequential programming and a branch-bound strategy [38].

Table 2. Computational complexities of the proposed, SDP and KKT-based algorithms.

Algorithms	Complexity Comparisons
SDP algorithm [30]	$\mathcal{O}(N_T^5 N_R^{3.5} N_E^3 \log(\epsilon))$
KKT-based algorithm [36]	$\mathcal{O}(N_T^2 (N_R N_E)^3 \log(\epsilon_q \epsilon_o))$
Proposed algorithm	P2-A: $\mathcal{O}(N_E (N_T N_R)^2 \log(\epsilon_q))$
	P3-A: $\mathcal{O}(N_T N_E^2 N_R^3 \log(\epsilon_o))$

5. Conclusions

This paper investigates the trade-offs between secrecy capacity–harvested energy for the MIMO SWIPT system under two dynamic transmission modes in the presence of a multi-antenna eavesdropper. The system is modeled for co-located EH/ID receivers under the PS scenario. The optimization problems are formulated to compute the transmit power matrix and the power sharing ratio. The formulated problem has multiple variables in the product form; the problem is subdivided and solved individually to obtain respective solutions. Numerical results show the relationship between the secrecy capacity and the harvested energy. The performance of the proposed solution is studied for two transmission scenarios and compared with the algorithms available in the literature. The proposed solution shows a significant performance gain in terms of the secrecy capacity values for both transmission scenarios. Further research can be carried out with multiple multi-antenna legitimate receivers in the presence of eavesdropper(s). In addition, we can also explore resource allocation strategies to secure the SWIPT network in the presence of an active eavesdropper (thus extending the study beyond the passive eavesdropping scenario) for a more robust security framework for MIMO-SWIPT systems in vehicular networks.

Author Contributions: Conceptualization of the model: R.R. and T.O.; Methodology and formulation description of the optimization problem: V.G. and R.R.; Investigation: V.G. and R.R.; Visualization of results: R.R. and T.O.; Software: V.G.; Validation: R.R. and T.O.; Writing—original draft preparation: V.G.; Writing— Review and editing, R.R. and T.O.; Supervision: R.R. and T.O. All authors have read and agreed to the published version of the manuscript.

Funding: This research received no external funding.

Institutional Review Board Statement: Not applicable.

Informed Consent Statement: Not applicable.

Data Availability Statement: Data sharing not applicable.

Conflicts of Interest: The authors declare no conflict of interest.

Appendix A

The relation between the optimization variable Q and the channel fading state between Alice and Bob determines the secrecy capacity constraint and its variation over time. From constraint C4 in problems $P1 - A(7)$ and $P1 - B(8)$, we have the following secrecy constraint (for brevity, we discussed only $P1 - A$):

$$S_{C_C} = \log_2 \left\| I_{N_R} + \frac{(I_{N_R} - \Omega_p)^{\frac{1}{2}} H_1 Q H_1^H (I_{N_R} - \Omega_p)^{\frac{1}{2}}}{\hat{\sigma}_i^2} \right\|$$

$$- \log_2 \left\| I_{N_E} + \frac{H_E^1 Q H_E^{1H}}{\hat{\sigma}^2} \right\| \quad (A1)$$

From (A1), we can observe that the S_{C_C} depends on the variables Q and Ω_p as well as on the channel fading states H_1 and H_E^1. Thus, the positive values of the secrecy capacity are determined by the change between the states of Q and the channel states. Since Eve is relatively closer to Alice, we can assume that $\hat{\sigma}_i^2 \geq \hat{\sigma}^2$. Let us assume

$$\alpha_1 = \frac{\|H_1\|}{\|H_E^1\|}, \tag{A2}$$

and,

$$\alpha_{scc} = \frac{\left\|\hat{\sigma}^2 (I_{N_R} - \Omega_p)^{\frac{1}{2}} H_1 Q H_1^H (I_{N_R} - \Omega_p)^{\frac{1}{2}}\right\|}{\left\|\hat{\sigma}_i^2 H_E^1 Q H_E^{1\,H}\right\|} \tag{A3}$$

The variation of the values of α_1 and α_{scc} determines the "relaxation zone" of the secrecy capacity for the considered dynamic system. To achieve a positive secrecy capacity, it is always better to keep α_{scc} greater than one. However, to achieve high values of harvested energy for the co-located SWIPT setup, it is imperative to have high values of Ω_p. This reduces Bob's information capacity, resulting in a value between 0 and 1.

References

1. Zhang, R.; Ho, C.K. MIMO Broadcasting for Simultaneous Wireless Information and Power Transfer. *IEEE Trans. Wirel. Commun.* **2013**, *12*, 1989–2001. [CrossRef]
2. Asheer, S.; Kumar, S. A Comprehensive Review of Cooperative MIMO WSN: Its Challenges and the Emerging Technologies. *Wirel. Netw.* **2021**, *27*, 1129–1152. [CrossRef]
3. Xiong, K.; Wang, B.; Liu, K.J.R. Rate-Energy Region of SWIPT for MIMO Broadcasting Under Nonlinear Energy Harvesting Model. *IEEE Trans. Wirel. Commun.* **2017**, *16*, 5147–5161. [CrossRef]
4. Xu, Y.; Gui, G.; Ohtsuki, T.; Gacanin, H.; Adebisi, B.; Sari, H.; Adachi, F. Robust Resource Allocation for Two-Tier HetNets: An Interference-Efficiency Perspective. *IEEE Trans. Green Commun. Netw.* **2021**, *5*, 1514–1528. [CrossRef]
5. Allahzadeh, S.; Daneshifar, E. Simultaneous Wireless Information and Power Transfer Optimization via Alternating Convex-Concave Procedure with Imperfect Channel State Information. *Signal Process.* **2021**, *182*, 107953. [CrossRef]
6. Gharavol, E.A.; Liang, Y.C.; Mouthaan, K. Robust Downlink Beamforming in Multiuser MISO Cognitive Radio Networks with Imperfect Channel-State Information. *IEEE Trans. Veh. Technol.* **2010**, *59*, 2852–2860. [CrossRef]
7. Khisti, A.; Wornell, G.W. Secure Transmission with Multiple Antennas—Part II: The MIMOME Wiretap Channel. *IEEE Trans. Inf. Theory* **2010**, *56*, 5515–5532. [CrossRef]
8. Kumar, M.S.; Ramanathan, R.; Jayakumar, M. Keyless Physical Layer Security for Wireless Networks: A Survey. *Eng. Sci. Technol. Int. J.* **2022**, *35*, 101260.
9. Porambage, P.; Gür, G.; Osorio, D.P.M.; Liyanage, M.; Gurtov, A.; Ylianttila, M. The Roadmap to 6G Security and Privacy. *IEEE Open J. Commun. Soc.* **2021**, *2*, 1094–1122. [CrossRef]
10. Ju, Y.; Chen, Y.; Cao, Z.; Liu, L.; Pei, Q.; Xiao, M.; Ota, K.; Dong, M.; Leung, V.C.M. Joint Secure Offloading and Resource Allocation for Vehicular Edge Computing Network: A Multi-Agent Deep Reinforcement Learning Approach. *IEEE Trans. Intell. Transp. Syst.* **2023**, *24*, 5555–5569. [CrossRef]
11. Vu, T.H.; Nguyen, T.V.; Kim, S. Cooperative NOMA-Enabled SWIPT IoT Networks with Imperfect SIC: Performance Analysis and Deep Learning Evaluation. *IEEE Internet Things J.* **2022**, *9*, 2253–2266. [CrossRef]
12. Zhao, R.; Wang, Y.; Xue, Z.; Ohtsuki, T.; Adebisi, B.; Gui, G. Semisupervised Federated-Learning-Based Intrusion Detection Method for Internet of Things. *IEEE Internet Things J.* **2023**, *10*, 8645–8657. [CrossRef]
13. Ouamri, M.A.; Barb, G.; Singh, D.; Adam, A.B.M.; Muthanna, M.S.A.; Li, X. Nonlinear Energy-Harvesting for D2D Networks Underlaying UAV with SWIPT Using MADQN. *IEEE Commun. Lett.* **2023**, *27*, 1804–1808. [CrossRef]
14. Alladi, T.; Gera, B.; Agrawal, A.; Chamola, V.; Yu, F.R. DeepADV: A Deep Neural Network Framework for Anomaly Detection in VANETs. *IEEE Trans. Veh. Technol.* **2021**, *70*, 12013–12023. [CrossRef]
15. Zhao, H.; Tang, J.; Adebisi, B.; Ohtsuki, T.; Gui, G.; Zhu, H. An Adaptive Vehicle Clustering Algorithm Based on Power Minimization in Vehicular Ad-hoc Networks. *IEEE Trans. Veh. Technol.* **2022**, *71*, 2939–2948. [CrossRef]
16. Sun, C.; Xiong, X.; Ni, W.; Ohtsuki, T.; Wang, X. Max-Min Fair 3D Trajectory Planning for Solar-Powered UAV-Assisted Data Collection. In Proceedings of the 2022 IEEE/CIC International Conference on Communications in China (ICCC), Foshan, China, 11–13 August 2022; pp. 610–615.
17. Zhu, X.; Zeng, W.; Xiao, C. Precoder Design for Simultaneous Wireless Information and Power Transfer Systems with Finite-Alphabet Inputs. *IEEE Trans. Veh. Technol.* **2017**, *66*, 9085–9097. [CrossRef]

18. Nam, J.; Caire, G.; Debbah, M.; Poor, H.V. Capacity Scaling of Massive MIMO in Strong Spatial Correlation Regimes. *IEEE Trans. Inf. Theory* **2020**, *66*, 3040–3064. [CrossRef]
19. Kumar, D.; Alcaraz López, O.L.; Joshi, S.K.; Tölli, A. Latency-Aware Multi-Antenna SWIPT System with Battery-Constrained Receivers. *IEEE Trans. Wirel. Commun.* **2023**, *22*, 3022–3037. [CrossRef]
20. Le, T.; Mayaram, K.; Fiez, T. Efficient Far-Field Radio Frequency Energy Harvesting for Passively Powered Sensor Networks. *IEEE J. Solid-State Circuits* **2008**, *43*, 1287–1302. [CrossRef]
21. Liang, Y.; Li, B.; Zhang, R.; Li, H.; Zhao, S. Distributed Beamforming for Energy-Harvesting Relaying in Vehicular Networks. *J. Commun. Inf. Netw.* **2020**, *5*, 160–167. [CrossRef]
22. Yang, C.; Lu, W.; Huang, G.; Qian, L.; Li, B.; Gong, Y. Power Optimization in Two-way AF Relaying SWIPT-based Cognitive Sensor Networks. In Proceedings of the 2020 IEEE 92nd Vehicular Technology Conference (VTC2020-Fall), Victoria, BC, Canada, 18 November–16 December 2020; pp. 1–5.
23. Li, Q.; Zhang, Q.; Qin, J. Secure Relay Beamforming for SWIPT in Amplify-and-Forward Two-way Relay Networks. *IEEE Trans. Veh. Technol.* **2016**, *65*, 9006–9019. [CrossRef]
24. Tashman, D.H.; Hamouda, W.; Moualeu, J.M. On Securing Cognitive Radio Networks-Enabled SWIPT Over Cascaded $\kappa - \mu$ Fading Channels with Multiple Eavesdroppers. *IEEE Trans. Veh. Technol.* **2022**, *71*, 478–488. [CrossRef]
25. Telatar, E. Capacity of Multi-antenna Gaussian Channels. *Eur. Trans. Telecommun.* **1999**, *10*, 585–595. [CrossRef]
26. Nair, A.R.; Kirthiga, S. Impact of Total Harmonic Distortion in SWIPT Enabled Wireless Communication Networks. In Proceedings of the 2021 Smart Technologies, Communication and Robotics (STCR), Sathyamangalam, India, 9–10 October 2021; pp. 1–5.
27. Nair, A.R.; Kirthiga, S. Analysis of Energy Harvesting in SWIPT using Bio-inspired Algorithms. *Int. J. Electron.* **2022**, *110*, 291–311. [CrossRef]
28. Vieeralingaam, G.; Ramanathan, R.; Jayakumar, M. Convex Optimization Approach to Joint Interference and Distortion Minimization in Energy Harvesting Wireless Sensor Networks. *Arab. J. Sci. Eng.* **2020**, *45*, 1669–1684. [CrossRef]
29. Memon, S.; Memon, K.A.; Uqaili, J.A.; Soothar, K.K.; Uqaili, R.S.; Cengiz, K. Joint Optimal Power Splitting and Relay Selection Strategy under SWIPT. *Wirel. Netw.* **2021**, *27*, 5385–5395. [CrossRef]
30. Alageli, M.; Ikhlef, A.; Alsifiany, F.; Abdullah, M.A.M.; Chen, G.; Chambers, J. Optimal Downlink Transmission for Cell-Free SWIPT Massive MIMO Systems with Active Eavesdropping. *IEEE Trans. Inf. Forensics Secur.* **2020**, *15*, 1983–1998. [CrossRef]
31. Bereyhi, A.; Asaad, S.; Müller, R.R.; Schaefer, R.F.; Poor, H.V. Secure Transmission in IRS-assisted MIMO Systems with Active Eavesdroppers. In Proceedings of the 2020 54th Asilomar Conference on Signals, Systems, and Computers, Pacific Grove, CA, USA, 1–4 November 2020; pp. 718–725.
32. Ji, B.; Li, Y.; Cao, D.; Li, C.; Mumtaz, S.; Wang, D. Secrecy Performance Analysis of UAV Assisted Relay Transmission for Cognitive Network with Energy Harvesting. *IEEE Trans. Veh. Technol.* **2020**, *69*, 7404–7415. [CrossRef]
33. Ma, R.; Wu, H.; Ou, J.; Yang, S.; Gao, Y. Power Splitting-Based SWIPT Systems with Full-Duplex Jamming. *IEEE Trans. Veh. Technol.* **2020**, *69*, 9822–9836. [CrossRef]
34. Ng, D.W.K.; Schober, R. Secure and Green SWIPT in Distributed Antenna Networks with Limited Backhaul Capacity. *IEEE Trans. Wirel. Commun.* **2015**, *14*, 5082–5097. [CrossRef]
35. Gorski, J.; Pfeuffer, F.; Klamroth, K. Biconvex Sets and Optimization with Biconvex Functions: A Survey and Extensions. *Math. Methods Oper. Res.* **2007**, *66*, 373–407. [CrossRef]
36. Kim, B.; Kang, J.M.; Kim, H.M.; Kang, J. Joint Channel Estimation, Training Design, Tx Power Allocation, and Rx Power Splitting for MIMO SWIPT Systems. *IEEE Commun. Lett.* **2021**, *25*, 1269–1273. [CrossRef]
37. Boyd, S.; Boyd, S.P.; Vandenberghe, L. *Convex Optimization*, 1st ed.; Cambridge University Press: Cambridge, UK, 2004.
38. Cormen, T.H.; Leiserson, C.E.; Rivest, R.L.; Stein, C. *Introduction to Algorithms*, 4th ed.; MIT Press: Cambridge, MA, USA, 2022.

Disclaimer/Publisher's Note: The statements, opinions and data contained in all publications are solely those of the individual author(s) and contributor(s) and not of MDPI and/or the editor(s). MDPI and/or the editor(s) disclaim responsibility for any injury to people or property resulting from any ideas, methods, instructions or products referred to in the content.

Article

Modeling and Performance Analysis of LBT-Based RF-Powered NR-U Network for IoT

Varada Potnis Kulkarni * and Radhika D. Joshi

Department of Electronics and Telecommunication Engineering, COEP Technological University, Formerly College of Engineering Pune, Wellesley Road, Shivajinagar, Pune 411005, India; rdj.extc@coeptech.ac.in
* Correspondence: varadapotnis@gmail.com; Tel.: +91-808-782-5460

Abstract: Energy harvesting combined with spectrum sharing offers a promising solution to the growing demand for spectrum while keeping energy costs low. New Radio Unlicensed (NR-U) technology enables telecom operators to utilize unlicensed spectrum in addition to the licensed spectrum already in use. Along with this, the energy demands for the Internet of Things (IoT) can be met through energy harvesting. In this regard, the ubiquity and ease of implementation make the RF-powered NR-U network a sustainable solution for cellular IoT. Using a Markov chain, we model the NR-U network with nodes powered by the base station (BS). We derive closed-form expressions for the normalized saturated throughput of nodes and the BS, along with the mean packet delay at the node. Additionally, we compute the transmit outage probability of the node. These quality of service (QoS) parameters are analyzed for different values of congestion window size, TXOP parameter, maximum energy level, and energy threshold of the node. Additionally, the effect of network density on collision, transmission, and energy harvesting probabilities is observed. We validate our model through simulations.

Keywords: LBT; Markov chain; NR-U; green; cellular IoT; RF-EH; RF-powered

Citation: Potnis Kulkarni, V.; Joshi, R.D. Modeling and Performance Analysis of LBT-Based RF-Powered NR-U Network for IoT. *Sensors* **2024**, *24*, 5369. https://doi.org/10.3390/s24165369

Academic Editors: Onel Luis Alcaraz López and Katsuya Suto

Received: 15 April 2024
Revised: 16 August 2024
Accepted: 18 August 2024
Published: 20 August 2024

Copyright: © 2024 by the authors. Licensee MDPI, Basel, Switzerland. This article is an open access article distributed under the terms and conditions of the Creative Commons Attribution (CC BY) license (https://creativecommons.org/licenses/by/4.0/).

1. Introduction

Communication among humans, machines, and between humans and machines is growing exponentially. The ever-increasing number of things connected to the internet and the corresponding rise in exciting and useful applications pose further challenges to the design and development of communication networks. The Internet of Things (IoT) is evolving into the Internet of Everything (IoE). The number of IoT devices will reach more than USD 39 billion by 2033, marking more than a twofold increase over the next nine years, as reported by https://www.statista.com (accessed on 8 August 2024) [1].

Also, the number of IoT devices per person was 3.6 on a worldwide average in the year 2023 according to a survey [2]. To cater to the increased demand for data rates and the number of devices connected to the network, fifth-generation (5G) and beyond fifth-generation (B5G) communication technologies are being standardized. Novel techniques are being developed to simultaneously support high bandwidth, low latency, high reliability, and fairness among incumbent technologies.

To meet the growing demand for data rates and to support the massive number of connections, it is essential to utilize all the spectrum types as described by Qualcomm Technologies [3]. There are mainly three spectrum types: (a) licensed spectrum (exclusive use, e.g., over 40 bands of long-term evolution used globally), (b) shared spectrum with new paradigms (e.g., 2.3 GHz in Europe and 3.7 GHz in the USA), and (c) unlicensed spectrum (shared use, 2.4 GHz/5–7 GHz/57–71 GHz). This paper focuses on New Radio Unlicensed (NR-U) technology, which involves the use of an unlicensed spectrum by cellular operators. It represents a fifth-generation (5G) extension to License-Assisted Access (LAA) used in 4G communications and standardized by the Third-Generation Partnership

Project (3GPP) [4]. NR-U coexists with other technologies using unlicensed spectra such as Wi-Fi and Bluetooth. It includes a Listen Before Talk (LBT) mechanism for multiple access, which is similar to the Carrier Sense Multiple Access (CSMA) used in Wi-Fi [5]. In the LBT protocol, nodes willing to transmit data 'sense' the carrier before sending the data and adjust their backoff depending on the congestion level. NR-U is widely used in various sectors involving IoT networks, as described by Y. Liu et al. [6]. The use of NR-U in the oil and gas industry is discussed by A.H.S et al. [7].

IoT networks with high data rates and large numbers of users require significant energy for their operations. The energy harvesting feature helps maintain the sustainability of the network, as described by N Ansari et al. [8]. In energy harvesting communication networks, various types of harvesting technologies can be incorporated. The base stations can be equipped with renewable power supplies such as solar power or wind power. Smartphones and other portable devices can harvest energy from thermal changes in the environment, piezoelectric effects, and RF energy present in the ambiance, as described by the authors [9]. RF energy is widely available and can be harvested cost-effectively, hence, RF-powered IoT networks are popular, as mentioned by M. A. Abd-Elmagid et al. [10]. These networks are used in applications such as smart farming [11], museum ambiance control [12], and segments such as civil infrastructure and manufacturing [13].

There are various ways in which a wireless network can be RF-powered. Inductive and magnetic coupling are useful for very short distances, ranging up to a few meters. For longer distances up to a few kilometers, the far-field region of the propagation wave can be used, as described by Niyato et al. [14]. In this technique, RF power from cellular base stations (BS), wireless access points, as well as radio and television transmitters is used. Due to the ubiquity of cellular systems, this work considers a base station-powered RF network. There are different ways in which a base station can power IoT nodes, e.g., using the duty cycle method for data and energy transfer, polling nodes for energy harvesting, or employing CSMA/LBT-based mechanisms. In order to combine the benefits of spectrum sharing and cellular systems, we consider an RF-powered IoT network that uses NR-U protocol for communication in which LBT is mandated for transmission. Thus, LBT-based RF power transfer is a natural choice.

In this paper, the base station is powered by traditional non-energy harvesting methods, and other nodes in the network harvest RF energy from BS during the data transfer. In particular, we mathematically model the LBT-based RF-powered network and analyze the performance of the nodes in the network. Using the expressions provided by the model, we analyze the node throughput, delay, and outage probability. The node throughput is specifically useful for enhanced multimedia broadband (eMBB) applications of 5G and 6G networks, and the mean delay of a node is significant in the context of ultra-reliable low latency communications (URLLC).

This paper is organized as follows: in Section 2, we present the related work regarding energy harvesting communication networks as well as a literature survey of mathematical modeling attempts. We describe the system model in Section 3. In Section 4, we present the performance analysis of the quality of service (QoS) parameters. We check the validity of the mathematical model with simulations in Section 5 and present the conclusions in Section 6. In Section 7, we discuss future work.

2. Literature Survey

The algorithms and protocols designed for wireless networks work differently when nodes harvest energy and work under energy constraints. Also, at the same time, the design should meet the criteria of 5G/6G networks. In [15], Huang et al. proposed an architecture for IoT applications in 5G communications that incorporates software-defined network management and energy harvesting nodes with a mobile charger. The paper describes the management of both uplink and downlink energy transfers in multi-hop wireless sensor networks, assisted by a mobile charger capable of replenishing energy for nodes. The authors also note that next-generation communication systems will likely be a combination of advanced technologies, highlighting the need to integrate these technologies and optimize them for better efficiency. In another study, the eHealth application involving wearable devices is modeled using the Poisson cluster process and analyzed for the probability that a wearable device correctly notifies the message along with other QoS parameters [16] by P-V Mekikis et al. The wearable devices are wirelessly charged.

There are relatively few papers on the modeling and analysis of energy harvesting solutions for 5G and B5G systems. In [17], Di Zhai et al. proposed two multi-user scheduling schemes for IoT networks with wireless-powered nodes and a hybrid access point capable of transferring wireless power. The first policy attempts to achieve maximum system throughput in the absence of channel state information, considering residual energy at nodes. The authors consider the accumulate and transmit protocol. The second policy [17] attempts to obtain fairness for nodes with poor channel quality. The delay constraints as well as spectrum sharing are not considered. The authors in [18] presented a Markovian model for wireless sensor networks where nodes can be chosen between unlicensed and licensed bands. They also provided a Markov decision process formulation to derive an optimal transmission scheduling policy. In this model, energy harvesting is not considered, as the nodes are powered by traditional methods.

We briefly describe the analytical models for CSMA, which are used by LBT. The homogenous coexistence consisting of only NR nodes in the absence of Wi-Fi in the NR-U network was considered [19] by S. Muhammad. The throughput, delay, and collision probability were calculated for different priorities of nodes in a multiclass environment. The model was applied to the 5G NR-U-enabled ICU hospital setup. The nodes were non-energy harvesting (non-EH). The modeling and analysis of WLAN with RF-powered nodes were performed [20] by Y. Zhao et al. The access point (AP) was grid-powered while the stations harvested RF energy from the AP when the AP transmitted data to other nodes. The network was modeled using a three-dimensional Markov chain. The uplink and downlink throughput were calculated for AP as well as a station. In another attempt, for general energy harvesting nodes without dependence on AP, modeling and analysis were performed [21]. The stations harvested energy, modeled as the Poisson process. As the number of slots required for sufficient charging of stations was assumed to be large compared to the average backoff duration, the throughput and delay calculated were independent of other parameters such as congestion window size and the number of stations. The Markov chain analysis performed in [19–21] is based on the classic Bianchi model [22]. The literature presents models for RF-powered WLAN to compute throughput and delay. In addition to these parameters, the outage probability is an important metric for energy-harvesting wireless networks. Most of the work in the literature is dedicated to finding outage probability, considering the signal-to-interference noise ratio (SINR) at the receiver's end, e.g., by R Vaze [23]. Reference [24] computed and analyzed the transmit outage probability.

This paper considers an IoT network that is RF-powered by BS. The network operates under the NR-U protocol using the LBT feature and is modeled using a three-dimensional Markov chain. Closed-form expressions for the collision probability of BS and nodes, as well as transmission probability, energy harvesting probability, and throughput, are calculated. The expression for the mean delay of a packet at a node is also derived. The outage probability of the node is computed. These QoS parameters are analyzed for the size of the network, the size of the congestion window, the TXOP value, the maximum energy level, and the energy threshold of the node. While there are numerous attempts in the literature to model RF-powered WLAN, to the best of our knowledge, this is the first attempt to model and analyze an RF-powered NR-U network for IoT communication. The model helps us to analyze the key QoS parameters for NR-U-specific parameters such as TXOP. In [19], the NR-U network with non-EH nodes is analyzed for collision probability, throughput, and mean delay. In this work, we analyze the RF-powered NR-U network for collision, transmission, and energy harvesting probabilities, as well as for throughput and delay. Also, we compute the transmit outage probability of RF-powered LBT, which adds novelty to the existing work. The authors' contribution is to envision the NR-U network with RF energy harvesting and comprehensively analyze it for QoS parameters, which include throughput, delay, and transmit outage.

3. System Model

This section is organized as follows: Section 3.1 describes the model and the assumptions. In Section 3.2, the introduction to the LBT protocol is provided. Section 3.3 describes the Markov chain modeling. In Section 3.4, the resultant formulae of QoS parameters are provided.

3.1. Model and Assumptions

In this paper, we consider an NR-U network where nodes/user equipment (UE) harvest RF energy from the access point or BS. The IoT nodes function equivalently to UE. The nodes harvest energy when the BS sends data to other nodes. As standardized in 3GPP, the nodes and the BS follow the LBT protocol for random access [4]. For simplification, we only consider the NR-U nodes and do not consider the existence of Wi-Fi networks and other incumbent technologies such as Bluetooth. This can also be referred to as homogenous coexistence [19]. It applies to scenarios where other technologies do not exist in the area considered, or some form of multiplexing—either spatial or temporal—is used to avoid interference among these different technologies.

The node is equipped with an RF energy harvester. A typical node with an RF energy harvester is shown in Figure 1.

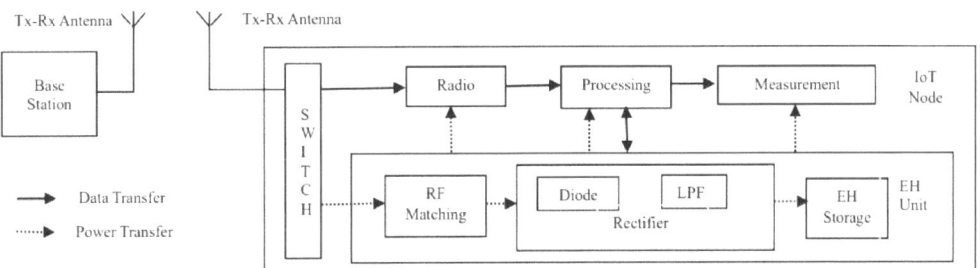

Figure 1. A typical node with an RF energy harvester.

The RF energy harvester consists of an antenna, a matching circuit, a rectifier, and a load or energy storage device [25]. Various types of RF energy harvesters—according to bandwidths—are available, namely single-band, multi-band, and wide-band [26]. For the system model under consideration, single-band RF energy harvesters for industrial, scientific, and medical (ISM) bands are suitable. There is substantial literature available on rectenna design for ISM bands (2.4 GHz and 5.8 GHz) [26]. For example, Khan et al. designed a single-band rectenna that generates 2.75 V at a distance of 10 meters [27] by Khan et al.

We consider a base station with a transmit power of Q watts located at the center of a circular area with a radius of R meters. The system model is shown in Figure 2.

The nodes are randomly present in the circular area. The nodes harvest sufficient energy to execute the NR-U protocol. The base station is powered by the grid and is always ON while the nodes are considered RF-powered. The nodes harvest RF power from BS when the BS transmits data to other nodes. The energy is stored in the supercapacitor. It is assumed that in one successful frame transmission of the BS, nodes harvest one unit of energy. For the transmission of one frame, the entire stored energy is used.

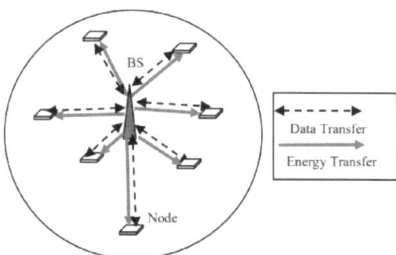

Figure 2. System model.

3.2. Introduction to LBT in RF-Powered NR-U

The flowchart of the LBT procedure followed by an RF-powered node is provided in Figure 3.

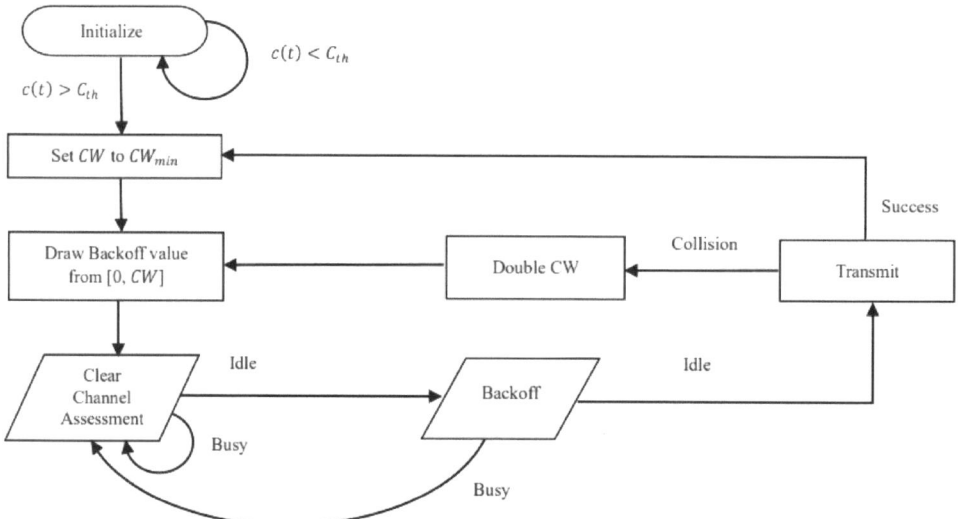

Figure 3. LBT procedure of RF-powered NR-U node.

When the nodes and BS have data to transmit, they wait for the clear channel assessment (CCA) duration. The nodes detect the energy level in the channel. According to 3GPP TR38.889 [4], the energy detection threshold is −73 dBm. If the channel is sensed to be free and the energy of the node is sufficient (i.e., greater than the threshold C_{th}), then the node draws a random value for backoff from the congestion window size, denoted by CW. The notations used in the paper are summarized in Table 1. The minimum value of the congestion window is denoted by CW_{min} and the maximum value is denoted as CW_{max}. The node with the least backoff counter transmits first, and others try in the next frame duration. If a node's energy is less than the threshold, then the nodes wait and will transmit when the energy becomes sufficient. The BS always has the energy to transmit as it is powered by the grid. If two or more nodes transmit in the same frame duration, a collision occurs. A collision can also happen when a node transmits simultaneously with the BS. The collision window is doubled each time a collision occurs.

Table 1. Notations.

Notation	Description
Q	The transmit power of the BS
R	The radius of the circular area
C_{th}	Threshold for energy level above which a node can transmit
C_{max}	Maximum value of the energy level of the supercapacitor
CW	Congestion window size
CW_{min}	Minimum value of the congestion window
CW_{max}	Maximum value of the congestion window
N	Number of stations
W_0	Initial congestion window size
W_m	Congestion window size after m^{th} retransmission attempt
M	Maximum number of times the congestion window can be doubled
$m(t)$	Number of times the congestion window doubled until time t
$b(t)$	Backoff counter value at time t
$c(t)$	Charging level of supercapacitor at time t
P_e	Probability that a node successfully harvests the energy from BS
τ_N	Probability that a node attempts a transmission
P_{c-N}	Probability that a node collides with another node
P_{c-BS}	Probability that BS encounters a collision
τ_{BS}	Transmission probability of BS
P_e	Energy harvesting probability of the node
P	Probability transition matrix of the Markov chain
θ_N, θ_{BS}	Normalized saturation throughput of nodes and BS, respectively
P_{Ts}	Probability of the success of a node
P_{Tc}	Probability of the collision of a node
$\overline{BD}$	Average backoff duration
δ	One slot duration
$E[D]$	Expected delay of a node
N_p	Mean number of packets waiting for transmission at the front of the queue in the network
θ_p	Throughput in terms of the mean of packets per second
T_s	Transmission time in case of success
P_o	Probability of outage of a node

The potential protocol stack of the envisioned system is shown in Figure 4.

The layers in the protocol stack of the NR-U base station and the IoT node using the NR-U framework are shown. Both the NR-U base station and the RF-powered node include an LBT manager in addition to other modules similar to New Radio 5G. A detailed description of the 5G New Radio architecture is available in 3GPP documentation [28]. Also, the variations in the protocol stack for NR-U are provided [29] by M. Hirzallah et al. The PHY and MAC layer modules perform functions for physical layer aspects and medium access control, respectively. They communicate with the scheduler in the NR-U base station, which in turn communicates with the LBT manager for channel sensing and decision-making based on the signal level. The scheduler is specific to the BS for coordinating the nodes in the network. The radio link control (RLC) layer ensures reliable data delivery. The packet data convergence protocol (PDCP) provides data compression for user and control data. The service data adaption protocol (SDAP) layer is responsible for the QoS framework of 5G. The functions of the radio resource control (RRC) layer include the establishment, maintenance of radio resources, and security management. In addition to these NR-U protocol layers, a power manager is essential in the architecture of an RF-powered IoT node, which performs the following functions: if the energy is available above the threshold, then signaling—like non-EH NR-U—takes place. If the energy is less than the threshold, then the node waits until its energy level is above the threshold by recharging itself with BS power.

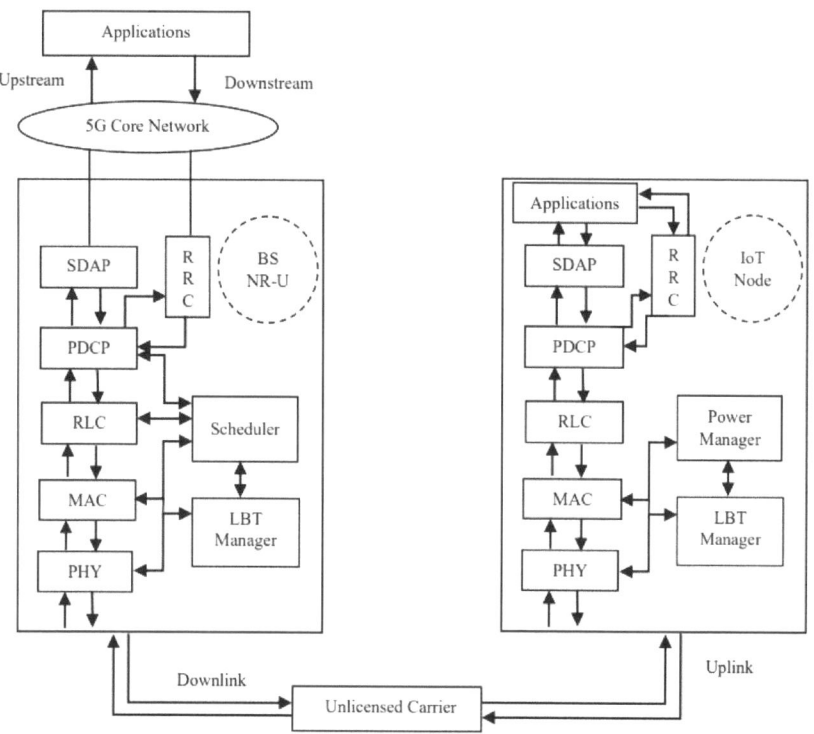

Figure 4. Protocol stack of the RF-powered NR-U system.

3.3. Markov Chain Modeling

Markov chain modeling is a powerful mathematical tool used to compute steady-state probabilities, leading to the analysis of QoS parameters in communication networks. Here, we model the sequence of events in the LBT procedure followed by the RF-powered NR-U node using a discrete-time Markov chain (DTMC).

Let the number of stations be denoted as N. The initial congestion window size is W_0. The maximum number of times the window size can be doubled is denoted as M. The maximum charging level of the supercapacitor is denoted by C_{max}. Let $m(t)$ denote the number of times the congestion window doubled (retransmission attempts), $b(t)$ denote the backoff counter value, and $c(t)$ denote the charging level of supercapacitor at time t. Then $\{m(t), b(t), c(t)\}$ is a discrete-time stochastic process with the state space described in Figure 5. Let P_e denote the probability that a node successfully harvests energy from the BS. Let τ_N be the probability that a node attempts a transmission, and P_{c-N} denote the probability that a node collides with another node. Let P_{c-BS} denote the probability that the BS encounters a collision and let the transmission probability of the τ_{BS}. The five categories of state space are described as follows:

- Category 1: In this category, the backoff counter $b(t)$ is decremented by one in every slot, but it does not reach zero, so the node cannot transmit. It harvests energy with the probability P_e and the charging level is increased by one.
- Category 2: As the charging level of the node is full, the only change in the state space is that the backoff counter is decremented by one for this set of states.
- Category 3: In this category, the backoff counter is zero, and the node harvests energy with probability P_e when BS transmits.
- Category 4: The node transmits as the backoff counter reaches zero. The collision occurs with the probability P_{c-N} and the next value of the backoff counter is chosen

with uniform probability from the congestion window size W_{m+1}. If the transmission is successful, then the congestion window is reset to size W_0.
- Category 5: The transitions are similar to that of Category 4. The difference is that the retransmission attempts reach the maximum value of M.

From the state space transitions, it can be seen that $\{m(t), b(t), c(t)\}$ is a DTMC. Here, it is important to consider three variables simultaneously so that the sequence of events under consideration follows the Markov property [30]. Therefore, the Markov chain is three-dimensional.

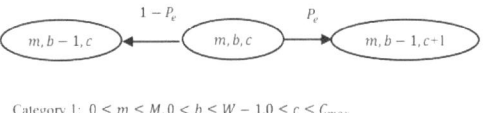

Category 1: $0 \leq m \leq M, 0 < b \leq W - 1, 0 \leq c \leq C_{max}$

Category 2: $0 \leq m \leq M, 0 < b \leq W - 1, c = C_{max}$

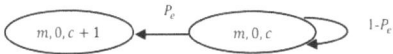

Category 3: $0 \leq m \leq M, b = 0, 0 < c \leq C_{max}$

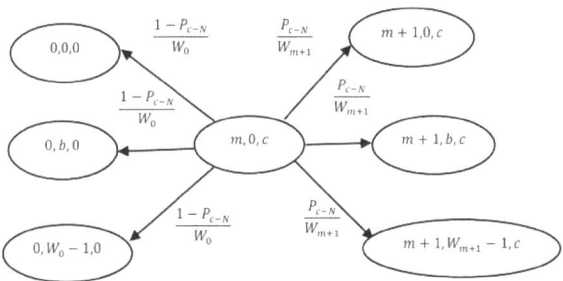

Category 4: $0 \leq m < M, b = 0, C_{th} \leq c \leq C_{max}$

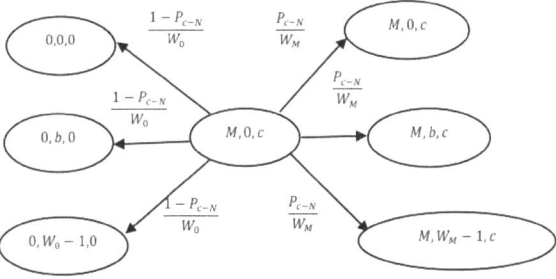

Category 5: $m = M, b = 0, C_{th} \leq c \leq C_{max}$

Figure 5. State space diagram.

The Markov chain under consideration is too complex to obtain closed-form expressions by solving $\pi = \pi P$ where P is the transition probability matrix. The steady-state distribution can be obtained either by solving numerically or using the decoupling technique [20]. The closed-form expressions after using the decoupling approximation are enlisted in Section 3.4.

3.4. QoS Parameters

The transmission probability of a node (τ_N), the collision probability of a node (P_{c-N}), and the energy harvesting probability of a node (P_e) can be found using the Markov chain model from Section 3.3. These probabilities depend on the collision probability and transmission probability of the BS. The expressions for P_{c-BS} and τ_{BS} are obtained from the analysis involving non-EH nodes initially derived [22] and used later in other works, including [20]. The collision of a BS occurs when at least one node transmits along with the BS. Hence, P_{c-BS} is expressed in (1) as follows:

$$P_{c-BS} = 1 - (1 - \tau_N)^N. \tag{1}$$

The transmission probability of BS (τ_{BS}) in terms of P_{c-BS} is given as follows:

$$\tau_{BS} = \frac{2}{1 + W + P_{c-BS} W \sum_{i=0}^{M-1} (2 P_{c-BS})^i}. \tag{2}$$

Note that the expressions (1) and (2) serve as the starting point to compute QoS parameters for the Markov chain of RF-powered nodes. Equation (1), in practice, provides an upper bound on the collision probability of the BS involving RF-powered IoT nodes. Similarly, (2) provides a lower bound on the transmission probability of the BS. This is due to the fact that in a network with EH nodes, the total number of nodes participating in the competition is less than the N as some of them might not have sufficient energy.

The transmission probability of a node (τ_N), the collision probability of a node (P_{c-N}), and the energy harvesting probability of a node (P_e) are expressed as simultaneous Equations (3)–(5), along with (1) and (2). The QoS parameter expressions for an IoT node are obtained by simplifying the expressions provided [20] after the decoupling approximation. The transmission probability of a node (τ_N) is expressed as follows:

$$\tau_N = \frac{t_2 - \sqrt{t_2^2 - 8 t_1 P_e}}{2 t_1}, \tag{3}$$

where

$$t_1 = \left((W-1)P_e + P_{c-N} W \sum_{i=0}^{M-1} (2 P_{c-N})^i\right) C_{th}(1 - P_{c-N}),$$

$$t_2 = (W+1)P_e + P_{c-N} P_e W \sum_{i=0}^{M-1} (2 P_{c-N})^i + 2 C_{th}(1 - P_{c-N}).$$

The collision of a node occurs if some other node or the BS transmits when the node transmits. Hence, the collision probability of a node (P_{c-N}) is expressed as follows:

$$P_{c-N} = 1 - (1 - \tau_{BS})(1 - \tau_N)^{(N-1)}. \tag{4}$$

The energy harvesting probability depends on the transmission and collision probability of the BS and is expressed as follows:

$$P_e = \frac{N-1}{N} \tau_{BS} (1 - \tau_N)^{(N-1)}. \tag{5}$$

These expressions are used further to derive the remaining QoS parameters in Section 4.

4. Performance Analysis

In this section, we derive the important performance indices for a node in the network. The structure of an NR-U frame is discussed in Section 4.1. The closed-form expressions for the normalized saturated throughput of nodes and the BS are derived in Section 4.2. The mean delay at a node is calculated in Section 4.3. The outage probability is derived in Section 4.4.

4.1. The NR-U Frame Structure

The frame structure of the NR-U node with sufficient energy to transmit is depicted in Figure 6.

Figure 6. Frame Structure of the NR-U Node.

Short interframe spacing (SIFS) is similar to that in the IEEE 802.11b standard [5], where the node waits at the beginning of the frame to avoid collisions due to delayed acknowledgments (ACKs). In CCA duration, the node senses the channel for occupancy. This duration is typically 1 to 7 slots according to [4]. Next, if the channel is found to be free, the node draws a random value from the congestion window (CW_{min}, CW_{max}) as the backoff duration, denoted as BD. At the end of the backoff, the node transmits. If two or more nodes transmit, then it results in a collision; otherwise, it is a success. The transmission duration in the case of success (T_s) is assumed to be the TXOP duration. The collision duration (T_c) is assumed to be equal to one slot. Let δ be one slot duration. The average backoff duration ($\overline{BD}$) in the network required for the computation of the throughput can be computed as follows.

In NR-U, the collision window is doubled after a collision, making the backoff process binary exponential, similar to carrier sense multiple access with collision avoidance (CSMA/CA). This can be modeled using exponential random variables, as described in [5]. Let $b_1, b_2, \ldots, b_N$ be exponential random variables with mean β. The idle period in the network ends when one of the backoffs from N nodes finishes. This period is a minimum of N i.i.d. exponential random variables with mean β. This is, again, an exponential random variable with a mean of $\frac{\beta}{N}$. As the backoff, a node is uniformly distributed over (CW_{min}, CW_{max}), and the mean value of the backoff duration of a node (β) can be expressed as follows:

$$\beta = \frac{CW_{min} + CW_{max}}{4}. \tag{6}$$

Hence,

$$\overline{BD} = \frac{CW_{min} + CW_{max}}{4N} \times \delta. \tag{7}$$

4.2. The Normalized Saturation Throughput

Next, we provide the analysis of the normalized saturation throughput. The normalized saturation throughput of nodes is denoted as θ_N and that of BS as θ_{BS}. The throughput is saturated as each node always has a packet to transmit. Let P_{Ts} be the probability of success and P_{Tc} be the probability of collision. They are expressed as follows:

$$P_{Ts} = \tau_{BS}(1 - \tau_N)^N + N\tau_N(1 - \tau_{BS})(1 - \tau_N)^{(N-1)}, \tag{8}$$

$$P_{Tc} = 1 - P_{Ts}. \tag{9}$$

Then, the normalized saturation throughput of nodes, which is the ratio of the expected time spent in successful transmission to the expected total time duration, is given as follows:

$$\theta_N = \frac{N\tau_N(1-\tau_{BS})(1-\tau_N)^{N-1}T_s}{P_{T_s}T_s + (1-P_{T_s})T_c + T_{CA}}, \quad (10)$$

where

T_s = Time spent in successful transmission;
T_c = Time spent in collision;
$T_{CA} = SIFS + CCA + \overline{BD}$.

Next, θ_{BS} can be expressed as follows:

$$\theta_{BS} = \frac{\tau_{BS}(1-\tau_N)^N T_s}{P_{T_s}T_s + (1-P_{T_s})T_c + T_{CA}}. \quad (11)$$

θ_N and θ_{BS} are simulated and analyzed for different parameters in Section 5.

4.3. Mean Delay of a Packet at a Node

In order to find the mean delay of a packet at an IoT node, the network with RF-powered NR-U nodes—except BS—is visualized as a queue sending data to BS. The queue's service rate in terms of the number of packets transmitted per second is denoted as $s(t)$ and the number of packets waiting in the queue at time t is denoted as $N(t)$. Let D_i be the delay of ith packet. We consider that a node transmits one packet whenever it has an opportunity during a successful transmission time T_s. As the Markov chain of Section 3.3 is a positive recurrent DTMC, it is stationary and ergodic. Hence, $s(t)$ and D_i are mean ergodic processes. Therefore,

$$\frac{\int_0^T s(t).d(t)}{T} = \theta_p \text{ as } T \to \infty,$$

$$\frac{1}{n}\sum_{i=1}^{n} D_i = E[D] \text{ as } n \to \infty.$$

where θ_p is throughput in terms of the mean number of packets transmitted per second. Then using Little's theorem [31], we have the following:

$$E[N(t)] = \theta_p E[D]. \quad (12)$$

Under the saturated network condition, $E[N(t)]$ is N. Hence,

$$E[D] = \frac{N}{\theta_p}. \quad (13)$$

θ_p is $\theta_N \backslash T_s$. Therefore,

$$E[D] = \frac{NT_s}{\theta_N}. \quad (14)$$

The derived results in this section are validated using simulations in Section 5.

4.4. Outage Probability of the Node

The charging level of the supercapacitor, i.e., the energy level of the node c(t) can be represented by the Markov chain shown in Figure 7.

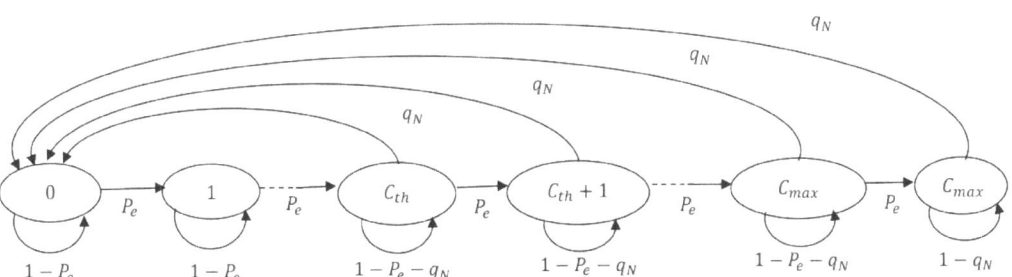

Figure 7. Markov chain of the charging level of the node.

The charging level is incremented by one from 0 to C_{max} with probability P_e. The energy of the node reduces to zero whenever a successful transmission occurs for states with energy greater than or equal to C_{th}. The probability of successful transmission for a node denoted as q_N is $\tau_N(1 - P_{c-N})$. The stationary distribution of the Markov chain in Figure 7 can be computed by solving $\pi = \pi P_E$ where P_E is the transition matrix of the Markov chain.

The outage probability of a node is the probability that the node does not have the energy to transmit the packet even though the packet is available at the node. Due to the assumption of saturated conditions, the outage probability of the node is the probability that the energy level of the node is less than C_{th}. Let P_o be the outage probability of the node, which is represented as follows:

$$P_o = \sum_{i=0}^{C_{th}-1} Pr(c(t) = i). \tag{15}$$

The outage probability is computed and analyzed in Section 5.

5. Simulations

In this section, the simulation results for the QoS parameters of the RF-powered NR-U network described in Sections 3 and 4 are presented. To check the validity of the mathematical model, a Monte Carlo simulator involving essential functions of an RF-powered LBT network is built in the programming language C. The transmit power of the base station (Q) is considered 1 W. The radius of the circular area is considered to be 9 m. The distances and power of the base station are considered according to the energy harvesting requirement of the nodes similar to [20] and also the 3GPP regulatory requirements on transmit power [4]. These distances are suitable for ISM bands (2.4 GHz and 5–7 GHz). For the mmWave band, the distances will be shorter, nonetheless, the model will be applicable. The values of simulation parameters are listed in Table 2. The number of nodes varied from 1 to 100 according to the requirements of different IoT applications, e.g., an agricultural IoT network has a size of around 100 nodes. The congestion window size combinations and TXOP parameter values are chosen according to the 3GPP guidelines provided [4].

Table 2. Simulation parameters.

Parameters	Figures 8 and 9	Figure 10	Figure 11	Figure 12	Figure 13	Figure 14	Figure 15–19
CW_{max}	128	128, 1028	128	128	128	128	128
TXOP (in ms)	8	8	4, 6, 8	8	4, 6, 8	8	8
C_{max} (in Volts)	8	8	8	8	8	8, 14, 20	8
C_{th} (in Volts)	4	4	4	4	4	4	3, 4, 5, 6
CW_{min}				16			
No. of Nodes				1–100			
SIFS (in μs)				16			
CCA (in μs)				63			
δ (in μs)				9			

In Figure 8, the collision probability and the transmission probability of the node are plotted with respect to the number of nodes. As the number of nodes increases, the collision probability of the node increases, and the transmission probability decreases exponentially as more nodes compete simultaneously.

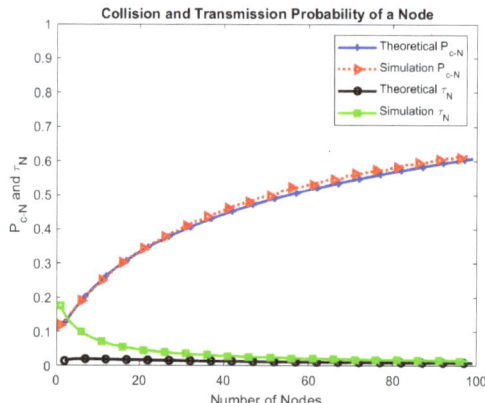

Figure 8. Collision and transmission probabilities of a node.

In Figure 9, the energy harvesting and outage probabilities of the IoT node are plotted. The energy harvesting probability of the node depends on the transmission probability of the BS, hence, it decreases with the increase in the number of nodes. As the number of nodes increases further, there is no significant change in the energy harvesting probability and the outage probability, indicating the energy sustainability of the network. The gap between theoretical and simulation values for fewer than 40 nodes is due to the fact that BS expressions (1) and (2) provide bounds instead of exact values. As the number of nodes increases, this effect reduces.

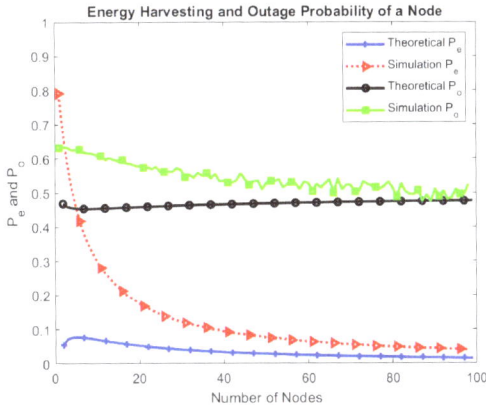

Figure 9. Energy harvesting and outage probabilities of a node.

The total throughput of nodes for different congestion window sizes is shown in Figure 10.

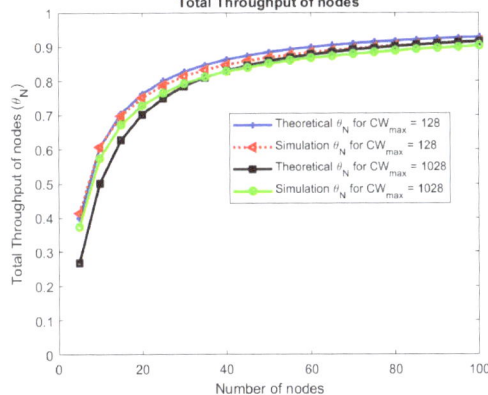

Figure 10. Total throughput of nodes w.r.t. CW_{max}.

The analytical model matches well with the simulations. It can be seen that there is no change in the throughput for the congestion window size.

In Figure 11, the throughput for nodes, as well as the BS for different values of the TXOP parameter, are plotted. The simulation values of throughput for TXOP values 4 ms, 6 ms, and 8 ms are shown for the congestion window sizes (16, 128). Changing the TXOP value has very little effect on the throughput.

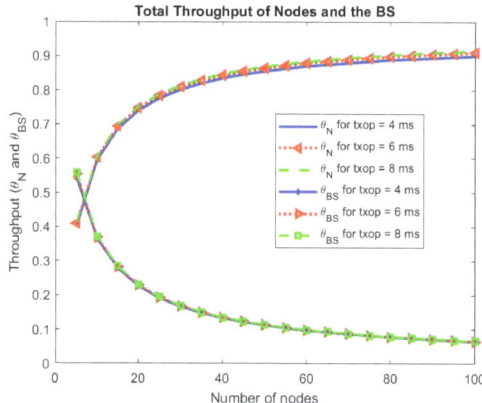

Figure 11. Total throughput of nodes and BS w.r.t. TXOP.

The mean delay of a packet at a node is shown for different values of the number of nodes in the network in Figure 12.

As the number of nodes increases, the competition among nodes increases. Hence, nodes have a lesser chance to transmit, increasing the mean delay of a packet at a node linearly. The change in the mean delay at a node for the TXOP parameter is shown in Figure 13. As the TXOP parameter value increases, the mean delay increases.

The effect of changing C_{max}, which is the energy storage capacity of the node, is observed in Figure 14. As seen in the theoretical analysis of Section 3, the performance of the node is independent of the C_{max}.

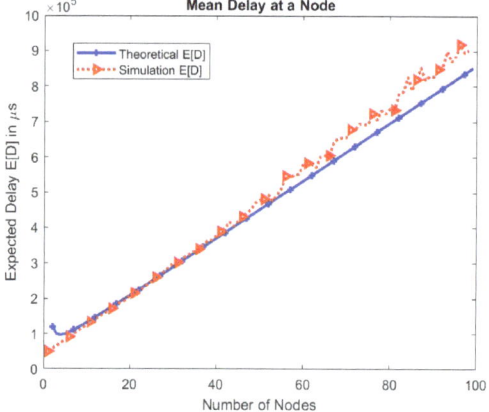

Figure 12. Mean delay of a node.

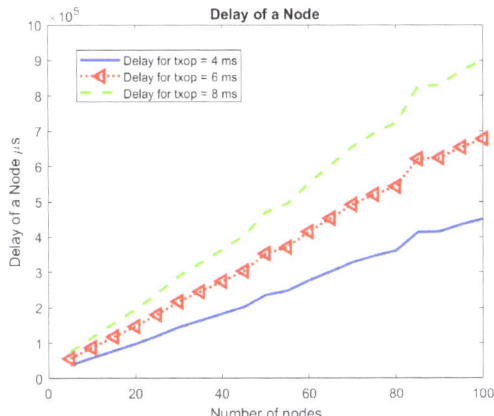

Figure 13. Mean delay at a node w.r.t. TXOP.

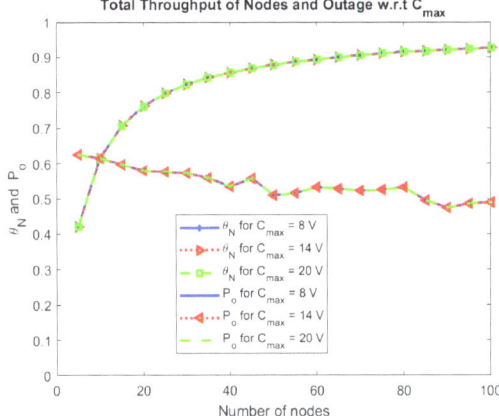

Figure 14. Total throughput of nodes and outage w.r.t. C_{max}.

The effect of changing the energy threshold C_{th} on the throughput of nodes and the mean delay is observed in Figure 15 and Figure 16, respectively. The throughput of nodes decreases, while the mean delay of the node remains unchanged with an increase in the minimum energy required for operation.

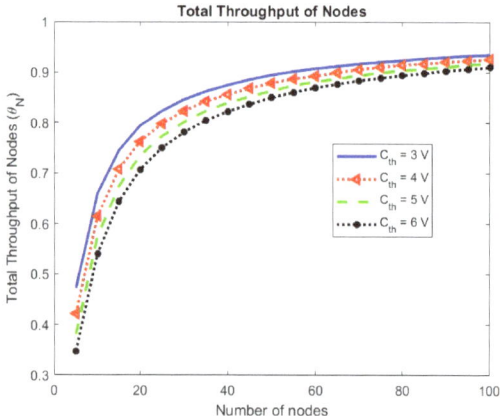

Figure 15. Total throughput of nodes w.r.t. C_{th}.

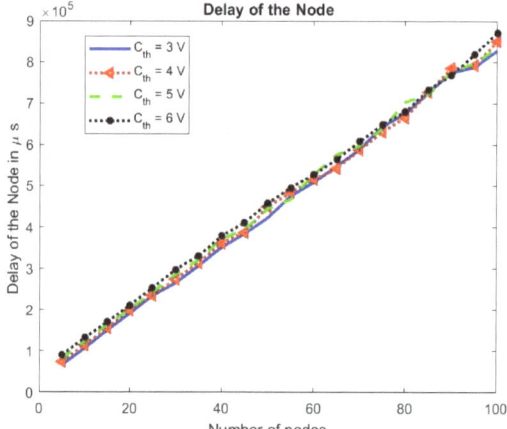

Figure 16. Delay at a node w.r.t. C_{th}.

The outage probability of the node increases with an increase in the energy threshold C_{th} as seen in Figure 17.

The effect of change in C_{th} on the collision and transmission probabilities of the node and the BS is shown in Figure 18 and Figure 19, respectively. It is seen that the transmission probability of the node remains the same for change in C_{th} while the transmission probability for the BS increases. This is because fewer nodes become eligible for participation due to increased C_{th} giving more chance to BS. An increase in the BS transmissions results in more harvested energy, compensating for the increased value of the minimum energy required for operation. Consequently, the transmission probability of the node remains unchanged. The collision probability for both the node and the BS decreases.

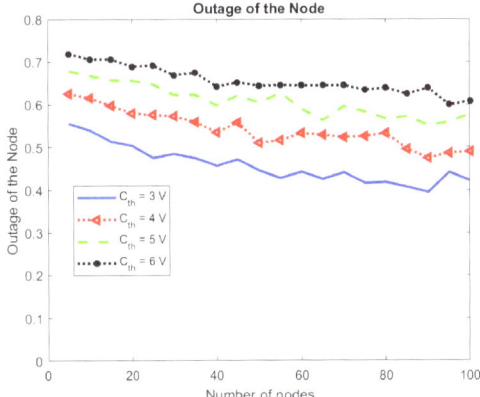

Figure 17. Outage probability of node w.r.t. C_{th}.

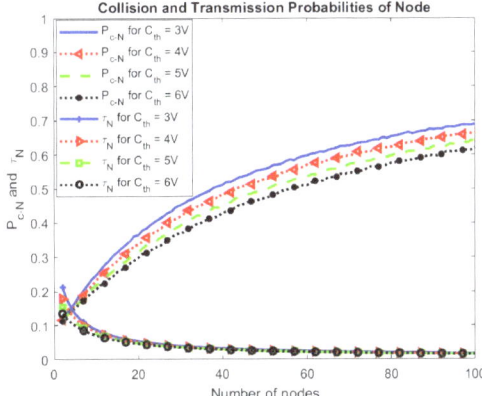

Figure 18. Collision and transmission probabilities of node w.r.t. C_{th}.

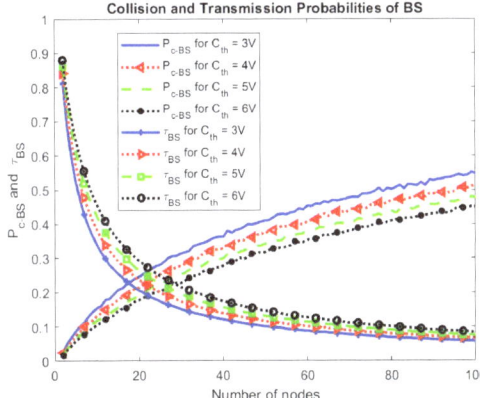

Figure 19. Collision and transmission probabilities of BS w.r.t. C_{th}.

6. Conclusions

We modeled an NR-U network with RF energy harvesting nodes in this work using a three-dimensional Markov chain. The closed-form expressions for the normalized saturated throughput of the IoT node and the BS, as well as the mean delay of a packet at the node, are derived. In addition, the node's transmit outage probability is computed. The collision and transmission probabilities of the node and the BS, along with the energy harvesting probability of the node, are analyzed based on the network's density. The theoretical values for QoS parameters align well with simulations. The collision probability increases and the transmission probability decreases for both the node and the BS as the number of nodes increases. The node's energy harvesting probability decreases with an increase in the number of nodes. The throughput of the BS decreases, while the total throughput of nodes increases with an increase in the number of nodes. The mean delay is observed to increase linearly with the density of the network. The outage probability of the node remains the same even though the number of nodes increased.

The effects of different congestion window sizes on the throughput and delay are observed. The change in TXOP value does not have a significant impact on the throughput but the mean delay changes linearly. As seen in the theoretical expressions and corresponding simulations, the throughput, delay, and outage are independent of the maximum energy level of a node. The throughput and outage probability depend on the energy threshold rather than the maximum energy storage level of the node. The delay is independent of the energy threshold. Thus, a theoretical framework for an RF-powered IoT network following the LBT protocol is provided, and QoS parameters are analyzed. The closed-form expressions presented in this work can be used to optimize the QoS parameters for system variables of interest. This framework is useful for analyzing the performance of LBT-based RF-powered IoT networks used in diverse areas, including agriculture, industrial, and healthcare.

7. Future Work

The model can be generalized to consider different EH requirements of each node. Also, different energy storage capacities and thresholds can be incorporated into the model. In this work, homogeneous coexistence is considered. The model can be extended to take into account the existence of Wi-Fi (either RF-powered or traditional), although the complexity will increase.

Author Contributions: Conceptualization, software, V.P.K.; methodology, V.P.K.; validation, V.P.K. and R.D.J.; formal analysis, investigation, V.P.K.; writing—original draft preparation, V.P.K.; writing—review and editing, V.P.K. and R.D.J.; supervision, R.D.J. All authors have read and agreed to the published version of the manuscript.

Funding: This research received no external funding.

Institutional Review Board Statement: Not applicable.

Informed Consent Statement: Not applicable.

Data Availability Statement: The original contributions presented in the study are included in the article.

Conflicts of Interest: The authors declare no conflicts of interest.

References

1. Statista. Available online: https://www.statista.com/statistics/1183457/iot-connected-devices-worldwide (accessed on 8 August 2024).
2. Statista. Available online: https://www.statista.com/statistics/1190270/number-of-devices-and-connections-per-person-worldwide (accessed on 8 August 2024).
3. Qualcomm. Available online: https://www.qualcomm.com/news/onq/2016/11/5g-spectrum-sharing-brings-new-innovations (accessed on 8 August 2024).

4. 3GPP Technical Specification Group Radio Access Network. *Study on NR-Based Access to Unlicensed Spectrum (Release 16)*; Technical Report; 3GPP TR 38.889 v1.1.0; 3GPP: Sophia Antipolis, France, 2018.
5. Kumar, A.; Manjunath, D.; Kuri, J. *Wireless Networking*; The Morgan Kaufmann Series in Networking; An imprint of Elsevier: Burlington, MA, USA, 2008.
6. Liu, Y.; Deng, Y.; Nallanathan, A.; Yuan, J. Machine Learning for 6G Enhanced Ultra-Reliable and Low-Latency Services. *IEEE Wirel. Commun.* **2023**, *30*, 48–54. [CrossRef]
7. Arpitha, H.S.; Anand, K.R.; Gullapalli, B. Digital Transformation of Oil and Gas Fields Architecting Multi-Services Digital Private Network on 5G NR-U Model. In Proceedings of the 2022 IEEE Wireless Antenna and Microwave Symposium (WAMS), Rourkela, India, 5–8 June 2022; pp. 1–5. [CrossRef]
8. Ansari, N.; Han, T. *Green Mobile Networks: A Networking Perspective*; IEEE Press: Piscataway, NJ, USA; Wiley: Hoboken, NJ, USA, 2017.
9. Shirvanimoghaddam, M.; Shirvanimoghaddam, K.; Abolhasani, M.M.; Farhangi, M.; Barsari, V.Z.; Liu, H.; Dohler, M.; Naebe, M. Paving the Path to a Green and Self-Powered Internet of Things. *arXiv* **2017**, arXiv:1712.02277.
10. Abd-Elmagid, M.A.; Dhillon, H.S.; Pappas, N. Online Age-Minimal Sampling Policy for RF-Powered IoT Networks. In Proceedings of the 2019 IEEE Global Communications Conference (GLOBECOM), Waikoloa, HI, USA, 9–13 December 2019; pp. 1–6. [CrossRef]
11. Liu, Y.; Li, D.; Du, B.; Shu, L.; Han, G. Rethinking Sustainable Sensing in Agricultural Internet of Things: From Power Supply Perspective. *IEEE Wirel. Commun.* **2022**, *29*, 102–109. [CrossRef]
12. Eltresy, N.A.; Dardeer, O.M.; Al-Habal, A.; Elhariri, E.; Hassan, A.H.; Khattab, A.; Elsheakh, D.N.; Taie, S.A.; Mostafa, H.; Elsadek, H.A.; et al. RF Energy Harvesting IoT System for Museum Ambience Control with Deep Learning. *Sensors* **2019**, *19*, 4465. [CrossRef] [PubMed]
13. Sanislav, T.; Mois, G.D.; Zeadally, S.; Folea, S.C. Energy Harvesting Techniques for Internet of Things (IoT). *IEEE Access* **2021**, *9*, 39530–39549. 2021.3064066. [CrossRef]
14. Niyato, D.; Kim, D.I.; Maso, M.; Han, Z. Wireless Powered Communication Networks: Research Directions and Technological Approaches. *IEEE Wirel. Commun.* **2017**, *24*, 88–97. [CrossRef]
15. Huang, X.; Yu, R.; Kang, J.; Gao, Y.; Maharjan, S.; Gjessing, S.; Zhang, Y. Software Defined Energy Harvesting Networking for 5G Green Communications. *IEEE Wirel. Commun.* **2017**, *24*, 38–45. [CrossRef]
16. Mekikis, P.-V.; Antonopoulos, A.; Kartsakli, E.; Passas, N.; Alonso, L.; Verikoukis, C. Stochastic modeling of wireless charged wearables for reliable health monitoring in hospital environments. In Proceedings of the 2017 IEEE International Conference on Communications (ICC), Paris, France, 21–25 May 2017; pp. 1–6. [CrossRef]
17. Zhai, D.; Chen, H.; Lin, Z.; Li, Y.; Vucetic, B. Accumulate Then Transmit: Multiuser Scheduling in Full-Duplex Wireless-Powered IoT Systems. *IEEE Internet Things J.* **2018**, *5*, 2753–2767. [CrossRef]
18. Ghazaleh, H.A.; Alfa, A.S. Optimal Scheduling in Cognitive Wireless Sensor Networks with Multiple Spectrum Access Opportunities. In Proceedings of the 2017 IEEE Wireless Communications and Networking Conference (WCNC), San Francisco, CA, USA, 19–22 March 2017; pp. 1–6. [CrossRef]
19. Muhammad, S.; Refai, H.H.; Al Kalaa, M.O. 5G NR-U: Homogeneous CoexistenceAnalysis. In Proceedings of the GLOBECOM 2020-2020 IEEE Global Communications Conference, Taipei, Taiwan, 7–11 December 2020; pp. 1–6.
20. Zhao, Y.; Hu, J.; Diao, Y.; Yu, Q.; Yang, K. Modelling and Performance Analysis of Wireless LAN Enabled by RF Energy Transfer. *IEEE Trans. Commun.* **2018**, *66*, 5756–5772. [CrossRef]
21. Yang, G.; Lin, G.; Wei, H. Markov chain performance model for IEEE 802.11 devices with energy harvesting source. In Proceedings of the 2012 IEEE Global Communications Conference (GLOBECOM), Anaheim, CA, USA, 3–7 December 2012; pp. 5212–5217.
22. Bianchi, G. Performance analysis of the IEEE 802.11 distributed coordination function. *IEEE J. Sel. Areas Commun.* **2000**, *18*, 535–547. [CrossRef]
23. Vaze, R. Transmission capacity of wireless ad hoc networks with energy harvesting nodes. In Proceedings of the 2013 IEEE Global Conference on Signal and Information Processing, Austin, TX, USA, 3–5 December 2013; pp. 353–358. [CrossRef]
24. Li, Y.; Zhao, R.; Deng, Y.; Shu, F.; Nie, Z.; Aghvami, A.H. Harvest-and-Opportunistically-Relay: Analyses on Transmission Outage and Covertness. *IEEE Trans. Wirel. Commun.* **2020**, *19*, 7779–7795. [CrossRef]
25. Li, Y.; Rajendran, J.; Mariappan, S.; Rawat, A.S.; Sal Hamid, S.; Kumar, N.; Othman, M.; Nathan, A. CMOS Radio Frequency Energy Harvester (RFEH) with Fully On-Chip Tunable Voltage-Booster for Wideband Sensitivity Enhancement. *Micromachines* **2023**, *14*, 392. [CrossRef]
26. Halimi, M.A.; Khan, T.; Nasimuddin; Kishk, A.A.; Antar, Y.M.M. Rectifier Circuits for RF Energy Harvesting and Wireless Power Transfer Applications: A Comprehensive Review Based on Operating Conditions. *IEEE Microw. Mag.* **2023**, *24*, 46–61. [CrossRef]
27. Khan, N.U.; Khan, F.U.; Farina, M.; Merla, A. RF energy harvesters for wireless sensors, state of the art, future prospects and challenges: A review. *Phys. Eng. Sci. Med.* **2024**, *47*, 385–401. [CrossRef] [PubMed]
28. 3GPP. *NR and NG–RAN Overall Description–Stage 2*; no. 3GPP TS 38.300 v15.4.0; 3GPP: Sophia Antipolis, France, 2018.
29. Hirzallah, M.; Krunz, M.; Kecicioglu, B.; Hamzeh, B. 5G New Radio Unlicensed: Challenges and Evaluation. *IEEE Trans. Cogn. Commun. Netw.* **2021**, *7*, 689–701. [CrossRef]

30. Ross, S. *Stochastic Processes*, 2nd ed.; Wiley India Pvt. Ltd.: New Delhi, India, 1996.
31. Papoulis, A.; Pillai, S.U. *Probability, Random Variables, and Stochastic Processes*, 4th ed.; Tata McGraw-Hill Publishing Company Limited: New Delhi, India, 2002.

Disclaimer/Publisher's Note: The statements, opinions and data contained in all publications are solely those of the individual author(s) and contributor(s) and not of MDPI and/or the editor(s). MDPI and/or the editor(s) disclaim responsibility for any injury to people or property resulting from any ideas, methods, instructions or products referred to in the content.

Article

Efficient Multi-Hop Wireless Power Transfer for the Indoor Environment

Janis Eidaks [1], Romans Kusnins [1], Ruslans Babajans [1,*], Darja Cirjulina [1], Janis Semenjako [1] and Anna Litvinenko [2]

[1] Institute of Microwave Engineering and Electronics, Riga Technical University, Azenes St. 12, LV-1048 Riga, Latvia; janis.eidaks@rtu.lv (J.E.); romans.kusnins@rtu.lv (R.K.); darja.cirjulina@rtu.lv (D.C.); janis.semenjako@rtu.lv (J.S.)

[2] SpacESPro Lab, Riga Technical University, Azenes St. 12, LV-1048 Riga, Latvia; anna.litvinenko@rtu.lv

* Correspondence: ruslans.babajans@rtu.lv

Abstract: With the rapid development of the Internet of Things (IoT) and wireless sensor networks (WSN), the modern world requires advanced solutions for the wireless powering of low-power autonomous devices. The present study addresses the wireless power transfer (WPT) efficiency problem by exploiting a multi-hop concept-based technique to increase the received power at the end sensor node (ESN). The current work adopts efficient multi-hop technology from the communications field to examine its impact on WPT performance. The investigation involves power transfer modeling and experimental measurements in a sub-GHz frequency range, chosen for being capable of providing a greater distance to transmit power. The paper proposes a multi-hop (MH) WPT concept based on signal amplification and demonstrates the fabricated multi-hop node (MHN) prototype. The experimental verification of the MHN is performed in the laboratory environment. The present paper examines two WPT scenarios: line-of-sight (LoS) and non-line-of-sight (NLoS). The turn-on angle of 90 degrees on MHN is used for the NLoS case. The received power and RF-DC converted voltage on the ESN are measured for all investigated scenarios. Moreover, the paper proposes an efficient simulation approach for the performance evaluation of MH WPT technology, providing an opportunity to analyze and optimize wireless sensor nodes' spatial distribution to increase the received power.

Keywords: wireless power transfer; multi-hop; internet of things; wireless senor networks; RF-DC conversion efficiency

1. Introduction

The growing number of smart interconnected devices and systems that make the Internet of Things (IoT) and the underlying wireless sensor networks (WSN) majorly impact various branches of the industry, creating smart interconnected environments with many low-power smart sensors that autonomously collect and exchange a huge amount of data [1,2]. The Ericsson Mobility Report [3] estimates cellular, wide-area, and short-range IoT connection to be 13.2 billion in 2022 and is expected to increase to 34.7 billion by 2028. The increasing integration of IoT is a significant driving force of economic development [4,5]. In 2013, the McKinsey Global Institute estimated the IoT economic impact to be between USD 2.7 and USD 6.2 trillion [6,7]; in 2021, the estimation was USD 5.5, being expected to rise to USD 12.6 trillion by 2030 [7]. The increasing number of autonomous devices and sensor nodes (SNs) poses a challenge to powering the devices employed in the network. The cases where the power delivery infrastructure cannot power an SN are usually covered by battery power. However, using batteries to power SNs poses a challenge in maintaining SNs by changing or charging batteries. The placement of the SNs can complicate maintenance or make it impossible. The increasing number of sensors and the need to maintain the integrity of the sensor network only amplifies the problem.

Radio frequency (RF) wireless power transfer (WPT) has developed over the last several years [8,9]. This technology not only benefits the powering solution but also facilitates control over the power consumption of an individual SN. WPT initially branched from the topic of RF energy harvesting (EH), which focused on powering individual autonomous SNs. The focus of RF EH was on developing solutions to convert the ambient RF signals to DC power. With the rapid increase in information exchange, RF EH approaches had to be complemented to be integrated into the WSN, thus moving from ambient EH to WPT. The simple structure of WPT thus consists of a power transmitter and a power receiver. The transmitter generates a power-carrying signal that the power receiver then converts to a DC voltage. The power receiver is an RF–DC converter paired with an antenna and an energy storage device (battery or a capacitor) connected to the SN. The main focus of current studies in the field is to enhance WPT system performance, with the most attention paid to the design of the rectenna (antenna paired with an RF–DC converter), studying possible RF–DC topologies, antenna designs, and their implementation [10–13].

The above-mentioned development of the WPT topic was single-device-scaled. An emerging concept of simultaneous wireless information and power transfer (SWIPT) has also received considerable attention, enabling backscatter technology [14–16]. The continuous development of WPT now shifts towards network-scale solutions, incorporating and combining new technologies to increase the performance of WPT on a network scale. The inspiration for such development is from the solutions developed for wireless communications.

In any wireless propagation, the wireless channel between the transmitter and receiver limits the system's performance [17,18]. In the case of WPT, the goal is adaptive power-signal forming matched to the channel and RF–DC converter of the harvesting node with a goal to increase the DC converted power [19–21]. Utilizing the propagation environment in a more efficient way might result in better performance of the WPT system not only in the case of a single power transmitter and receiver but also on a network scale [22]. The work [23] demonstrates that theoretically, channel adaptive waveforms might increase the delivered DC power by 100%, and in [24], the experiments demonstrate that joint waveform and beamforming can boost output DC power by 100% and increase the WPT range. However, the design of the most efficient power-carrying waveforms may be computationally complex, and thus a compromise must be made between the achievable efficiency and application possibilities [23].

Progress in RF micro-electromechanical systems (MEMS), as well as the field of artificial materials, influenced a number of emerging transmission technologies, such as intelligent reflecting surfaces (IRSs) [25,26]. IRSs are a promising approach to enhancing the performance of wireless sensor networks, as they may, in the foreseeable future, replace their active counterparts, extensively employed in wireless communications. Recent studies have demonstrated that a further reduction in the energy consumption of IRS-assisted WSNs [27] as well as reduced transmit power in the case of satellite-terrestrial relay networking [28], can be achieved through passive beamforming optimization. This approach can be followed to achieve a higher SWIPT efficiency [29,30]. Beamforming itself is an essential extension of WPT aimed at increasing the power transfer range and steering the power-carrying signal direction to enhance performance [31]. Using IRSs for WPT can increase the distance of power transfer as well as improve the performance in the non-line-of-sight (NLoS) energy transfer case. The work [32] experimentally demonstrates a 20 dB gain in received power when utilizing IRS for WPT. However, for the optimal deployment of IRSs in WPT, the placement and number of surfaces must be considered [33].

Multi-hop routing is another promising technique that can enhance the power efficiency of wireless communication systems, WPT systems, or even SWIPT [34,35]. Applying this wireless communications technique to WPT simplifies the overall infrastructure by reducing the number of power transmitters (power beacons, PB). The multi-hop node (MHN) will have enough power for the connected SN and transmit some energy to another SN not covered by the PB. This approach can slightly extend the coverage of the WPT without extra PB and also improve the performance in the NLoS propagation.

The work [36] experimentally demonstrated that the voltage gain on the end node powered using a multi-hop approach could increase up to 32% depending on the configuration. However, unlike the energy-aware routing developed in conventional wireless networks, the routing protocols in multi-hop WPT systems must consider the RF energy propagation and the circuit design of network nodes. Thus, the power-efficient multi-hop nodes must be supplemented with power-efficient routing protocols [37,38]. This limits the potential scale of the deployed multi-hop WPT system.

Each of the techniques described has its limitations and specific deployment, and thus developing an efficient WPT system should not be limited to any single one of the listed techniques [19]. Combining the aforementioned approaches can lead to efficient WPT and a further reduction in the energy consumption of WSNs based on WPT. The current project's goal is to verify the impact of advanced techniques of WPT on energy transfer performance. This paper is devoted to studying the impact of multi-hop on WPT performance using the designed multi-hop node (MHN). The main focus of previous works on multi-hop wireless power transfer was theoretical and simulation studies to optimize the WPT network. The work [39] determined the efficiency of the energy transfer verses the number of hops, while work [40] presented an optimization model to determine the number of required PBs. The development of MHNs and experimental validation of MHN's performance in the indoor environment has received much less attention. The work [36] experimentally validated the MHN prototype that generates the power-carrying signal in the form of data packets with a 915 MHz carrier frequency. The system uses off-the-shelf evaluation boards; however, the distance between the PB and the end node is only 30 cm, and only the line-of-sight (LoS) is explored. Another aspect not covered is modeling the performance of the MHN node. Developing models with acceptable accuracy and low resource use is challenging. However, the theoretical models that can estimate the performance of WPT under real-life scenarios have high applicability in the system's development.

The current paper contributes to many aspects of MHN-based WPT development. First, the work proposes a new concept of MHN based on signal amplification. The presented approach results in a simpler design that can provide greater distance than the previously mentioned signal generation-based approach. The carrier frequency of the WPT system in the study is 865.5 MHz, as this frequency was used in our prior research [41] on RF–DC converter designs for WPT purposes. The emphasis on the sub-GHz range (primarily the ISM bands) is due to the increased distance for power transfer with compatibility with IoT and LORA. For example, commercially-available wireless power transfer-based products by Powercast specifically target these frequencies and applications [42]. It is important to point out that commercially-available WPT products do not utilize the multi-hop approach as of yet. Second, the paper discusses the challenges of MHN design and proposes a method of MHN development. Third, the validation of the MHN is performed using laboratory measurements and simulations emulating real-life WPT scenarios: LoS energy transfer and NLoS energy transfer. A turn-on angle of 90 degrees for MHN is required to mimic a non-direct scenario in a typical indoor environment, like an "L" shaped corridor. Fourth, the paper addresses the challenges of developing simulation models for the provided WPT scenarios and suggests an algorithm based on using different computational methods: the finite element full-wave method for acquiring the far-field antenna patterns, the ray-tracing method for estimating the power transfer efficiency, and the harmonic balance method for estimating the converted DC power.

The paper is organized as follows: Section 2 describes the multi-hop wireless power transfer concept. Section 3 details the developed multi-hop WPT node-design process and parameters. Section 4 presents experimental verification of multi-hop WPT technology, while Section 5 discusses the multi-hop WPT model and proposes an efficient simulation approach for performance evaluation. The final section concludes the work.

2. Concept of Multi-Hop Wireless Power Transfer

This section elaborates more on the multi-hop concept in the architecture of WPT and discusses possible implementations of the multi-hop node. The WPT architecture comprises a power transmitter called a power beacon (PB) that transmits the power-carrying signals to the surrounding SNs. This scenario is depicted in Figure 1a. The figure demonstrates that one PB cannot provide sufficient power to all SNs, as some are out of the cover range of the PB. Hence, additional PBs are required. This is not an optimal strategy, as an additional PB may be required to power only a single SN. Additionally, several PBs may interfere in situations where their coverage overlaps. An alternative architecture is illustrated in Figure 1b; it exploits multi-hop nodes to slightly extend the coverage to reach the SNs that otherwise would require an additional PB. This mitigates the interference while simplifying the overall architecture of the WPT network.

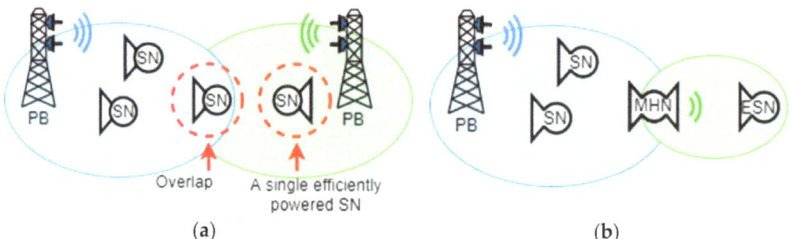

Figure 1. WPT architecture based on individual power beacons (**a**) and WPT architecture based on multi-hop power transfer (**b**).

The conceptual design of the MHN is defined by how the "hop" is implemented. Figure 2 demonstrates an MHN based on signal amplification. In this method, the RF-DC converted power is stored in a capacitor or a battery to power the connected SN. When the energy storage has sufficient energy for the SN, some is used to power the power amplifier (PA) to amplify the received power-carrying signal and send it further when the PA is active. Then, the received signal is partially amplified and transmitted, with some part of the signal rectified to charge the capacitor. This concept was adopted due to encouraging results reported in one of the previous contributions [43], showing it as a more viable approach for MHN. However, this technique possesses several shortcomings to be addressed. To be more precise, a proper MHN design necessitates balancing multiple requirements, e.g., high output power, low current consumption, high gain, and high RF–DC power conversion efficiency. Additionally, switching between the amplifying and charging modes must be optimized.

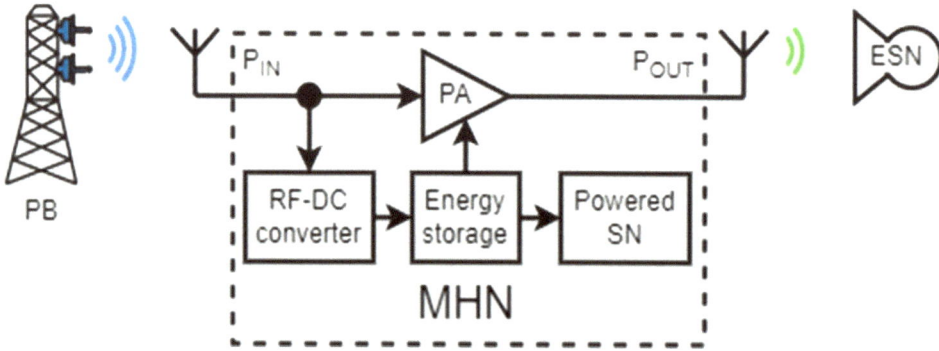

Figure 2. Multi-hop node based on signal amplification.

3. Multi-Hop Node's Design and Parameters

This section elaborates on the development process of the multi-hop node, outlining the necessary development stages and presenting the final design based on the signal amplification approach. The section also discusses limitations and compromises introduced at each stage of development. The section concludes by demonstrating the fabricated MHN prototype.

3.1. Workflow of the Multi-Hop Node's Development

The MHN based on signal amplification development steps, shown in Figure 3, starts with determining the node's output power level from the amplifiers. The second step requires analyzing and determining the approximate input power level range either with simulations or experimentally, as these parameters will determine the requirements of the amplifiers and the RF–DC converters used in the design. The third step of development determines the duration of the node's active (amplification) state. The active node amplifier's current consumption and storage element's capacity determines how long the node will be active; the expected input power level will allow for estimating how long the node will charge the storage element to the required threshold for it to operate. After the initial selection of design specifications is finished, the fourth step is devoted to selecting the topology and components and testing the components' performance. The fifth and final step is the prototyping and validating the complete MHN.

Figure 3. The MHN development process's block diagram.

3.2. Multi-Hop Node's Design Limitations and Final Structure

As mentioned previously, the development of the MHN introduced design compromises, as the MHN must satisfy numerous requirements in order to fulfill a useful and efficient solution. The MHN relaying power performance depends on each component's performance, and the following discussion demonstrates this and how the development steps influence one another.

One of the hurdles introduced in the development is matching the node's input impedance, as there is an input mismatch between the active state and the state of capacitor charging. The initial design solution was to use solid-state RF switches or RF relays to direct the received signal fully to the rectifier or amplifier circuitry. However, the mismatch when the switch is in an undefined state due to battery storage being in "cold" start conditions introduces additional issues in the matching system for the "cold" startup state and active state. When developing a prototype with discrete components, matching becomes a pronounced problem due to changes in different system operating modes, for example, switching the input signal to the rectifier in the system "cold" start conditions when the system storage voltage level is low or empty. This problem could be partially solved using the power of the storage system; however, it will always require a minimal charge to generate control signals for switches and discharge them. For this reason, the final design does not employ RF switches or RF relays; thus, this creates a compromise between better RF–DC matching and avoiding the uncertainty state which may brick the node. In the current MHN design, the input signal port is split into two: one path to the RF–DC rectifier and the other to the RF amplifier cascade. Input matching is performed specifically for the MHN charging mode, as this state will mostly occur relative to the active MHN state; therefore, with better matching, the more energy the system will harvest. There will be losses due to mismatch when the MHN is in a charging state due to reflections from the

PA's input in the MHN off state. There will also be a mismatch when the MHN becomes active and works in transmitting mode. However, it will have a negligible overall impact on the total harvested amount of energy, as such a state will be relatively short compared to the MHN charging state.

As the amount of the harvested energy depends on the received signal power, it will directly translate to the time required for storing energy. In order to improve the transmission mode duty cycle of the MHN, it is required to improve the RF–DC converter's efficiency as much as possible and reduce the amplifier's power consumption while still being capable of further transmitting a considerable amount of power. RF–DC converter matching and tuning were selected for the simulated input power range from modeling; the input power range and the required output power range requirements also narrowed down the chip selection for the amplifier and also introduced a lot of limitations due to requirements for the amplifier: The amplifier must have a considerable gain and produce a high output power at the same time, consuming less current than the counterparts. In this endeavor and comparison of amplifier datasheets, it was concluded that a two-stage amplifier design is required to bridge the necessary gain and output power level with reasonable current consumption to satisfy low power consumption. The PA's efficiency directly impacts the overall MHN system's performance, as most of the harvested energy is consumed by the power amplifier stage.

The power amplifiers require a stable supply when in active mode and a higher voltage than the RF–DC converter can provide. For this, an additional DC–DC boost converter chip must be introduced. The DC–DC converter must have a maximum power point tracking (MPPT) function to maximize the harvested power at the optimal operation point. The DC–DC converter will work in the set MPPT operation point, where the maximum power can be extracted from the RF–DC converter. Additionally, the DC–DC converter's output switching and amplifier turn-on speed must be fast in comparison to the active MHN state duration.

A final MHN structure was formed by performing the MHN development steps and considering the mentioned limitations. As a result, the final MHN in Figure 4 consists of several parts that must fulfill the required functions, such as the RF–DC converter, DC–DC boost converter, a storage element, and a two-stage amplifier consisting of PA_1 and PA_2. The RF–DC converter is used to convert the RF signal to the DC signal. The DC–DC boost converter is used to charge the external capacitor from the RF–DC converter output, boost the output voltage, and supply the voltage to the power amplifier cascade through external transistors. The two-stage amplifier cascade is used to amplify the signal from the input of the MHN system and relay it to the ESN. A storage element is required, in this case, a capacitor, to store the harvested energy from the RF–DC converter.

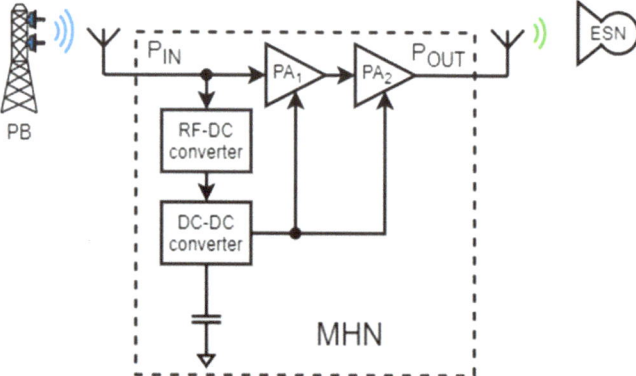

Figure 4. The structure of the multi-hop node.

Theoretically, the node can also supply some low-power IoT sensors or WSN nodes; however, it will impact the transmission period of the MHN system, depending on the received power level from PB and the current consumption of the sensor. Such implementation was not reviewed in this paper.

3.3. Multi-Hop Node's Prototype Implementation

After the structure of the MHN was finalized and the requirements for each component were set, the next step was aimed at selecting the components and prototyping the MHN. The schematic in Figure 5 shows the MHN design layout and component selection. The chosen RF–DC converter is based on the well-known and extensively used voltage doubler topology with two Schottky diodes SMS7630 [41]. These diodes are well-suited for high-frequency applications, owing to the low voltage drop they exhibit. The RF–DC converter output voltage is boosted using a DC–DC switching converter BQ25570 and stored in a 1 mF capacitor connected externally. As for the amplifiers, the first stage is the Qorvo RF2878 amplifier, while the second stage amplifier is Guerrilla RF GFR5020.

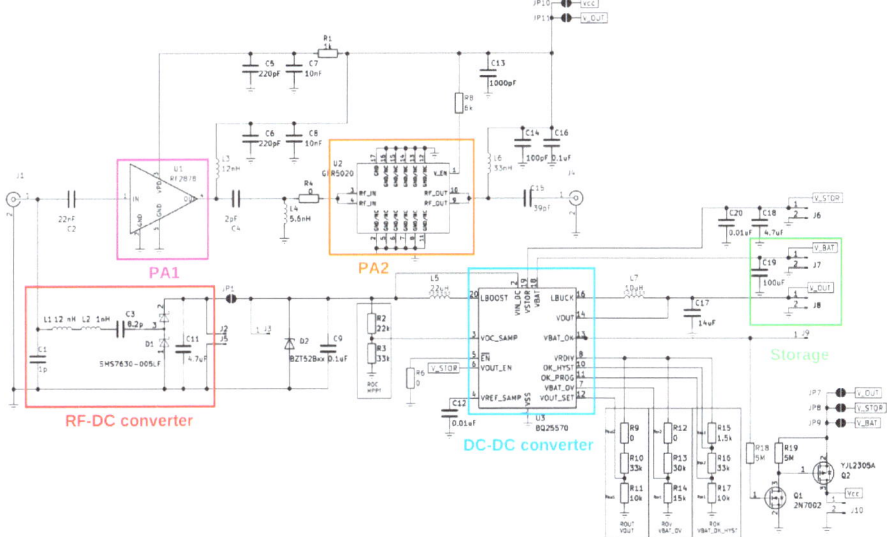

Figure 5. Schematic of the multi-hop node prototype.

The fabricated MHN prototype is shown in Figure 6. The prototype is a compact layout on a 1.6 mm FR4 base, with the board itself being 5 by 5 cm. The node's prototype was constructed so that most of the individual parts' performance could be measured, which is elaborated more in our previous work [43].

Figure 6. Photo of the multi-hop node prototype.

4. Experimental Verification of the Multi-Hop Wireless Power Transfer Technology

To prove the validity and demonstrate the benefits of the proposed concept, two indoor wireless power transfer scenarios are examined: LoS energy transfer and NLoS energy transfer. In the LoS scenario outlined in Figure 7, the PB's, MHN's, and ESN's antennas are all lined up. This experimental study compares the received (rectified) power level on the ESN with the MHN to the case without the MHN (only PB and ESN).

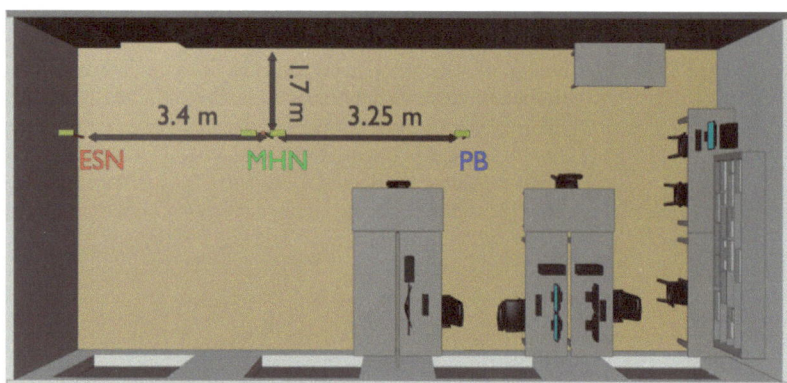

Figure 7. Layout of the indoor environment for studying the LoS MHN-aided WPT.

All antennas employed are four-element Yagi–Uda designed to operate at 865.5 MHz. The antenna gain at this frequency is 9.18 dBi. The Yagi–Uda antennas are employed due to high gain and narrow beam, reducing the multipath propagation's impact. The distance between the antennas is measured from the first dipole, which is approximately 10 cm from the edge of the antenna. The distance between the PB's transmitting antenna and the MHN's receiving antenna is set to 3.25 m. The distance between the MHN's transmitting antenna and the ESN's antenna changes from 1.15 m to 3.4 m. All antennas are located at the same height of 1.07 m.

In the NLoS scenario outlined in Figure 8, the imaginary straight line passing through the PB's antenna and the receiving antenna of the MHN subtends a right angle with that passing through the transmitting antenna of the MHN node and ESN's antenna. The antennas and the distances are similar to the LoS scenario. The key difference in this study is that the turn angle between the PB and ESN is 90 degrees; thus, there is no direct line of sight between these nodes. Due to the NLoS and the narrow beam of the antennas, this experimental study evaluates the performance only in the case of the MHN.

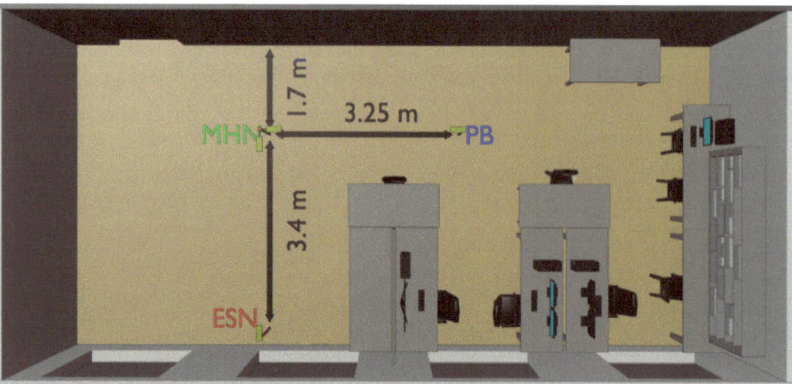

Figure 8. Layout of the indoor environment for studying the NLoS MHN-aided WPT.

The next subsection elaborates more on the ESN's RF–DC converter prototype used to evaluate the impact of MHN on the WPT performance, followed by the measurements setup used for both WPT scenarios and the analysis of the acquired results.

4.1. End Sensor Node's RF–DC Converter Prototype

The ESN prototype's RF–DC converter used in this study is the voltage doubler-based design shown in Figure 9. The design is based on the Skyworks SMS7630 diodes. The fabricated prototype in Figure 10 is a compact solution fabricated on a 1.6 mm FR4 base. This design was explored in our works [41,44] and we proved its efficiency. The latest iteration of this design uses low-DC-resistance inductors from Coilcraft, allowing for reaching significantly higher power conversion efficiency (PCE).

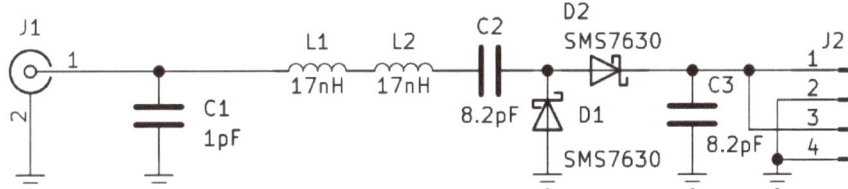

Figure 9. Schematic of the end node's RF–DC converter.

Figure 10. Photo of the end node's RF–DC converter.

The PCE of the RF–DC converter is also highly dependent on the load resistance. Figure 11 demonstrates the dependence of the PCE on the load resistance of this converter at various input power levels. The following formula is used to calculate the PCE:

$$\text{PCE} = \frac{P_{rec}}{P_{in}} \times 100\%, \tag{1}$$

where P_{rec} is the rectified power on the output of the RF–CD converter, and P_{in} is the input power of the RF–CD converter.

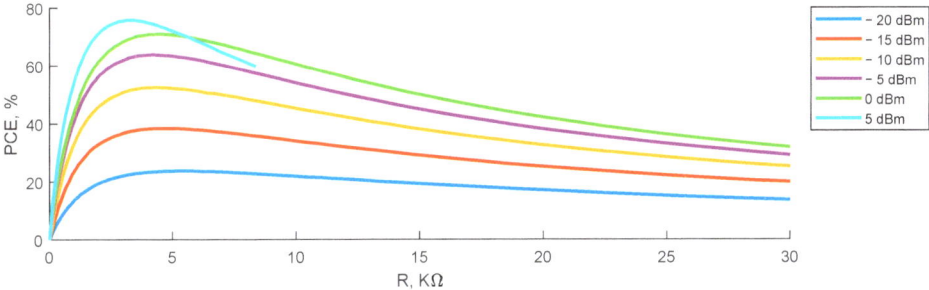

Figure 11. The PCE of the voltage doubler rectifier (end sensor node rectifier) depending on load resistance at different input power levels.

At the input power level of −20 dBm, the power conversion efficiency reaches 23.7% with 5 kΩ load resistance, efficiency at −15 dBm input power at 5 kΩ load resistance reaches 38.4%, efficiency at −10 dBm input power at 4.3 kΩ load resistance reaches 52.5%, efficiency at −5 dBm input power at 4.2 kΩ load resistance reaches 63.8%, efficiency at 0 dBm input power at 4.4 kΩ load resistance reaches 70.9%, and efficiency at 5 dBm input power at 3.3 kΩ load resistance reaches 75.7%. For the use cases with the MHN, the rectifier will be used in the power range from 0 to 5 dBm; therefore, the optimal load resistance is 3 kΩ. After defining this crucial parameter of the ESN's RF–DC converter, the ESN can be used in the experimental study.

4.2. Experimental Measurements Setup

The experimental study on the LoS propagation scenario is performed with the setups depicted in Figures 12 and 13. Figure 13 demonstrates the setup used to measure the received power level and rectified voltage at the ESN and its dependence on the distance between the MHN and ESN. As the setup shows, the 865.5 MHz power-carrying signal on the PB is generated by the R&S SMC100A generator through the power amplifier (PA) MMG3006N by NXP, resulting in the 26 dBm output power. The signal waveform from the ESN with active MHN is saved with the Tektronix DPO72004 oscilloscope (Salt Lake City, UT, USA) and later processed to calculate the average power absorbed into the oscilloscope's 50 Ω input termination, considering the SMA cable attenuation. The voltage across the RF–DC converter with 3 kΩ load resistance at the end node is measured with Digilent Analog Discovery 2 (AD2). In this case, the cable attenuation to the oscilloscope is also considered in the calculations. The setup in Figure 13 is similar but without the MHN, and the distance is measured between the PB and ESN antennas.

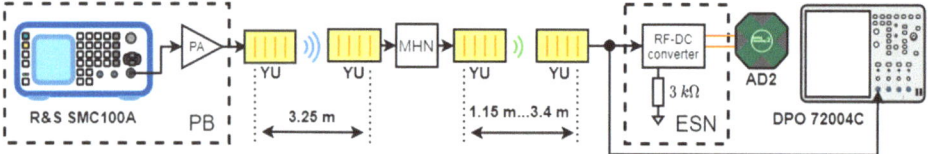

Figure 12. Experimental setup for measuring the received power and rectified voltage at the ESN depending on the distance between the MHN's and the edge sensor node's antennas.

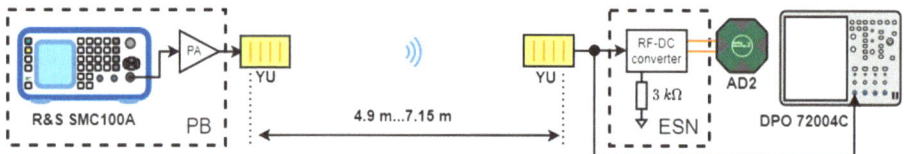

Figure 13. Experimental setup for measuring the received signal power and rectified voltage at the ESN from the transmitter depending on the distance between the antennas (without MHN).

For the NLoS scenario, only the experimental setup demonstrated in Figure 13 is used for measuring the received power and rectified voltage at the ESN, as the NLoS scenario requires the MHN for operation.

4.3. Experimental Results Analysis

The results of the experimental study are compiled in Figures 14 and 15. Figure 14 demonstrates the dependence of the ESN's received power level and rectified voltage on the distance between the PB and ESN with and without MHN. The case of LoS with MHN shows a steady power level decrease on the ESN from 11.66 dBm to 1.67 dBm in the distance range from 4.9 m to 6.25 m. However, from the distance of 6.25 m to 7 m, the received power level shows a reverse trend, as the power level increases up to 3.42 dBm at the 7 m distance and a sharp decline at the 7.15 m with the received power level of 2 dBm. The rectified voltage level follows the same trend as the received power level: a steady decrease from 5 V to 1.74 V across the 3 kΩ load resistance in the distance range from 4.9 m to 6.25 m, then a slight increase in the rectified voltage level from 6.25 m onwards, reaching the 2 V voltage level at 7 m and decreasing at 7.15 m with a 1.6 V level. Thus, the use of MHN, even in the presence of the line-of-sight propagation path, has been shown to sufficiently increase the amount of received power. The use of MHN in LoS conditions in the active state increases the received power level from 5.5 dB to 19.16 dB depending on the position from the PB, where the lowest difference in power levels is measured at 6.25 m and the largest at a distance of 6.85 m. The difference in the rectified voltage at the ESN with MHN in active state is significant; the voltage increase relative to the rectified voltage without MHN is from 203% to 3189%. MHN allows the receiving of significantly higher power levels and, therefore, higher rectified voltage compared to non-MHN measurement scenarios.

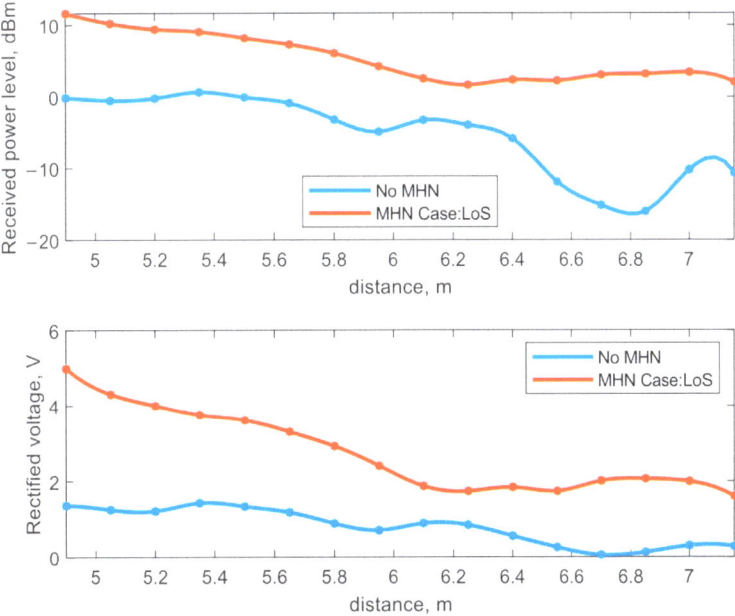

Figure 14. Measurement of received power level and rectified voltage with voltage doubler at the ESN antenna depending on distance between the PB position and ESN in LoS scenario.

Figure 15 presents the measurement results for the NLoS scenario. The received power level follows the same trend as in the previous scenario, where the received power level decreases linearly from 11.21 dBm to 0.92 dBm in the distance range from 4.9 m to 6.55 m. From 6.55 m to 7 m, the received power level fluctuates around the same value and, in the end, drops to a −0.34 dBm power level. The rectified voltage level follows the same trend as the received power level: the voltage level linearly decreases from 4.84 V to 1.52 V in the distance range from 4.9 m to 6.55 m, and from 6.55 m to 7 m fluctuates in the same voltage range and drops at the last measurement point to 1.32 V.

The ratio of the ESN harvested energy and MHN transmitted power also depends on distance. In the LoS case, the ratio is 3.7% at 4.9 m and 0.4% at 7.15 m. In the NLoS case, the ratio is 3.3% at 4.9 m and 0.2% at 7.15 m.

The acquired results not only show the benefits of using MHN in the two different WPT scenarios but also demonstrate that the received power is highly dependent on the ESN's position. When developing a WPT network, figuring out the optimal positions of the nodes via experimental measurements is costly. This highlights the need to develop

WPT models that sufficiently match real-life measurements while demanding the least computational time and resources. This is the next step of this research and is described in the next section.

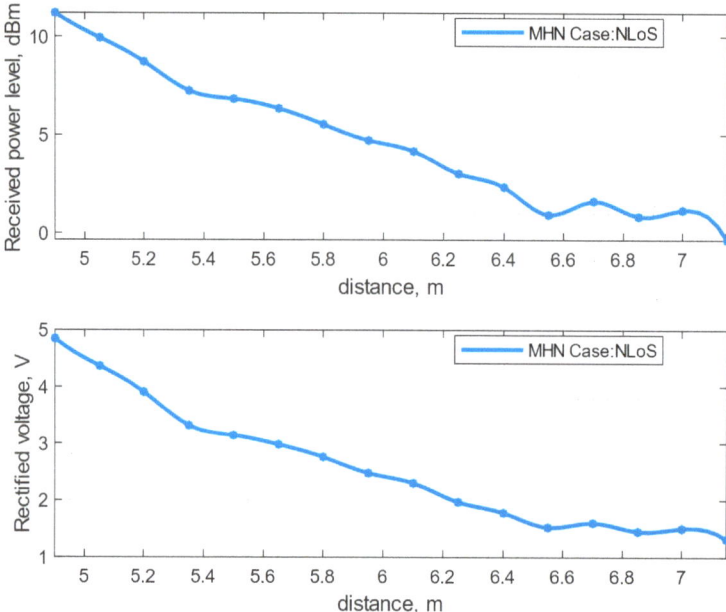

Figure 15. Measurement of received power level and rectified voltage with voltage doubler at the ESN antenna depending on distance between the PB position and ESN in NLoS scenario with MHN.

5. Model Development for Multi-Hop Wireless Power Transfer

Fast and accurate PB-to-ESN power transfer estimation is of primary concern in WPT systems antenna positioning optimization seeking such configurations of a WPT system deployed in some indoor environment that minimizes the amount of wasted power. The numerical modeling of power transfer channels obviates the need for experiments that, in some cases, may require expensive instrumentation.

While in sparse outdoor environments it is possible to estimate power transfer efficiency based on the well-known Friis transmission equation [45] or some refinements and extensions thereof [46–48], in indoor environments, even those with relatively simple geometry, the power they use of it may lead to dramatic discrepancies between the theoretical results and those obtained through measurements. The main source of the discrepancies is the fact that this equation does not consider multiple reflections off the walls, as well as other elements of the indoor environment under study. Furthermore, the highly approximate theory based on which the Friis equations are derived completely ignores various diffraction mechanisms that may contribute considerably to the discrepancy. This is particularly the case in environments featuring a number of objects with sharp edges [49].

The presence of walls, ceiling, floor, and windows, as well as plenty of different cabinets and other office elements, make the estimation process quite a challenging problem that is not easy to address using the existing methods, especially when the wavelength at a working frequency is not ten or more times smaller than the environment (e.g., an office room) dimensions but is comparable with those of objects located in it and furniture and decoration elements, such as a suspended ceiling. Multipath propagation resulting from multiple reflections of the transmitting (TX) antenna radiated waves from the walls and various objects present in the environment result in interference that, in turn, leads to variations in the received amount of power with the antenna position. In some cases, the variations may be so significant that a slight receiving (RX) or TX antenna shift might affect the received power.

In such a case, the modeling accuracy can be improved by applying some empirical or theoretical corrections to the Friis model. Although introducing corrections to the Friis model may seem simple to implement at first glance, a closer examination of this approach reveals that it may not always be the most optimal way. A more accurate approach to modeling indoor environments is the one based on the use of numerical techniques exploiting high-frequency approximations of wave equations, such as the geometrical optics approximation (GO, also commonly known as the ray-tracing method) [50–52] and the physical optics approximation (PO) [53,54]. However, this approach may fail to accurately predict the behavior of real-life models because of their approximate nature, albeit being very fast.

In an attempt to improve accuracy while maintaining an acceptable computational burden when analyzing large-scale real-life problems, various extensions and combinations of both PO and GO, as well as other methods, have been developed over the last few decades [55], e.g., the shooting and bouncing rays method (SBR) [56,57] and the hybrid PO–SBR–PTD Method [58]. Both plain GO (ray tracing) and PO methods have been successfully applied to solving a wide variety of practically important problems, such as radio wave propagation channel parameter estimation in both outdoor and indoor environments, e.g., tunnels [59–61].

However, in the case of environments involving objects with sharp edges, even refined high-frequency methods may still fail to provide acceptable accuracy when the diffraction effect is pronounced. To mitigate this issue, a few formulations that approximately describe the underlying diffraction mechanisms have been developed [62]. The geometrical theory of diffraction, originally developed by Keller [63] and subsequently extended by several researchers, mitigates the inability of the geometrical optics approximation to describe the fields in the region of geometrical optics' shadow, which fails to describe the fields resulting from the diffraction effect at edges, tips, vertices, and other object surface discontinuities. Furthermore, the original Keller's theory, as well as its extensions, are capable of ensuring sufficient field approximation accuracy relatively far from the objects the incident field interacts with due to the fact that they are derived from approximate closed-form expressions under the high-frequency assumption. The method is based on approximating the diffracted fields in terms of a set of rays consisting of two straight lines meeting at points of a sharp edge the incident rays hit or other geometrical discontinuities. However, it was shown that the GTD fails to ensure uniformity in the transition regions separating the reflection shadow and diffraction shadow regions. Later, in an attempt to overcome this issue, an improved formulation was proposed by Kouyoumjian and Pathak [62], which is nowadays commonly known as the uniform theory of diffraction. Similar to Keller's GTD, the UTD also approximates diffracted fields as a bunch of rays obeying an extended version of the Fermat principle. While the UTD can provide a more consistent approximation of diffracted fields, the ray field coefficients (diffraction coefficients) in this case are also derived under the assumption that the observation point is many wavelengths away from the object under consideration. At smaller distances, both the GTD and UTD may provide highly inaccurate results. Therefore, one cannot expect that the two approximate high-frequency techniques will ensure acceptable field computation accuracy when a model being studied involves many relatively closely-spaced fine elements with edges.

An alternative diffraction field approximation technique applies a correction to the physical optics approximation results to improve the field calculation accuracy. The technique is termed the physical theory of diffraction (PTD) in the scientific literature [64,65], and the correction to the surface currents used to compute the scattered field is derived using the same reasoning as the one used to derive the GTD diffraction coefficients. As the correction is deduced based on the same reasoning as in the case of the GTD and the UTD, the PTD has the same shortcoming; namely, it is not guaranteed to provide an adequate diffracted field approximation when a diffraction-causing object (e.g., suspended ceiling fixture) and other objects, such as the ceiling in the present study are not well separated. To summarize, in the case of indoor environments with complicated structures, the above-mentioned methods may fail to provide adequate results due to their approximate nature. In such a situation, the last resort is to employ general-purpose full-wave analysis methods, e.g., the finite element method (FEM) [66,67] and the method of moments (MoM) [68]. The latter is also known as the integral equation (IE) method. Despite their remarkable flexibility and capability to handle structures with arbitrary geometry no matter how complicated it may be, they are, unfortunately, tremendously computationally expensive, and the CPU time grows very quickly as the dimensions of the environment model increase, for example, when applied to models whose dimensions are multiples of a wavelength [69,70].

It is worth noting that, in some cases, the computational burden can be reduced by the use of semi-analytical methods. These methods can be applied when the environment can be decomposed into several regions, some of which have simple geometry that enables one to exploit field description in terms of the entire domain basis function well-suited to the given geometry, such as cylindrical functions, spherical, etc. [71].

Considering the benefits and pitfalls of high-frequency and full-wave techniques, in the present paper, the antenna-to-antenna power transfer is analyzed using two approaches, and the obtained results are compared with each other, as well as against the measurement data acquired in a real-life indoor environment (an office room). This comparative study's main objective is to determine how close the numerical results are to the experimental ones for a given indoor WPT system model.

The ray-tracing method-based model of the indoor environment is examined first, then the received power level for an indoor environment, the laboratory room where the measurements of the MHN-assisted WPT systems take place. Then, the same model is analyzed using Ansys HFSS, which solves Maxwell's equations employing general-purpose numerical methods, including the FEM or MoM.

While the ray-tracing method is approximate, the wavelength at the operating frequency is approximately 34.6 cm, which is much smaller than the room dimensions, which implies that the simulation results may provide a reasonably good approximation of the actual field distribution while requiring much less CPU resources compared to the full-wave analysis based power level estimation, as the full-wave simulation for such an electrically (relative to the wavelength) large model would lead to an amount of CPU time prohibitive for off-the-shelf computers.

5.1. Ray-Tracing Method-Based Analysis

As far as the ray-tracing method is concerned, in the numerical studies described herein, not only the line-of-sight path but also reflections from the room side walls, ceiling, and floor are taken into account, including multiple reflections; a maximum of two reflections from the model sides are considered. Namely, only the waves that have undergone no more than two reflections from the room sides before arriving at the receiving antenna are assumed to contribute to the total received wave, which is nothing but a superposition of different multipath components.

The ray-tracing-based modeling procedure comprises two stages. In the first stage, the far-field pattern for a four-element Yagi–Uda antenna is computed. Additionally, the ray-tracing method requires far-field data for the antenna-to-antenna power transfer calculation. The corresponding HFSS model is constructed for the antenna, and the far-field pattern is calculated using the finite element full-wave solver incorporated into the software.

By performing multiple antenna simulations for different parameter value sets, the optimal variation from the maximum gain point of view has been found, and four identical antenna prototypes were fabricated. These antenna prototypes were utilized to validate the multi-hop node-assisted power transfer systems. Both the E-plane and H-plane radiation patterns for the gain, which factors in the antenna input reflection, were computed and are illustrated in Figure 16.

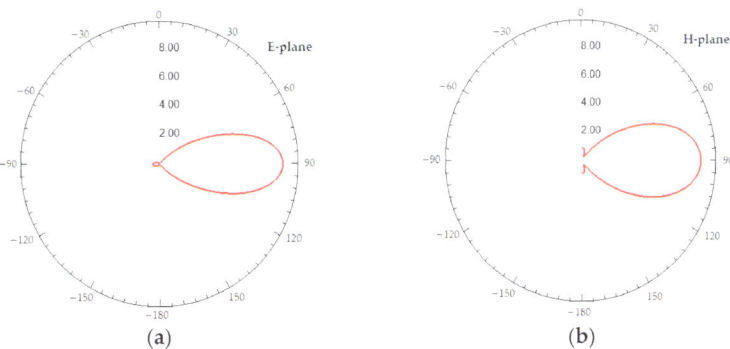

Figure 16. The calculated radiation pattern (calculated using Ansys HFSS) of the Yagi–Uda antenna in the E-plane (**a**) and H-plane (**b**).

Once the far-field patterns are found, they are exported from HFSS into MATLAB for the indoor environment under study ray-tracing. In the second stage, the transferred power from the transmitting antennas to the receiving one is calculated using the GO approximation. Specifically, the field radiated by a transmitting antenna is treated as a set of rays emerging from a point source representing the antenna. The parameters of the ray-tracing model are summarized in Table 1. In this model, the floor and ceiling are assumed to be infinite in extent; in this case, such an assumption does not significantly affect the results.

Table 1. Room model parameters for the ray-tracing method.

Parameter	Value
Antenna to ceiling distance	2000 mm
Antenna to floor distance	1070 mm
Antenna to the closest wall distance.	17,000 mm
Antenna to the farthest wall distance	3300 mm
Wall material (concrete) dielectric constant	4.69
Wall material (concrete) loss tangent	0.176
Floor length	∞
Floor width	∞
Ceiling length	∞
Ceiling width	∞

The receiving antenna is also assumed to be a point receiver. It is worth noting that the aperture area is calculated as the product of the isotropic antenna aperture and the antenna gain in a specific direction, e.g., the direction of arrival of an incident wave (ray). For each multipath component, the path is determined based on the location of the phase centers of both the receiving and transmitting antenna and reflection points that can be readily found using simple geometrical methods. Each time a ray bounces off the model surface, its phasor is multiplied by the corresponding reflection coefficient calculated based on the Fresnel transmission equations.

The received power levels against the separation between the antennas computed using Ansys HFSS (full-wave) and the ray-tracing method are presented in Figure 17. The curve obtained for free space is plotted on the same graph as the ones obtained using Ansys HFSS and the ray-tracing method. It is worth noting that the effect of the suspended ceiling is not accounted for by the ray-tracing method, as outlined above. Both the full-wave and the ray-tracing models treat the ceiling and the floor as conducting planes with sufficiently high conductivity. In this study, the conductivity of iron is assumed.

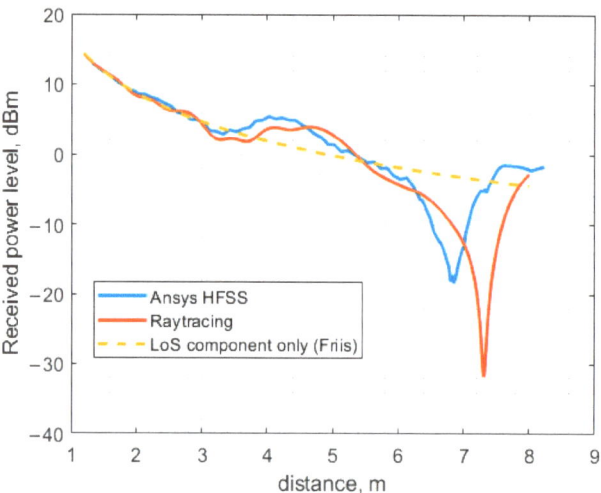

Figure 17. The received power level as a function of the distance between the antennas calculated using three different methods.

It should be noted that, in the theoretical analysis, the modeling of the amplifying node is not carried out. The effect of this node on the rest of the system is incorporated through the node gain and the total phase shift between the sine waves at the node's input and output. While the importance of the former node parameter is obvious, the role of the latter one may seem not essential at first; however, it is not the case. Specifically, the phase determines the shape of the received power curve in the case of the LoS scenario, as the total power received by the ESN is not determined by the power radiated by the MHN node and the separation between the nodes. Still, it is the power of the superposition of two waves, one being the wave due to the MHN node, whereas the other is the wave coming from the PB.

The power level of this direct component is smaller by an order of magnitude than the one received from the MHN. However, the phase difference between the waves depends on the distance between them because the direct wave is composed of several waves: the LoS one and the waves reflected from the ceiling, floor, and walls. Due to this difference, the shape of the curve varies with the phase delay introduced by the node; as a result, the curves obtained under LoS and NLoS scenarios may differ somewhat.

Another set of curves is shown in Figure 18. In this case, the full-wave model is the same as in Figure 17, but the ray-tracing model is slightly modified. More precisely, the effect of the suspended ceiling is approximately accounted for via the introduction of the effective antenna axis to ceiling distance, approximately 11.5 cm shorter than in the case without a ceiling. The main idea of this approach is based on the fact that the metallic grid of the suspended ceiling exhibits a non-zero transmission coefficient at incident angles that are not sufficiently large. Consequently, some fraction of a wave impinging on the grid passes through it, and then it undergoes multiple bounces between the actual ceiling surface and the grid, which results in multiple reflected waves with different amplitudes and phases that form the total reflected wave. The amplitude of the total reflected wave is still close to unity if the ceiling is assumed to be conducting, but the phase may differ considerably from the one reflected from the ceiling surface only. Moreover, its dependence on the angle of incidence differs from that governed by the classical Fresnel equations. It is possible to achieve the required phase for a specific angle by changing the distance between the ceiling surface and the antenna axis. However, the change should not be large to not affect the wave amplitude due to the variation in the wave travel distance.

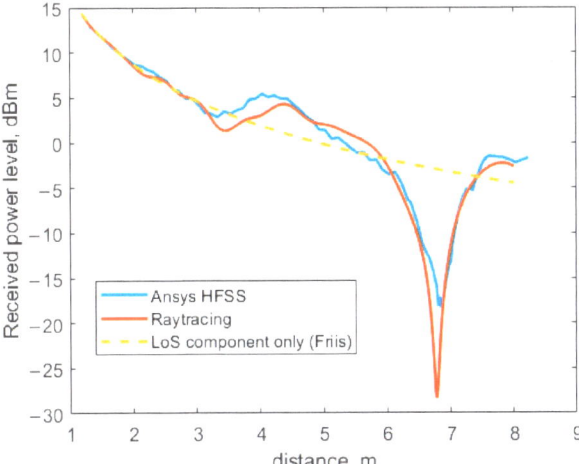

Figure 18. The received power level as a function of the distance between the antennas calculated using three different methods.

As evidenced in Figure 18, the effective distance approach provides better results; the large minima in both curves coincide in this case. However, the curve shape differs considerably between the minimum and the maximum, occurring at approximately 3.5 m and 4.2 m, respectively. The discrepancy is likely due to the dependence of the effective distance on the incidence angle.

In general, finding an effective antenna axis to ceiling distance in many cases may not be a trivial task, and it is not guaranteed that the approach based on the effective distance concept will ensure acceptable power level estimation accuracy; moreover, it is beyond the scope of the present investigation, and a more in-depth treatment of this approach will likely be provided in one of our forthcoming papers.

5.2. Full-Wave Analysis

To perform a full-wave analysis of the indoor environment under study, Ansys HFSS is employed. The main advantage of the full-wave analysis is that it takes into account the effect of the suspended ceiling metallic elements (metallic grid) on the wave propagation directly without using any physical or geometrical optics or other approximations. Ignoring this effect might lead to a dramatic discrepancy between the estimation and measured results, as was shown in the previous subsection on the ray-tracing analysis. However, direct full-wave analysis is highly computationally intensive. Therefore, the indoor environment model examined in this study is significantly simplified to reduce the number of model discretization elements, thus reducing CPU time. To be more precise, only the ceiling and floor, including the fixtures of the suspended ceiling (metallic grid), are accounted for in the model, while the effect of the wall is completely ignored. This simplification is justified by a simple investigation of the radiation pattern of the narrow beam antenna used in the study and the behavior of the waves impinging on the floor, ceiling, and side walls, which is governed by the Fresnel equations. According to the Fresnel equations for the parallel incident plane wave polarization (TE case), the waves impinging on the side walls with a small grazing angle (due to the narrow antenna beam) have a small reflection coefficient. Thus, the contribution of the wave reflected from the side walls can be ignored, and therefore, they can be removed from the model. The flowchart of the entire procedure for full-wave analysis of the indoor environment is depicted in Figure 19.

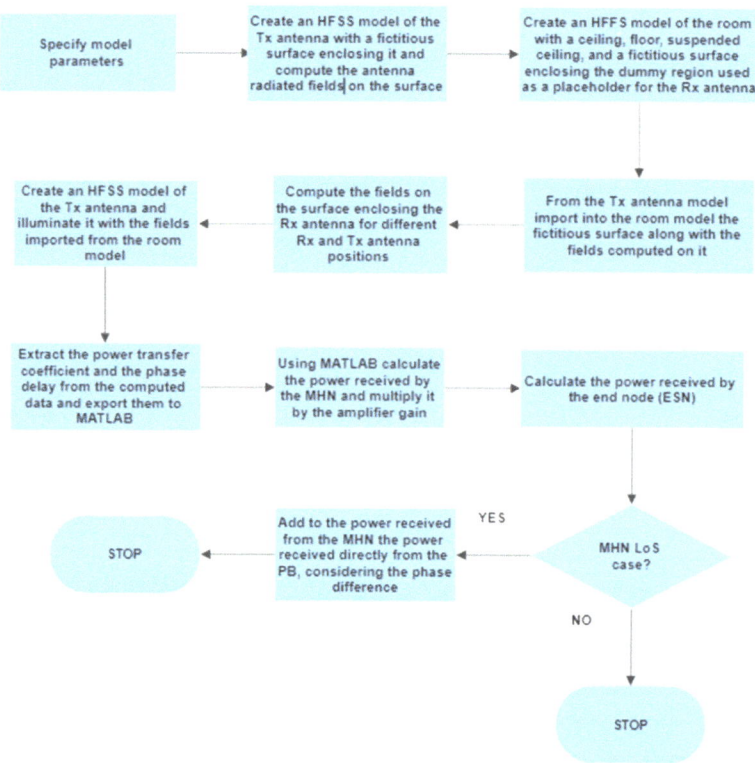

Figure 19. The flowchart of the transferred power estimation procedure using full-wave analysis.

To reduce the computational burden, the indoor environment model is divided into three parts: the receiving antenna model, the transmitting antenna model, and a model of the environment. Each of the three sub-models of the original model is treated separately. For the full-wave analysis of both receiving (RX) and transmitting (TX) antenna models, the finite element method solver is employed. The solution space is truncated using absorbing boundary conditions (ABC), which in the HFSS are available as the radiation conditions option. The conditions imitate the absorption of the antenna-radiated waves. These boundary conditions are less computationally expensive than their PML and Hybrid FEM–IE counterparts. However, it comes at the expense of higher reflection from the absorbing boundary when the absorbing surface is not sufficiently far from the antenna model itself. The dimensions of the box to the surface of which the above-mentioned boundary conditions are assigned are as follows: width—390 mm, length—890 mm, and height—200 mm. These dimensions were determined empirically by performing a series of numerical studies with gradually increasing the bounding box dimensions. Additionally, a fictitious surface enclosing the antenna model is created. No boundary conditions are assigned to the surface; it is intended to be utilized as an equivalent radiation surface in the second part of the model, which is intended to be linked with this one. First, the fields are calculated for the TX model, including the ones on the fictitious inner surface. Then, the fictitious surface, together with the field distribution on it computed by performing simulations of the TX antenna model, is transferred to the next model, namely, the room model, where it will be used as an excitation. The created antenna model of the Yagi–Uda antenna is depicted in Figure 20.

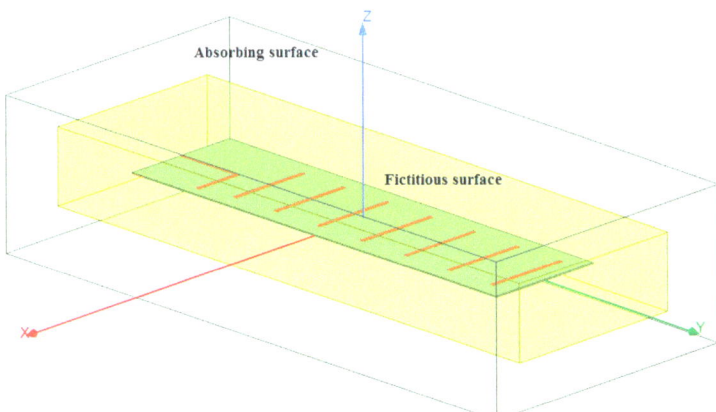

Figure 20. The Ansys HFSS model of a four-element Yagi–Uda antenna with a fictitious surface enclosing the antenna to compute near fields to be imported to another HFSS model.

The Ansys HFSS model of the indoor environment (room) is shown in Figure 21. In the room model, a dummy region of space where the TX antenna model would otherwise be located (TX antenna placeholder) is created and treated as being filled with the perfect electric conductor (PEC) to prevent the solver from seeking the fields within it.

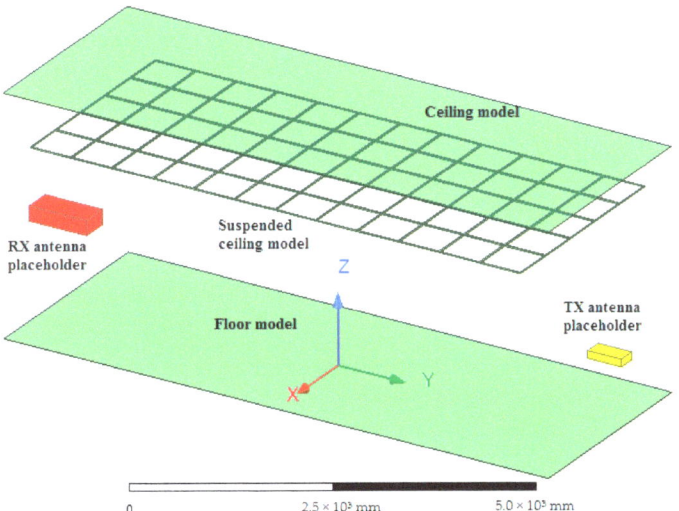

Figure 21. The Ansys HFSS model for the antenna-to-antenna power transfer analysis.

The floor and the ceiling are modeled as flat rectangular surfaces, which are assigned impedance boundary conditions and are defined as hybrid IE regions. Specifically, the ceiling and floor models are handled by an integral equation solver (IE solver) that employs the MoM for surfaces. The surface impedance value corresponds to that of iron. The model of the metal grid (suspended ceiling) is a solid 3D object on the surface of which other impedance boundary conditions are imposed. This object is also treated by the IE solver. As mentioned above, this room model does not contain the model of the TX antenna; in place of it, another fictitious surface is created whose dimensions are slightly larger than those of the fictitious surface used in the RX antenna model. The region of space inside the surface may be regarded as a placeholder for the TX antenna. The surface is necessary for computing fields on it.

Once the fields on the second fictitious surface are found, they are transferred through the link established between the room model and that of the RX antenna and used to illuminate the TX antenna model as if it were located in the room model. The RX model is identical to that of the TX antenna. The parameters of the full-wave model under study are presented in Table 2.

Table 2. Simplified Ansys HFSS room model parameters.

Parameter	Value
Ceiling to suspended ceiling grid distance	435 mm
Antenna to ceiling distance	2000 mm
Antenna to floor distance	1070 mm
Floor length	8000 mm
Floor width	3000 mm
Ceiling length	8000 mm
Ceiling width	3000 mm
Metal grid main (cross) tee width	590 mm
Metal grid main (cross) tee length	10 mm
Metal grid main (cross) tee height	50 mm

It is worth noting that the software exploits the surface equivalence principle stating that the antenna-generated fields outside the surface can be evaluated knowing the field on some surface enclosing it. This eliminates the need to discretize the room model part corresponding to the TX antenna. However, handling the indoor environment model in this way ignores the two-way interaction between the antenna and the rest of the model. Nevertheless, since, in the present case, the interaction is not strong, mainly affecting the waves reflected back to the antenna, which is not of concern in the present study, the partitioning of the original model will only slightly influence the final results.

The suspended ceiling used in most modern office facilities is a periodic metallic structure intended to hold stone wool slabs whose thickness typically ranges between 2 and 5 cm. The dielectric constant of the slab material in the indoor environment under study is approximately 1.2–1.5, and it has a negligible effect on the wave propagation, which was experimentally verified in the present study; placing a single slab between a pair of Yagi–Uda antennas does not affect the received power level under normal incidence. However, when two such stone wool slabs are stacked together, the amount of the received power is reduced by about 0.1 dB, which is still very small.

The calculated gain of the antenna is approximately 9.48 dBi, while the measured gain of the antenna prototype is in the vicinity of 9.17 dBi. Such a mismatch between the results might arise from the imperfections of the milling machine used to fabricate the prototype. Among other factors that might cause this may be the difference between the actual value of the substrate material dielectric constant and loss tangent from the ones assumed for the antenna design, some parasitic effects of the SMA connectors used, small inhomogeneous in the substrate material, etc. For the calculated results to be closer to the measured ones, the calculated antenna gain is rescaled so that the antenna model's maximum gain (in the boresight direction) matches the measured one.

Figure 22 displays the results obtained using the full-wave analysis and the experimentally measured ones in the case of direct antenna-to-antenna power transfer. The figure clearly shows that the results are in close agreement, which means that the full-wave analysis-based estimation, even with a simplified model such as the one used in the present study, may provide reliable power level estimation, enabling antenna placement optimization to increase the overall WPT systems efficiency.

The results calculated for the case of the direct power transfer and the MHN-assisted power transfer under the LoS scenario where all the antennas are lined up are indicated in Figure 23. The power in the LoS scenario is calculated by adding phasors of both the direct component from the PB and the one amplified and subsequently radiated by the

MHN. The phasors are computed using Ansys HFSS, while the phase delay due to the node circuitry and the cable connecting the node's antenna is established experimentally.

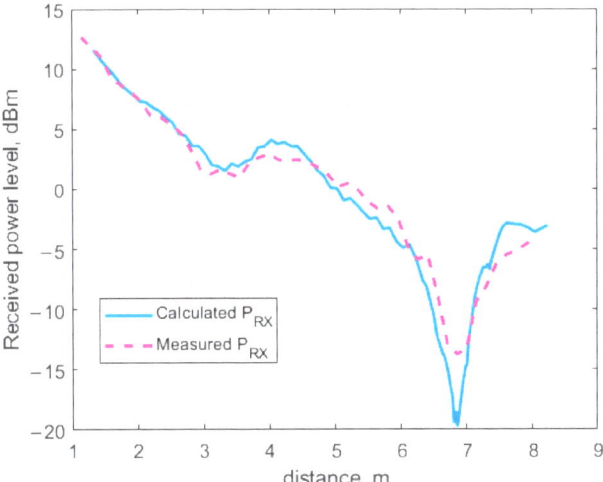

Figure 22. The calculated and the measured received power as a function of the distance between two four-element Yagi–Uda antennas with a fixed transmitted power of 26 dBm.

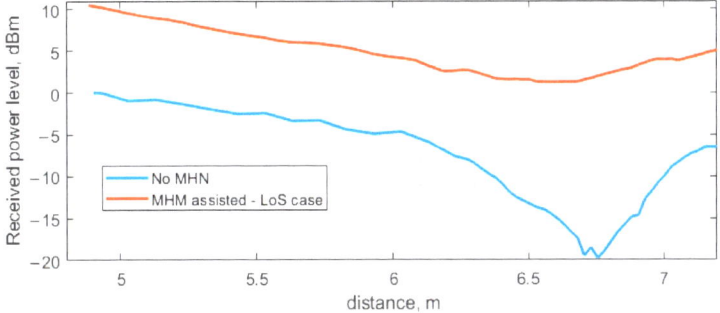

Figure 23. The power received at the ESN with and without the MHN when all the antennas are aligned (LoS scenario).

The full-wave simulation results obtained for the multi-hop assisted wireless power transfer in the case when there is no line-of-sight channel between the PB and ESN antenna are displayed in Figure 24. Similar to the previous case, two Yagi–Uda antenna pairs are used. The antennas are arranged so that the direct component due to the coupling between the PB antenna and the ESN antenna is eliminated. Alternatively, the two imaginary straight lines, one of which passes through the phase centers of the PB antenna and the receiving antenna of the MHN and the other one passes through the those of the MHN transmitting antenna and the one of the ESN, make a 90-degree angle. The received input power is calculated by combining the full-wave simulation results obtained using Ansys HFSS.

While it would be more favorable from the modeling accuracy perspective to perform analysis of an indoor environment model incorporating all four antennas, the equation system resulting from the model discretization would be huge, making the CPU time unacceptably large. Instead, in this study, the power transferred from one antenna to another is calculated for a two-antenna model, like in the case of the direct power transfer. Then, using the obtained data, the power received by the receiving antenna of the MHN is calculated, multiplied by the node amplifier gain, and eventually, the power transferred to the end node is found using the calculated amount of power transmitted by the MHN and the relation between the transmitted and the received power for the pair of antennas. The last step of the calculation procedure is identical to the first one, as the same data are used (calculated using HFSS). The effect of the antenna position relative to the suspended ceiling grid elements on the relation between the RX and TX power levels is found to be sufficiently little to neglect, which makes it possible to use the data acquired for only one antenna position.

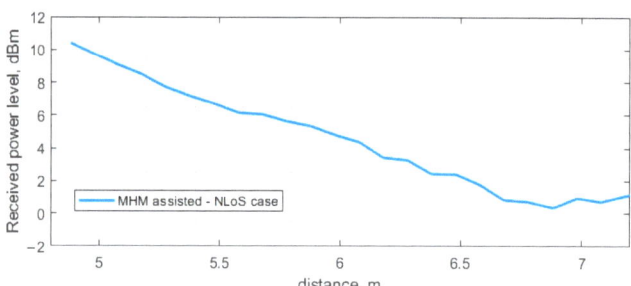

Figure 24. The power received at the ESN with the MHN in the case of a 90-degree turn (NLoS scenario).

5.3. Calculated and Measured Result Comparison

The numerical (modeling) results and the experimentally obtained ones discussed in the previous sections are presented on the same graphs for comparison. The results obtained under the LoS scenario are shown in Figure 25, while those corresponding to the NLoS scenario are observed in Figure 26.

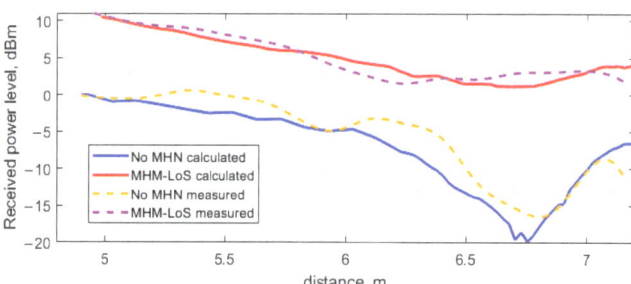

Figure 25. The calculated and measured power received at the ESN as a function of the distance between the PB and the ESN, obtained under the LoS scenario with and without the MHN assistance.

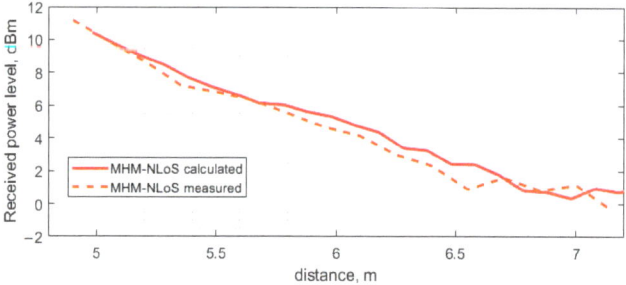

Figure 26. The calculated and measured power received at the ESN as a function of the distance between the PB and the ESN, obtained under the MHN-assisted NLoS scenario.

As can be seen, the discrepancy between the calculated and the measured results shown in Figures 25 and 26 is small, even though a simplified model was employed for the full-wave analysis of the indoor environment under study. The input power level of the transfer system under study was kept fixed at the level of 26.0 dBm throughout the investigation, regardless of the power transfer scenario. In the case of direct power transfer from the PB to the ESN, a distinct dip is observed at a distance of approximately 6.8 m. This received power blockage occurs due to the wave reflection off the floor. Since the beam of the antenna utilized in the present study is sufficiently narrow (about 60 degrees), the fluctuation of the received power level about the theoretical curve for free space calculated according to the well-known Friis transmission equations is not significant up to about 6.0 m where the basin of attraction of the minimum begins. It should be noted that if antennas having a much wider beam in the H-plane, e.g., an LDPA, dipole, or monopole antenna, were used, the minima and maxima would be appreciably more pronounced, thereby severely limiting the use of such antennas in indoor power transfer systems due to the antenna positioning complexity. Although this issue could be mitigated by exploiting the spatial diversity of MIMO or similar beamforming approaches, it would be much more expensive than deploying antenna systems based on the printed Yagi–Uda antenna equipped with a small autonomous power amplifier, as treated in the previous sections. The other two extrema, a minimum and a maximum occurring at 3 and 4 m, respectively, are much smaller than the large one at 6.8 m and should not considerably affect the power transfer system performance.

Regarding the difference between the theoretical and practical power level curves, Figure 25 shows that the largest deviation of the measured power level from the calculated one occurs in the vicinity of a maximum in the experimental curve at approximately 5.5 m and the adjacent maximum at 6.4 m. Interestingly, the theoretical curve exhibits no maxima over the antenna separation range from 4.8 to 6.4 m. The absence of the maxima may be attributed to the effect of some fine model elements that were not taken into account to simplify the model, thus reducing the amount of CPU time as described above. For instance, the presence of the lighting elements, including wires, fastening components and fixtures, was not accounted for by the model in the present study.

6. Conclusions

This research was devoted to the performance evaluation of multi-hop (MH) technology for wireless power transfer (WPT) at sub-GHz frequencies. The study presents an approach based on using an additional node for path loss compensation, a multi-hop node (MHN). The proposed approach was experimentally validated under real-life scenarios for line-of-sight and non-line-of-sight propagation conditions. However, to apply this prototype for powering IoT, LORA devices will need further optimization, like fine-tuning the time of active and charging modes to meet the power consumption of the powered device. Additionally, it was demonstrated that it is still possible to accurately estimate the power transfer efficiency in an indoor environment with a metallic grid (suspended ceiling) in a reasonable timeframe when the classical ray-tracing method fails to provide adequate modeling accuracy. The numerical analysis of the PB to ESN power transfer under different indoor propagation scenarios was carried out using a simplified full-wave model. The fields were computed using commercially available software Ansys HFSS. In addition, the results obtained using HFSS were compared with those obtained by means of the ray-tracing method. The far-field patterns required for the ray-tracing analysis were obtained using HFSS. It was found that even in the case of a simplified full-wave analysis model, the HFSS results are much closer to the experimental ones than those provided by the ray-tracing method.

Simulation and experimental study results of MHN-assisted WPT have shown that MH technology can sufficiently increase WPT efficiency. In the line-of-sight case, at a distance up to 7 m, the received power level at the end node was increased up to 19 dBm compared to the direct WPT without the MH. Furthermore, the results obtained under the non-line-of-sight scenario show that the MH approach enables sufficiently efficient power transfer even when the direct link between transmitters and the receiver antennas is absent.

Indeed, the proposed MH WPT simulation approach could be applied to optimizing the spatial distribution of MH WPT system elements in real-life scenarios.

Author Contributions: Conceptualization, A.L.; Formal analysis, R.K.; Funding acquisition, A.L.; Investigation, J.E.; Methodology, J.E. and R.K.; Project administration, A.L.; Software, J.E. and R.K.; Supervision, R.B., D.C., J.S. and A.L.; Validation, J.E. and R.K.; Visualization, R.B. and D.C.; Writing—original draft, J.E., R.K. and R.B.; Writing—review and editing, R.B., D.C., J.S. and A.L. All authors have read and agreed to the published version of the manuscript.

Funding: This research was funded by the Latvian Council of Science, grant No. lzp-2021/1-0170, "Advanced Techniques for Wireless Power Transfer".

Institutional Review Board Statement: Not applicable.

Informed Consent Statement: Not applicable.

Data Availability Statement: The experimental measurement data are available in the repository https://github.com/jaancis/Wireless_Power_Transfer_MHN (accessed on 27 July 2023).

Acknowledgments: This research was performed at Riga Technical University, Space Electronics and Signal Processing Laboratory–SpacESPro Lab.

Conflicts of Interest: The authors declare no conflict of interest.

References

1. Al-Fuqaha, A.; Guizani, M.; Mohammadi, M.; Aledhari, M.; Ayyash, M. Internet of Things: A Survey on Enabling Technologies, Protocols, and Applications. *IEEE Commun. Surv. Tutor.* **2015**, *17*, 2347–2376. [CrossRef]
2. Zhao, Y.; Wu, Y.; Hu, J.; Yang, K.; Clerckx, B. Energy Harvesting Modulation for Integrated Control State and Energy Transfer in Industrial IoT. *IEEE Wirel. Commun. Lett.* **2023**, *12*, 292–296. [CrossRef]
3. Ericsson Mobility Report November 2022. 2022. Available online: https://www.ericsson.com/en/reports-and-papers/mobility-report/reports (accessed on 16 August 2023).
4. Aboltins, U.; Novickis, J.; Romanovs, A. IoT Impact on Business Opportunities. In Proceedings of the 2020 61st International Scientific Conference on Information Technology and Management Science of Riga Technical University (ITMS), Riga, Latvia, 15–16 October 2020; pp. 1–6. [CrossRef]
5. Kanoun, O.; Khriji, S.; Naifar, S.; Bradai, S.; Bouattour, G.; Bouhamed, A.; El Houssaini, D.; Viehweger, C. Prospects of Wireless Energy-Aware Sensors for Smart Factories in the Industry 4.0 Era. *Electronics* **2021**, *10*, 2929. [CrossRef]
6. Disruptive Technologies: Advances That Will Transform Life, Business, and the Global Economy | McKinsey. Available online: https://www.mckinsey.com/capabilities/mckinsey-digital/our-insights/disruptive-technologies (accessed on 16 August 2023).
7. Where and How to Capture Accelerating IoT Value | McKinsey. Available online: https://www.mckinsey.com/capabilities/mckinsey-digital/our-insights/iot-value-set-to-accelerate-through-2030-where-and-how-to-capture-it (accessed on 16 August 2023).
8. Clerckx, B.; Kim, J.; Choi, K.W.; Kim, D.I. Foundations of Wireless Information and Power Transfer: Theory, Prototypes, and Experiments. *Proc. IEEE* **2022**, *110*, 8–30. [CrossRef]
9. Van Mulders, J.; Delabie, D.; Lecluyse, C.; Buyle, C.; Callebaut, G.; Van der Perre, L.; De Strycker, L. Wireless Power Transfer: Systems, Circuits, Standards, and Use Cases. *Sensors* **2022**, *22*, 5573. [CrossRef] [PubMed]
10. Sakai, N.; Noguchi, K.; Itoh, K. A 5.8-GHz Band Highly Efficient 1-W Rectenna With Short-Stub-Connected High-Impedance Dipole Antenna. *IEEE Trans. Microw. Theory Tech.* **2021**, *69*, 3558–3566. [CrossRef]
11. Chang, Y.T.; Claessens, S.; Pollin, S.; Schreurs, D. A Wideband Efficient Rectifier Design for SWIPT. In Proceedings of the 2019 IEEE Wireless Power Transfer Conference (WPTC), London, UK, 18–21 June 2019; IEEE: Piscataway, NJ, USA, 2019. Available online: https://ieeexplore-ieee-org.resursi.rtu.lv/document/9055666/ (accessed on 25 June 2023).
12. Gu, X.; Hemour, S.; Wu, K. Far-Field Wireless Power Harvesting: Nonlinear Modeling, Rectenna Design, and Emerging Applications. *Proc. IEEE* **2022**, *110*, 56–73. [CrossRef]
13. Shao, Y.; Yan, Z.; Mai, R. A High-Efficiency Wireless Power Transfer Converter with Integrated Power Stages. In Proceedings of the 2023 International Conference on Power Energy Systems and Applications (ICoPESA), Nanjing, China, 24–26 February 2023; pp. 155–160. [CrossRef]
14. Liu, P.; Gazor, S.; Kim, I.-M.; Kim, D.I. Energy Harvesting Noncoherent Cooperative Communications. *IEEE Trans. Wirel. Commun.* **2015**, *14*, 6722–6737. [CrossRef]
15. Gong, S.; Huang, X.; Xu, J.; Liu, W.; Wang, P.; Niyato, D. Backscatter Relay Communications Powered by Wireless Energy Beamforming. *IEEE Trans. Commun.* **2018**, *66*, 3187–3200. [CrossRef]
16. Song, C.; Ding, Y.; Eid, A.; Hester, J.G.D.; He, X.; Bahr, R.; Georgiadis, A.; Goussetis, G.; Tentzeris, M.M. Advances in Wirelessly Powered Backscatter Communications: From Antenna/RF Circuitry Design to Printed Flexible Electronics. *Proc. IEEE* **2022**, *110*, 171–192. [CrossRef]

17. Kumar, P.; Dhaka, K. Performance Analysis of Wireless Powered DF Relay System Under Nakagami-m Fading. *IEEE Trans. Veh. Technol.* **2018**, *67*, 7073–7085. [CrossRef]
18. Liu, Y.; Pan, Z.; Xiao, R.; Yang, H.; Yan, C. Throughput Analysis for WPT Networks Over Two-Way Log-Normal Fading Channels. In Proceedings of the 2019 IEEE/CIC International Conference on Communications in China (ICCC), Changchun, China, 11–13 August 2019; pp. 502–506. [CrossRef]
19. Shen, S.; Kim, J.; Clerckx, B. Closed-Loop Wireless Power Transfer with Adaptive Waveform and Beamforming: Design, Prototype, and Experiment. *IEEE J. Microw.* **2022**, *3*, 29–42. [CrossRef]
20. Eidaks, J.; Litvinenko, A.; Pikulins, D.; Tjukovs, S. The Impact of PAPR on the Wireless Power Transfer in IoT Applications. In Proceedings of the 2019 29th International Conference Radioelektronika (RADIOELEKTRONIKA), Pardubice, Czech Republic, 16–18 April 2019; pp. 1–5. [CrossRef]
21. Ayir, N.; Riihonen, T.; Heino, M. Practical Waveform-to-Energy Harvesting Model and Transmit Waveform Optimization for RF Wireless Power Transfer Systems. In Proceedings of the IEEE Transactions on Microwave Theory and Techniques, Piscataway, NJ, USA, 20 June 2023; pp. 1–17. [CrossRef]
22. Lin, Z.; Lin, M.; Zhu, W.-P.; Wang, J.-B.; Cheng, J. Robust Secure Beamforming for Wireless Powered Cognitive Satellite-Terrestrial Networks. *IEEE Trans. Cogn. Commun. Netw.* **2021**, *7*, 567–580. [CrossRef]
23. Clerckx, B.; Bayguzina, E. Low-Complexity Adaptive Multisine Waveform Design for Wireless Power Transfer. *IEEE Antennas Wirel. Propag. Lett.* **2017**, *16*, 2207–2210. [CrossRef]
24. Kim, J.; Clerckx, B. Range Expansion for Wireless Power Transfer Using Joint Beamforming and Waveform Architecture: An Experimental Study in Indoor Environment. *IEEE Wirel. Commun. Lett.* **2021**, *10*, 1237–1241. [CrossRef]
25. de Araujo, G.T.; de Almeida, A.L.F.; Boyer, R. Channel Estimation for Intelligent Reflecting Surface Assisted MIMO Systems: A Tensor Modeling Approach. *IEEE J. Sel. Top. Signal Process.* **2021**, *15*, 789–802. [CrossRef]
26. Fara, R.; Ratajczak, P.; Phan-Huy, D.-T.; Ourir, A.; Renzo, M.D.; De Rosny, J. A Prototype of Reconfigurable Intelligent Surface with Continuous Control of the Reflection Phase Modeling, Full-Wave Electromagnetic Characterization, Experi-mental Validation, and Application to Ambient Backscatter Communications. *IEEE Wirel. Commun.* **2022**, *29*, 70–77. [CrossRef]
27. Ahmed, M.F.; Rajput, K.P.; Venkategowda, N.K.D.; Mishra, K.V.; Jagannatham, A.K. Joint Transmit and Reflective Beamformer Design for Secure Estimation in IRS-Aided WSNs. *IEEE Signal Process Lett.* **2022**, *29*, 692–696. [CrossRef]
28. Lin, Z.; Niu, H.; An, K.; Wang, Y.; Zheng, G.; Chatzinotas, S.; Hu, Y. Refracting RIS-Aided Hybrid Satellite-Terrestrial Relay Networks: Joint Beamforming Design and Optimization. *IEEE Trans. Aerosp. Electron. Syst.* **2022**, *58*, 3717–3724. [CrossRef]
29. Wu, Q.; Guan, X.; Zhang, R. Intelligent Reflecting Surface-Aided Wireless Energy and Information Transmission: An Overview. *Proc. IEEE* **2022**, *110*, 150–170. [CrossRef]
30. Lee, W.; Yoon, Y.-K. Wireless Power Transfer Systems Using Metamaterials: A Review. *IEEE Access* **2020**, *8*, 147930–147947. [CrossRef]
31. Feng, Z.; Clerckx, B.; Zhao, Y. Waveform and Beamforming Design for Intelligent Reflecting Surface Aided Wireless Power Transfer: Single-User and Multi-User Solutions. *IEEE Trans. Wirel. Commun.* **2022**, *21*, 5346–5361. [CrossRef]
32. Tran, N.M.; Amri, M.M.; Park, J.H.; Kim, D.I.; Choi, K.W. Reconfigurable-Intelligent-Surface-Aided Wireless Power Transfer Systems: Analysis and Implementation. *IEEE Internet Things J.* **2022**, *9*, 21338–21356. [CrossRef]
33. Fu, M.; Mei, W.; Zhang, R. Multi-Active/Passive-IRS Enabled Wireless Information and Power Transfer: Active IRS Deployment and Performance Analysis. *IEEE Commun. Lett.* **2023**, *27*, 2217–2221. [CrossRef]
34. Mei, W.; Zhang, R. Cooperative Multi-Beam Routing for Multi-IRS Aided Massive MIMO. In Proceedings of the ICC 2021—IEEE International Conference on Communications, Montreal, QC, Canada, 14–23 June 2021; pp. 1–6. [CrossRef]
35. Nauryzbayev, G.; Rabie, K.M.; Abdallah, M.; Adebisi, B. On the Performance Analysis of WPT-Based Dual-Hop AF Relaying Networks in α–μ Fading. *IEEE Access* **2018**, *6*, 37138–37149. [CrossRef]
36. Kaushik, K.; Mishra, D.; De, S.; Basagni, S.; Heinzelman, W.; Chowdhury, K.; Jana, S. Experimental demonstration of multi-hop RF energy transfer. In Proceedings of the 2013 IEEE 24th Annual International Symposium on Personal, Indoor, and Mobile Radio Communications (PIMRC), London, UK, 8–11 September 2013; pp. 538–542. [CrossRef]
37. Lu, X.; Wang, P.; Niyato, D.; Kim, D.I.; Han, Z. Wireless Networks with RF Energy Harvesting: A Contemporary Survey. *IEEE Commun. Surv. Tutor* **2015**, *17*, 757–789. [CrossRef]
38. Nasir, A.A.; Zhou, X.; Durrani, S.; Kennedy, R.A. Relaying Protocols for Wireless Energy Harvesting and Information Processing. *IEEE Trans. Wirel. Commun.* **2013**, *12*, 3622–3636. [CrossRef]
39. Watfa, M.K.; AlHassanieh, H.; Selman, S. Multi-Hop Wireless Energy Transfer in WSNs. *IEEE Commun. Lett.* **2011**, *15*, 1275–1277. [CrossRef]
40. Rault, T.; Bouabdallah, A.; Challal, Y. Multi-hop wireless charging optimization in low-power networks. In Proceedings of the 2013 IEEE Global Communications Conference (GLOBECOM), Atlanta, GA, USA, 9–13 December 2013; pp. 462–467. [CrossRef]
41. Eidaks, J.; Kusnins, R.; Babajans, R.; Cirjulina, D.; Semenjako, J.; Litvinenko, A. Fast and Accurate Approach to RF-DC Conversion Efficiency Estimation for Multi-Tone Signals. *Sensors* **2022**, *22*, 787. [CrossRef]
42. G4developer. Powercast Releases Over-the-Air Wireless Charging Grip for Nintendo Joy-Con Controllers on Amazon. Powercast, 25 August 2020. Available online: https://www.powercastco.com/powercast-releases-over-the-air-wireless-charging-grip-for-nintendo-joy-con-controllers-on-amazon/ (accessed on 16 August 2023).

43. Eidaks, J.; Kusnins, R.; Kolosovs, D.; Babajans, R.; Cirjulina, D.; Krukovskis, P.; Litvinenko, A. Multi-Hop RF Wireless Power Transfer for Autonomous Wireless Sensor Network. In Proceedings of the 2022 Workshop on Microwave Theory and Techniques in Wireless Communications (MTTW), Riga, Latvia, 5–7 October 2022; pp. 51–56. [CrossRef]
44. Eidaks, J.; Kusnins, R.; Laksis, D.; Babajans, R.; Litvinenko, A. Signal Waveform Impact on RF-DC Conversion Efficiency for Different Energy Harvesting Circuits. In Proceedings of the 2021 IEEE Microwave Theory and Techniques in Wireless Communications (MTTW), Riga, Latvia, 7–8 October 2021; pp. 1–6. [CrossRef]
45. Bian, C.; Li, W.; Wang, M.; Wang, X.; Wei, Y.; Zhou, W. Path Loss Measurement of Outdoor Wireless Channel in D-band. *Sensors* **2022**, *22*, 9734. [CrossRef]
46. Friis, H. A Note on a Simple Transmission Formula. *Proc. IRE* **1946**, *34*, 254–256. [CrossRef]
47. Sun, S.; Rappaport, T.S.; Thomas, T.A.; Ghosh, A.; Nguyen, H.C.; Kovacs, I.Z.; Rodriguez, I.; Koymen, O.; Partyka, A. Investigation of Prediction Accuracy, Sensitivity, and Parameter Stability of Large-Scale Propagation Path Loss Models for 5G Wireless Communications. *IEEE Trans. Veh. Technol.* **2016**, *65*, 2843–2860. [CrossRef]
48. Maccartney, G.R.; Rappaport, T.S.; Sun, S.; Deng, S. Indoor Office Wideband Millimeter-Wave Propagation Measurements and Channel Models at 28 and 73 GHz for Ultra-Dense 5G Wireless Networks. *IEEE Access* **2015**, *3*, 2388–2424. [CrossRef]
49. MacCartney, G.R.; Deng, S.; Sun, S.; Rappaport, T.S. Millimeter-Wave Human Blockage at 73 GHz with a Simple Double Knife-Edge Diffraction Model and Extension for Directional Antennas. In Proceedings of the 2016 IEEE 84th Vehicular Technology Conference (VTC-Fall), Montreal, QC, Canada, 18–21 September 2016; pp. 1–6. [CrossRef]
50. Cotton, M.G.; Kuester, E.F.; Holloway, C.L. An investigation into the geometric optics approximation for indoor scenarios with a discussion on pseudolateral waves. *Radio Sci.* **2002**, *37*, 1–31. [CrossRef]
51. Remley, K.; Anderson, H.; Weisshar, A. Improving the accuracy of ray-tracing techniques for indoor propagation modeling. *IEEE Trans. Veh. Technol.* **2000**, *49*, 2350–2358. [CrossRef]
52. Numata, S.; Sato, H.; Yamane, T.; Chiba, H. Indoor propagation experiments in shielding TV studio at 2 GHz and analysis by geometrical optics method. In Proceedings of the 1999 International Symposium on Electromagnetic Compatibility (IEEE Cat. No.99EX147), Tokyo, Japan, 17–21 May 1999; pp. 432–435. [CrossRef]
53. De Coster, I.; Van Lil, E.; Van de Capelle, A. A PO approach for a more accurate calculation of the reflected field in indoor propagation. In Proceedings of the VTC '98. 48th IEEE Vehicular Technology Conference. Pathway to Global Wireless Revolution (Cat. No.98CH36151), Ottawa, ON, Canada, 21 May 1998. [CrossRef]
54. Perez, J.; Catedra, M. Application of physical optics to the RCS computation of bodies modeled with NURBS surfaces. *IEEE Trans. Antennas Propag.* **1994**, *42*, 1404–1411. [CrossRef]
55. Knott, E. A progression of high-frequency RCS prediction techniques. *Proc. IEEE* **1985**, *73*, 252–264. [CrossRef]
56. Tao, Y.; Lin, H.; Bao, H. GPU-Based Shooting and Bouncing Ray Method for Fast RCS Prediction. *IEEE Trans. Antennas Propag.* **2010**, *58*, 494–502. [CrossRef]
57. Ling, H.; Chou, R.-C.; Lee, S.-W. Shooting and bouncing rays: Calculating the RCS of an arbitrarily shaped cavity. *IEEE Trans. Antennas Propag.* **1989**, *37*, 194–205. [CrossRef]
58. Li, J.; Zhao, L.; Guo, L.-X.; Li, K.; Chai, S.-R. Hybrid PO-SBR-PTD method for composite scattering of a vehicle target on the ground. *Appl. Opt.* **2021**, *60*, 179–185. [CrossRef]
59. Choudhury, B.; Jha, R.M. A refined ray tracing approach for wireless communications inside underground mines and metrorail tunnels. In Proceedings of the 2011 IEEE Applied Electromagnetics Conference (AEMC), Kolkata, India, 18–22 December 2011; pp. 1–4. [CrossRef]
60. Gentile, C.; Valoit, F.; Moayeri, N. A raytracing model for wireless propagation in tunnels with varying cross section. In Proceedings of the 2012 IEEE Global Communications Conference (GLOBECOM), Anaheim, CA, USA, 3–7 December 2012; pp. 5027–5032. [CrossRef]
61. Li, M.; Zhou, L.; Zheng, Z. Analysis of electromagnetic wave propagation characteristics in rectangular tunnels based on the geometrical optics method. In Proceedings of the 2012 International Conference on Microwave and Millimeter Wave Technology (ICMMT), Shenzhen, China, 5–8 May 2012; Volume 2, pp. 1–3. [CrossRef]
62. Kouyoumjian, R.; Pathak, P. A uniform geometrical theory of diffraction for an edge in a perfectly conducting surface. *Proc. IEEE* **1974**, *62*, 1448–1461. [CrossRef]
63. Keller, J.B. Geometrical Theory of Diffraction*. *J. Opt. Soc. Am.* **1962**, *52*, 116–130. [CrossRef]
64. Ufimtsev, P.Y. *Fundamentals of the Physical Theory of Diffraction*, 2nd ed.Wiley-IEEE Press: Hoboken, NJ, USA, 2014; Volume 9781118753668, pp. 1–469. [CrossRef]
65. Ufimtsev, P.Y. Elementary Edge Waves and the Physical Theory of Diffraction. *Electromagnetics* **1991**, *11*, 125–160. [CrossRef]
66. Finite Element Method Electromagnetics: Antennas, Microwave Circuits, and Scattering Applications | Wiley. Available online: https://www.wiley.com/en-us/Finite+Element+Method+Electromagnetics%3A+Antennas%2C+Microwave+Circuits%2C+and+Scattering+Applications-p-9780780334250 (accessed on 26 July 2023).
67. The Finite Element Method in Electromagnetics, 3rd Edition | Wiley. Available online: https://www.wiley.com/en-us/The+Finite+Element+Method+in+Electromagnetics%2C+3rd+Edition-p-9781118571361 (accessed on 26 July 2023).
68. Davidson, D.B. *Computational Electromagnetics for RF and Microwave Engineering*, 2nd ed.; Cambridge University Press: Cambridge, UK, 2010. [CrossRef]

69. Zuo, S.; Donoro, D.G.; Zhang, Y.; Bai, Y.; Zhao, X. Simulation of Challenging Electromagnetic Problems Using a Massively Parallel Finite Element Method Solver. *IEEE Access* **2019**, *7*, 20346–20362. [CrossRef]
70. Shen, M.; Zuo, S.; Zhao, X.; Lin, Z. Analysis of Large Array Antenna Based on Parallel Finite Element Domain Decomposition Method. In Proceedings of the 2020 International Conference on Microwave and Millimeter Wave Technology (ICMMT), Shanghai, China, 20–23 September 2020; pp. 1–3. [CrossRef]
71. Bao, G.; Gao, J.; Lin, J.; Zhang, W. Mode Matching for the Electromagnetic Scattering from Three-Dimensional Large Cavities. *IEEE Trans. Antennas Propag.* **2012**, *60*, 2004–2010. [CrossRef]

Disclaimer/Publisher's Note: The statements, opinions and data contained in all publications are solely those of the individual author(s) and contributor(s) and not of MDPI and/or the editor(s). MDPI and/or the editor(s) disclaim responsibility for any injury to people or property resulting from any ideas, methods, instructions or products referred to in the content.

Article

Enabling Semantic-Functional Communications for Multiuser Event Transmissions via Wireless Power Transfer

Pedro E. Gória Silva [1,2], Nicola Marchetti [3], Pedro H. J. Nardelli [1,4,*] and Rausley A. A. de Souza [2]

[1] School of Energy Systems, Lappeenranta–Lahti University of Technology (LUT), 53850 Lappeenranta, Finland
[2] Department of Electrical Engineering, National Institute of Telecommunications (INATEL), Santa Rita do Sapucaí 37540-000, Brazil
[3] Connect Centre, Trinity College Dublin, D02 PN40 Dublin, Ireland
[4] 6G Flagship, University of Oulu, 90570 Oulu, Finland
* Correspondence: pedro.nardelli@lut.fi

Abstract: A central concern for large-scale sensor networks and the Internet of Things (IoT) has been battery capacity and how to recharge it. Recent advances have pointed to a technique capable of collecting energy from radio frequency (RF) waves called radio frequency-based energy harvesting (RF-EH) as a solution for low-power networks where cables or even changing the battery is unfeasible. The technical literature addresses energy harvesting techniques as an isolated block by dealing with energy harvesting apart from the other aspects inherent to the transmitter and receiver. Thus, the energy spent on data transmission cannot be used together to charge the battery and decode information. As an extension to them, we propose here a method that enables the information to be recovered from the battery charge by designing a sensor network operating with a semantic-functional communication framework. Moreover, we propose an event-driven sensor network in which batteries are recharged by applying the technique RF-EH. In order to evaluate system performance, we investigated event signaling, event detection, empty battery, and signaling success rates, as well as the Age of Information (AoI). We discuss how the main parameters are related to the system behavior based on a representative case study, also discussing the battery charge behavior. Numerical results corroborate the effectiveness of the proposed system.

Keywords: wireless power transfer; energy harvesting; semantic-functional communications; event-based communications

1. Introduction

Between the end of the 19th century and the beginning of the 20th century, Nikola Tesla envisioned, designed, and built an experimental station (Wardenclyffe Tower) to transmit energy over the air. Tesla's technique applied alternating electrical current to the ground so that a device touching the ground would close an electrical circuit through the atmosphere with the generator, and the gadget would thus be powered. Although Tesla did not address the use of radio waves, one can understand Tesla's experiment, the Wardenclyffe Tower, as a historic milestone initiating the radio frequency-based energy harvesting (RF-EH) field of research. Following the trend of miniaturization of Internet of Things (IoT) devices, including sensors, research focused on wireless charging devices at a distance has recently gained increasing relevance.

Wireless power transfer, a broader term than RF-EH, can be separated into two primary categories: near-field and far-field. This sub-classification arrives from the distinct behavior of electromagnetic waves due to the distance from the transmitting antenna. Overall, the area delimited by one wavelength from the electromagnetic wave source constitutes the near field. In general, this distance is no more than a few meters, thus drastically limiting the coverage area for possible Wireless power transfer harvesting (WPT) techniques with a near-field application; on the other hand, far-field methods are capable of covering a vast area at the cost of providing the device with reduced power compared to near-field ones [1–4].

In addition to energy solutions, one can highlight two other aspects relevant to the design of a system: Application and Communication. For simplicity, we group all works that address concepts related to a specific function, general objective, highest communication layer (e.g., Open Systems Interconnection (OSI) model), etc., inside the Application. In a way, the Application contemplates solutions solely dealing with the primary function of operating the system. The Communication umbrella covers works whose purpose is to assess the performance of information exchange; therefore, it contains methods, techniques, and analyses that study modulation performance, coding, channel capacity, different medium access control techniques, etc. Finally, we can add another class that includes work related to energy consumption, named Energy here. Energy refers to work that aims to understand and evaluate the system's energy consumption or expenditure.

An outline of how literature deals with these classes is described in Figure 1 together with Table 1. Note that there are intersection regions between the subsets in Figure 1 representing works that directly address two or three aspects listed above. Table 1 is directly related to Figure 1; each subset of Figure 1 (named #1, #2, #3, etc.) is exemplified through one or more papers from the literature in Table 1. Although one can extend the list of examples to subsets #1 through #6, to the best of the authors' knowledge, this is the first work to jointly consider aspects of Energy, Application, and Communication in the system design. Thus, in addition to contrasting with other works, it is not possible to draw a parallel or fairly compare our solution with other methods presented in the technical literature. In order to exemplify such a discrepancy, suppose a comparison with a Media Access Control (MAC) protocol. In this case, any clarification possibly acquired would refer exclusively to Communication, regardless of the specific method adopted (Time-Division Multiple Access (TDMA), Frequency-Division Multiple Access (FDMA), Code-Division Multiple Access (CDMA), etc.). Furthermore, one could propose a system covering the three classes (Energy, Application, and Communication) consisting of methods designed or evaluated specifically for one or two (never for all three) classes as a benchmark. This case would not be an adequate comparison either; as will become clear throughout this paper, the proposed solution jointly covers the three classes and cannot be understood as a combination of three dispersed methods.

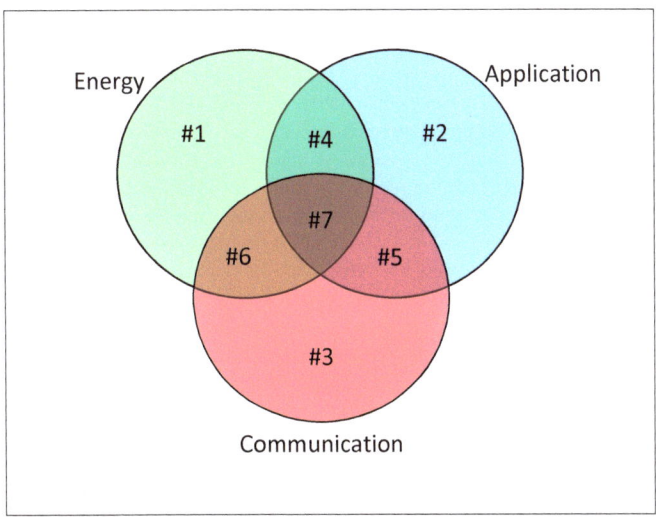

Figure 1. Classification of design approaches.

Table 1. Brief description and classification of works present in the literature.

Subset	Description	Ref.
#1	Theoretical analysis of RF-DC conversion	[5,6]
#2	Event-driven data acquisition for electricity metering Automatic fault detection of sensors	[7] [8]
#3	Performance of MPSK modulation over fading channel	[9,10]
#4	Boosting quantum battery via energy harvesting	[11]
#5	Semantic-functional communications for multiuser	[12]
#6	Simultaneous wireless information and power transfer	[1,13]
#7	Communications for multiuser event transmissions and RF-EH	This work

This paper contributes to the literature in the field (which will be reviewed in the sections to come; a complete landscape can be found in [14]) by combining wireless power transfer techniques and the semantic-function approach recently proposed in [12] that consider meaningful events that require explicit use of the communications channel to support the operation of a given cyber-physical system [15]. Our numerical results demonstrated that it is possible to have an effective recharge of batteries and optimized performance by merging RF-EH and the semantic-function approach.

The rest of this paper is divided as follows: Preliminary concepts are introduced in Section 2. The problem addressed is formulated in Section 3. The proposed system is introduced in Section 4. A case study is carried out in Section 5 in order to assess the performance of the system through simulations. Then, we present the conclusions of the paper in Section 6.

2. Background: WPT and Fundamentals of RF-EH

In this section, we will provide a brief overview of wireless power transfer and RF-based energy harvesting that will serve as the main theoretical background of this paper.

2.1. Near-Field WPT

The near-field or non-radiative region is the zone within one wavelength from the transmitting antenna. Inside the near-field area, one can power a device with up to a few tens of Watts by employing inductive coupling, resonant inductive coupling, or capacitive coupling in the range of tenths of Watts [4,16,17]. The near-field solutions, under certain conditions, reach more than 90% efficiency in the energy transfer process between the source and the powered device [17]. An advantageous feature of near-field methods is that power dissipation only occurs when we have an absorbing material (a device to be charged) within the actuation component. In other words, the transmitter in a near-field method drastically reduces its energy expenditure (energy consumed for its operation in addition to transfers when applicable) if there is no device to charge. This peculiarity provides an extra monitoring capacity in the system: the transmitter can identify the presence of nearby devices through its energy consumption. On the other hand, near-field techniques face an exponential decrease in the total power transmitted by expanding the gap between the source and load. Overall, we have smallish or no energy transfer when the distance between the source and load exceeds one wavelength.

2.1.1. Inductive Coupling

In inductive energy transfers for near-field WPT, the power transmitter recharges the devices through a magnetic field induced by coils. In this type of application, efficiency is as high as the proximity of the devices [18]. Inductive power transfer is supported theoretically by Ampere's law: an alternating electric current produces an oscillating magnetic field when passing through a coil or wire. Moreover, once the magnetic field generated by the power transmitter is close enough to the receiver coil, we have an alternating electric current in the receiver; then, the receiver recharges its battery as soon as it rectifies the obtained alternating current. A wide range of applications uses this technique. We can mention electric toothbrush stands at low frequencies (50/60 Hz) or consumer electronic devices such as laptops, cellular phones, and other portable devices, in addition to those used to charge electric automobiles at high frequencies [1]. In order to enhance inductive power transmission, one can add resonant circuits. In resonant inductive power transfer (RIPT), resonant circuits consist of a coil of wire connected to a capacitor or a resonator with internal capacitance. Like inductive power transfer techniques, magnetic fields convey energy from the transmitter to the receiver in RIPT; however, WPT based on resonant inductive power transfer can cover broader areas. In addition, there are several advantages of resonant RIPT over inductive power transfer, such as reduced electromagnetic interference, higher frequency of operation, and higher efficiency. Among the possible applications for RIPT, we can mention wireless power coverage, power lights, and recharging mobile batteries anywhere in a room without wired connections.

2.1.2. Capacitive Coupling

In capacitive energy transfers for near-field WPT, the power transmitter recharges the devices through an electric field between electrodes like metal plates. More precisely, the transmitting power source applies an alternating voltage across a metal plate, leading to electrostatic induction and an alternating electrostatic potential on the metal plates of the receiver. This receiver's metal plate is the alternating current source to recharge the device. Some fundamental factors for the efficiency of energy transfer are frequency, the square of the voltage, and the capacitance between the transmitter and receiver plates [19]. Another intriguing characteristic of capacitive coupling is the proportionality between the transferred power and the smallest area among the plates used in the system. Capacitive energy transfers involve limited power due to poisonous ozone gas produced by high voltage on the electrodes. Therefore, techniques of capacitive energy transfer are applied to low-power systems due to their hazardous property. However, capacitive couplings have advantages over inductive couplings, such as controlled interference, and the alignment between the transmitter and receiver can be more flexible. Recent work has pointed to capacitive coupling as a potential method for an electric double layer for wireless power transfer under seawater [20].

2.2. Far-Field WPT

The far-field or radiative region is the zone beyond one wavelength around the transmitting antenna. In this region, the electromagnetic field settles into ordinary electromagnetic radiation, i.e., we have transverse electric or magnetic fields with electric dipole characteristics. In addition, the radiated power is well-behaved, decreasing with the square of the distance, and the radiation absorption does not feed back to the transmitter. We can point out radio waves, microwaves, and laser beams as examples of electromagnetic radiations that are traditionally adopted for far-field WPT. In general, radio waves have a lower attenuation and are more efficient than microwaves or laser beams due to atmospheric attenuation. Given the intrinsic scattering of propagation of electromagnetic waves, as a rule of thumb, the efficiency of a far-field WPT does not exceed 50% [21]. On the other hand, for example, as long as there is a line-of-sight propagation, Ref. [22] has shown that more than 80% efficiency can be achieved.

The literature usually employs far-field WPT techniques to cover a wide area or charge devices far from the source. Some available power sources may not be dedicated to recharging devices and still be appropriated for the far-field WPT technique. In other words, a timely use can be made of the available frequency spectrum by harvesting the power traveling freely in the air. Therefore, it is reasonable to classify power sources into two classes: dedicated power sources and ambient power sources. Dedicated power sources can, for example, benefit from a licensed frequency band or even a license-free band to charge a remote device. On the other hand, an ambient power source is any transmitter whose transmitted electromagnetic waves reach the device. Although this distinction defines the power sources, it clarifies relevant aspects regarding the receiver design. In general, receivers that collect energy from ambient power sources must consider several medium statistics, while scenarios with dedicated sources are less susceptible to the environment. We shall see this in more detail throughout this work.

The literature has lately pointed to RF-EH as a disruptive technology that enables low-power portable devices and energy-constrained wireless networks to benefit from the electromagnetic energy available in the environment. Roughly speaking, the working principle of RF-EH consists of an electric-to-electromagnetic-to-electric conversion process in order to recharge wireless devices. It should be noted that the first process transformation, electric-electromagnetic, is not necessarily carried out by a source dedicated to the RF-EH system. Researchers have proposed several architectures to perform energy harvesting in order to coexist with traditional data transmission systems. Briefly, although there are different architectures for RF-EH, to the best of the authors' knowledge, the transmission method employed does not deviate from the more traditional modulation techniques. Figure 2 illustrates the main architectures for RF-EH regarding the arrangement of the data layer concerning the energy harvesting layer. We discuss each of them in more detail below.

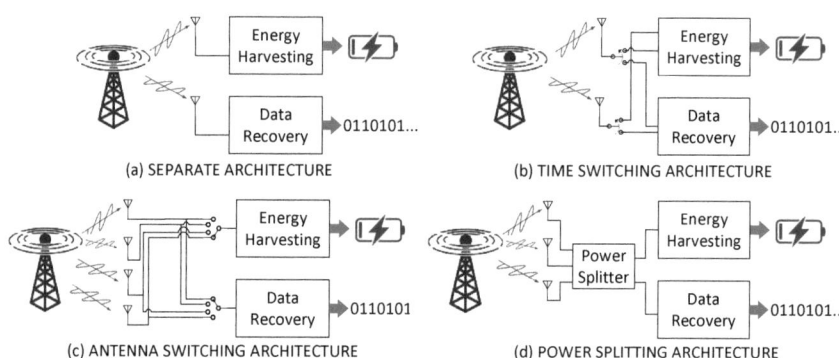

Figure 2. Layout of (**a**) separate, (**b**) time switching, (**c**) antenna switching, and (**d**) power splitting architecture for RF-EH.

2.2.1. Separate Architecture

As highlighted in Figure 2a, the separate architecture consists of two or more antennas and two independent receiver blocks. Each receiver block performs a different function: data decoding (data recovery) and electromagnetic wave conversion into the direct current (energy harvesting). More than one receiving antenna can be used for each block, depending on the technique employed. For example, multiple-input-multiple-output (MIMO) can be used for the data transmission system, or one can have antennas in different bands to harvest energy from various frequencies. Another example is [23], in which the authors propose a MIMO broadcast system for simultaneous wireless information and power transfer. In this type of architecture, the receiver can perform both the energy harvesting function and the data recovery continuously and concurrently.

2.2.2. Time Switching Architecture

A Time Switching architecture must have an antenna selector switch, a data receiver, and an energy harvesting block, as shown in Figure 2b. It is commonly assumed that all antennas feed exclusively one block at a time. The antenna selector switch can modify the receiving process of the system, i.e., it can select either energy harvesting operation or data recovery. In a nutshell, the working principle of time switching consists of periodically switching the beam between energy harvesting and data recovery blocks. Depending on the adopted data transmission method, it may be necessary to synchronize the operation mode selection. In other words, the receiver may not be able to receive the transmitted data because it is operating in energy harvesting mode. On the other hand, this architecture provides the highest peak value for the harvested energy compared to other architectures with equivalent systems. This advantage can be quite beneficial in environments with sporadic availability of energy. Ref. [24] studies the optimal design for simultaneous wireless information and power transfer in down-link orthogonal frequency-division multiplexing (OFDM) systems, in which the users apply either time switching or power splitting to coordinate the energy harvesting and data recovery processes.

2.2.3. Antenna Switching Architecture

A sketch of the antenna switching architecture is shown in Figure 2c. This architecture is constituted by two independent receiver blocks and two or more antennas. We have one data recovery block and an energy harvesting one. The Antenna Switching architecture is distinguished by using a low-complexity switch. Antennas can migrate from one block to the other under certain conditions. As the dynamic properties of antenna systems do not require high speed or periodic changes in the configuration of the antenna array of each block, one can claim that antenna switching involves a relatively low complexity and that this architecture is, somehow, more suitable for practical designs [1].

2.2.4. Power Splitting Architecture

The power splitting architecture has a peculiar characteristic: the division of the power collected by all the antennas into two streams of different power levels before the receiver performs signal processing. As with the separate architecture, the power splitting one operates with two independent receiver blocks. We have one receive block to recover the information and another to harvest energy, as depicted in Figure 2d. The number of antennas may vary according to the application and the frequency range of interest. This architecture allows the receiver to perform both the energy harvesting function and the data recovery continuously and concurrently. Some approaches, such as those proposed by [25], present the division of power between energy harvesting and data recovery blocks as dynamic. This characteristic is a significant advantage of this architecture, as it allows the receiver to operate with different reception rates according to the need for a recharge. Furthermore, one can design the power-splitting block like any other power splitter of conventional communication systems [26]. We can find a vast literature on applications to power division architectures, such as [27], which proposes a long-range optical wireless energy and information transfer system of high power and high rate.

2.3. RF Energy Sources and Statistics

Following the classification proposed earlier in this paper, Figure 3 summarizes the energy sources available to harvest energy. The first branch in Figure 3 concerns the energy emitter. As already discussed, sources emitting energy without the specific purpose of recharging a device—this certainly includes the forces of nature—are treated as ambient sources, and transmitters of electromagnetic waves to power a remote device are classified as dedicated sources.

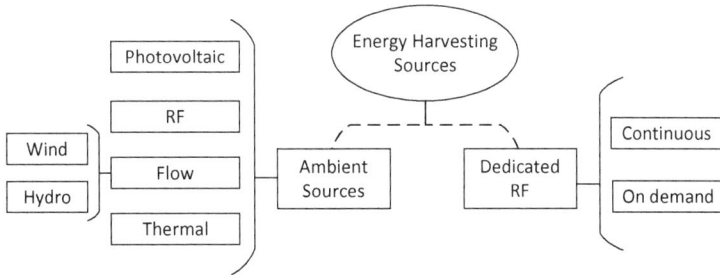

Figure 3. Sources to harvest energy.

We split ambient sources into sub-classes related to the nature of the energetic process. Photovoltaic covers sources that can be harvested by converting light into direct current; therefore, any light source belongs to the Photovoltaic class. Miscellaneous sources of electromagnetic waves outside the frequency of visible light are classified as RF class since they are not dedicated to power devices. For cases in which the energy harvesting process consists of a transformation of mechanical energy into electrical energy, we classify them as Flow; two sub-classes are highlighted for it: Wind and Hydro. The last class of Ambient Sources is related to thermal energies. For example, wearable sensors which can recharge themselves using human heat are classified as Thermal. Applications in the Thermal class require the device to handle thermoelectric phenomena.

Dedicated sources can handle the most diverse means of delivering power to the remote device. We could divide it into the same categories proposed for Ambient Sources; however, by and large, the literature suggests electromagnetic wave transmitters as the dedicated source. We then split the Dedicated RF class into Continuous and On Demand. The former refers to sources that operate and transmit uninterruptedly—note that this classification does not restrict the transmitter regarding the operating mode, i.e., the transmitter can transmit information continuously. The latter implies that the transmitter has some control and only transmits energy at opportune times, even if transmissions are periodic.

Since our focus in this work is RF-EH, we shall approach RF from Ambient Sources and Dedicated RF classes. The reader is referred to [2] to obtain further insights about other energy sources.

2.4. RF from Ambient Sources

We can highlight radio broadcasts, TV broadcasts, Wi-Fi, and mobile phone networks as the most typical sources for RF-EH in urban centers. Except for specific cases, such as the receiver being extremely close to a base station, the power available after harvesting from these sources does not exceed tens of mW. Due to this low power, the rectifier circuits and matching network can reach an efficiency of 70% at best [2]. However, an energy conversion efficiency of around 83% can be achieved, under certain conditions, by optimizing the antenna array and the rectifier circuit, employing the gradient method [28]. Due to the nature of the medium and sources, the amount of energy captured by the RF-EH method from ambient RF sources randomly varies over time. Some studies have focused on modeling several frequency ranges in a probability density function. For instance, a kernel distribution model is used to fit experimentally collected data on power density in some RF bands in [29]. Furthermore, the authors compare the data with the Rayleigh distribution. Ref. [30] assesses the broadband RF radiations in terms of power densities from 467.5 MHz to 3.5 GHz, in addition to graphically presenting the probability density function (pdf) and cumulative density function (cdf) of the power densities. Ref. [30] does not propose a theoretical pdf for the collected data; however, the authors indicate the mean and standard deviation of the samples. Since a trustworthy characterization of the medium is beyond the scope of this work, the knowledge available at present in the literature about random variations over time of the medium is sufficient for our exposition. In particular, we assume the following mathematical statement to characterize the power harvested from ambient sources.

Assumption 1. *Given an RF-EH receiver with efficiency η, the energy harvested from Ambient Sources after T seconds is given by*

$$\mathcal{E}_{Ab}(T) = \eta \int_T E_{Ab}(t)dt = T\eta p_{Ab}, \qquad (1)$$

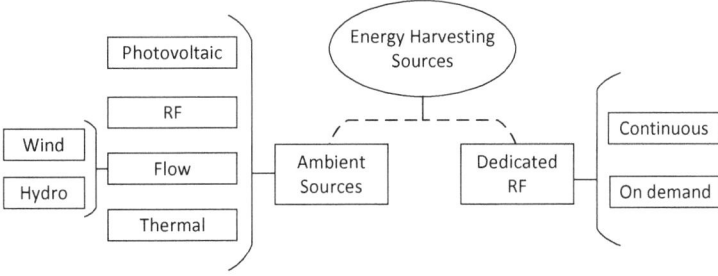

with $E_{Ab}(t)$ representing the the instantaneous energy density at the RF-EH receiver, and p_{Ab} follows a Rayleigh distribution as

$$f(p_{Ab}) = \frac{p_{Ab}}{\sigma_{Ab}^2} \exp\left(-\frac{p_{Ab}^2}{\sigma_{Ab}^2}\right). \tag{2}$$

The mean of p_{Ab} is given by $\sigma_{Ab}\sqrt{\pi/2}$.

2.5. Dedicated RF

As already discussed previously, we named as dedicated those RF sources that intentionally deliver energy in one or more RF frequency ranges to wireless sensor nodes or mobile devices. These sources are fully controllable, and thus, they are predictable. On the other hand, the system's unpredictability arises from the communication channel. Let us examine this issue more closely.

For a multipath channel, we have that the complex envelope of the received signal is given by [31]

$$r(t) = x(t)h(t) + n(t), \tag{3}$$

with $x(t)$, $h(t)$, and $n(t)$ being the complex envelope of the transmitted signal, channel coefficient, and additive white Gaussian noise (AWGN), respectively. Thereby, the RF-EH can harvest a certain amount of energy at the end of an exposition period T given by

$$\mathcal{E}_{De}(T) = \eta \int_T (r(t))^2 \, dt. \tag{4}$$

For an unmodulated carrier $x(t) = A_r$ and slow fading, and assuming a Rician fading channel, we have that energy harvested by RF-EH receiver is given by

$$\mathcal{E}_{De}(T) = \eta \int_T (A_r h(t) + n(t))^2 dt = \eta\left(T(A_r h)^2 + 2A_r hn + \int_T n(t)^2 dt\right), \tag{5}$$

in which n is zero-mean, σ^2-variance Gaussian variate, and h follows a Rician distribution with mean and variance given by $E[h] = H_m$ and $\text{VAR}[h] = \sigma_{De}$, respectively.

Precisely determining the pdf of $\mathcal{E}_{De}(T)$ is beyond the scope of this work. However, given the intrinsic characteristics of the RF-EH process (antenna, impedance matching cell, rectifier circuit, etc.), we assume that the RF-EH receiver's bandwidth is narrow enough such that $n(t)$ can be considered constant over a period of T seconds. Accordingly, we can rewrite (5) as

$$\mathcal{E}_{De}(T) = \eta T\left((A_r h)^2 + 2A_r hn + n^2\right). \tag{6}$$

Figure 4 presents three histogram examples for the energy harvested by RF-EH receiver $\mathcal{E}_{De}(T)$. We have $T = 1$, $\eta = 1$, $A_r = 1$, and $\sigma^2 = 0.01$ in all three histograms. A Rician fading with means of 2, 0.77, and 1 and variance of 0.1, 0.15, and 0.01 were assumed in blue, orange, and yellow, respectively.

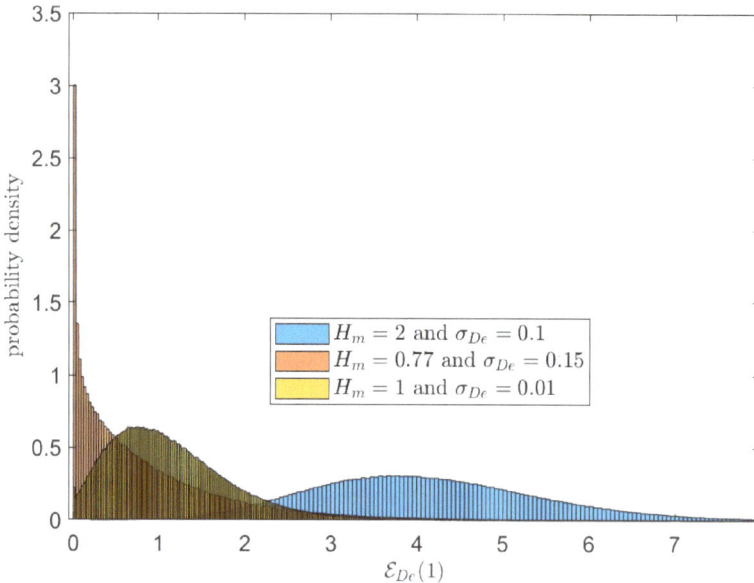

Figure 4. Histogram of energy harvested by RF-EH receiver $\mathcal{E}_{De}(T)$ for $T = 1$, $\eta = 1$, $A_r = 1$ and $\sigma^2 = 0.01$.

3. Problem Formulation

This section introduces the considered scenario and our system model. We discuss how our approach differs from traditional solutions by briefly presenting how the literature approaches the problem. Furthermore, we discuss how the sensors can harvest energy and receive information without jeopardizing the system's operation.

Feedback control demands communications as a way of assessing the current state of a physical entity. The feedback control design of any remote physical process must cover two main aspects: control theory and communication theory. From a control theory perspective, the decision-making unit (data fusion or control center) acquires information about the physical process through the sensor network, in order to predict which intervention should be performed by the physical entity. The key challenges are how and when to act (controller block) and what information to acquire from the physical system (sensors and feedback). By receiving meaningful events instead of periodic samples, the decision-maker can control the system adequately [32]. This approach is called event-triggered control (ETC) and may save network resources while ensuring system stability [7]. Concerning communication aspects, one can mention a Wireless Sensor Network (WSN) as a case of interest here. In a nutshell, a WSN consists of sensors that operate remotely and report their readings to the decision-maker through a wireless channel. For a WSN, event-driven communication (EDC), which consists of transmitting only when certain events occur, can work in harmony with ETC. The distinction between EDC and ETC is subtle, although they address different system parts—the former handles the communication layer, and the latter addresses the control part of the system.

The convergence of EDC and ETC in a specific event-driven communication system designed to perform event-triggered control is still an open problem. The work we proposed in [12], to the best of the author's knowledge, is the first to merge control and communication aspects in a dedicated communication system, taking advantage of implicit communications between transmitters (sensors) and receivers (controllers). The approach proposed in [12] determines the event (sensor that transmitted it) based on an identification code generated by a random map. Since the above-mentioned technique works with energy detection instead of a strict signal demodulation process, we will use it in conjunction with

RF-EH in order to ensure the operation of a monitoring system with sensors powered by RF. More details about the system and how it works are presented below.

4. System Model

Our scenario consists of N_s sensors spread over an area. This scenario can represent, for example, an electric power plant, an industry, a rural area, a farm, or any other environment that requires monitoring. In addition to the sensors, there are also N_u unmanned aerial vehicles (UAVs) randomly flying over the region in order to collect information from the sensors and forward it to the decision-maker. For comparison purposes, we also assume a scenario where the area is small enough so that the sensors communicate directly with the decision-maker. This work aims to investigate the behavior of the sensors regarding the battery charge and how they access the medium. Therefore, we consider an ideal channel between the UAVs and the decision-maker. Furthermore, we do not assess the control system's performance in terms of stability; only sensor network properties such as delay are investigated. We also assume that the system employs distinct frequency bands for Dedicated RF and Ambient Sources harvesting.

An operation diagram of sensor nodes is presented in Figure 5. The sensors remain in an idle state (deep sleep) as long as an event does not occur. Deep sleep is a widely used operation in embedded systems to save energy. In this state, micro-controllers and other components can operate under significantly low electrical currents, around 20 µA [33]. In addition, one can wake up the micro-controller due to external factors, in our case, as the result of an event. The rise of a signal (which can be, for example, the measure of temperature, pressure, or humidity) above a threshold may trigger what we treat as an event; more precisely, one can define a function event as general as possible as $f : \mathbb{R} \to \{0, 1\}$, such that $f(t) = 1$ represents the event's occurrence (the reader interested in more details about the event is referred to [12]). In other words, given a signal of interest, the event can be defined as required by the decision-maker. After identifying the event, the sensor node starts the transmission, which we discuss in more detail later. Immediately after the the end of the transmission, the sensor node saves the battery charge and accurately monitors the current from the RF-EH circuit. After W seconds of current monitoring, the sensor node decides whether to retransmit based on the behavior of the battery charge current over the last W seconds. Note that the acknowledgment (ACK) transmitted by one or more UAVs is not a conventional data packet; instead, our ACK is identified through the battery charging pattern in the sensor. Thus, the ACK simultaneously fulfills two roles: recharging the battery and confirming receipt of the message. Sporadic conditions, such as insufficient battery power, bring the sensor node back to deep sleep.

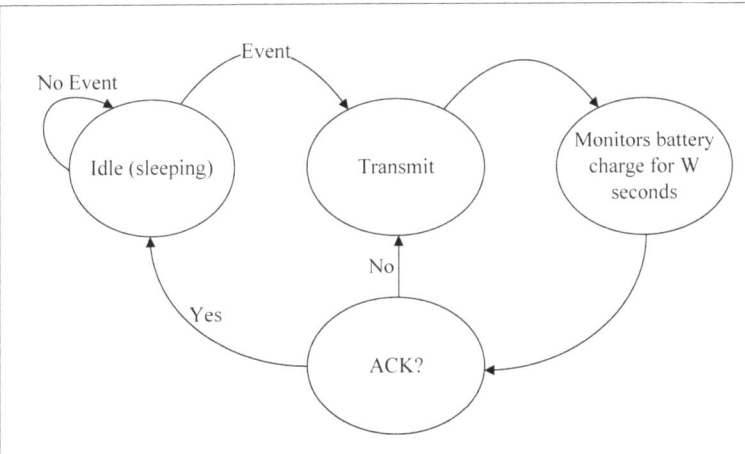

Figure 5. Sensor operation diagram.

Communication between sensors and UAVs in both directions is established through the semantic-functional communication (SFC) technique proposed in [12]. In a nutshell, each event obtains a sole transmission map that enables unambiguously determining the origin and the event at the receiver for SFC. This map guides the transmission of energy packets over time, called energy slots—to the best of the authors' knowledge, the concept of energy slot proposed in [12] differs from standard transmission techniques by not requiring a process of demodulation itself in addition to a distinctive methodology for controlling access to the medium; therefore, one should not understand it as a traditional method of transmission.

In our scenario, the UAV listens to the sensors through the SFC and then replies to them with an ACK again using the SFC. Thus, the UAV must look for patterns of energy variation over time in order to recover the transmitted information. Here, we will not thoroughly describe the UAV reception process; it suffices to say that SFC enables the UAV to always identify, under certain conditions, an event at the cost of a low rate (in many cases negligible) of false alarms. On the other hand, a sensor may be out of range of the UAV at a given time, leading to a failure in communication. We will assume that sensor transmissions have a success rate of R_s proportional to the number of UAVs per area.

The reception process at the sensor consists of determining if it hears an ACK transmission by scrutinizing the battery charging as follows. Let T_E and $T_E^{(k)} = \{t \in \mathbb{R} : (k-1)T_E < t - t_{TE}^{(n)} \leq kT_E\}$ be the duration of an energy slot and the time interval of the kth energy slot just after sensor transmission, respectively, with $t_{TE}^{(n)}$ being the time at which the nth transmission ended. Thus, after transmitting, the sensor node samples the charge (energy) of the battery every T_E seconds—for simplicity, we assume that this process is perfect, i.e., the amount of energy in a given instant is unequivocally read by the sensor. Let the battery charge at time t in the sensor $\mathcal{S}$ be given by $C_{\mathcal{S}}(t) \leq C_{\max}$. Thus, at the instant just after transmission, we have that the change in battery charge is given by

$$\Delta C[k] = C_{\mathcal{S}}\left(kT_E + t_{TE}^{(n)}\right) - C_{\mathcal{S}}\left((k-1)T_E + t_{TE}^{(n)}\right)$$
$$= P_{De}[k]\mathcal{E}_{De}\left(T_E^{(k)}\right) + \mathcal{E}_{Ab}\left(T_E^{(k)}\right) \ \forall k \in \{1,2,3,\ldots,K\} \quad (7)$$

where

$$P_{De}[k] = \begin{cases} 1, & \text{if UAV transmitted in } T_E^{(k)} \\ 0, & \text{otherwise} \end{cases}, \quad (8)$$

and $K \leq \lfloor W/T \rfloor$, with $\lfloor \cdot \rfloor$ rounding to the largest integer less than or equal to the argument. Let the threshold function $f : \mathbb{R} \to \{0,1\}$ be given by

$$f(a) = \begin{cases} 1, & \text{if } a > \delta \\ 0, & \text{otherwise} \end{cases}. \quad (9)$$

Then, by making $B_C[k] = f(\Delta C[k])$, the sensor node has a binary version of $\Delta C[k]$, and it can create a binary reception map $M_r = B_C[1], B_C[2], \ldots, B_C[k]$. If there is a binary sequence in M_r corresponding to its ACK, the sensor node decides that the UAV is aware of its transmission; otherwise, it retransmits if the battery has enough energy. One can draw an analogy with a traditional modulation process: $\mathcal{E}_{De}(T)$ represents the transmitted information while $\mathcal{E}_{Ab}(T)$ is the noise.

RF-EH can be upgraded by SFC because the battery charging can use the energy from the information signal. Note that this is a disruptive approach, as traditional communication methods make it impossible for the receiver to simultaneously demodulate and recharge the battery without splitting the received energy; in other words, the battery charging process can consume completely (not partially, as discussed in Section 2.2) the energy contained in the informational signal.

Harvesting and Consuming Energy

In order to extract the maximum available energy from the environment, the sensors keep the RF-EH circuit working even in deep sleep mode. In this way, the battery remains charged during the entire system operation. Therefore, the charge on the battery of sensor S at the instant t is given by

$$C_S(t) = \begin{cases} 0, & \text{if } c_S(t) < 0 \\ C_{\max}, & \text{if } c_S(t) > C_{\max}, \\ c_S(t), & \text{otherwise} \end{cases} \quad (10)$$

with

$$c_S(t) = \mathcal{E}_{\text{Ab}}(t) + E_{\text{De}}^{(t)} - nE_T. \quad (11)$$

$E_{\text{De}}^{(t)}$ is the sum of all energy packets (energy slots) sent in interval $(0, t)$ by UAV or other sensor nodes, E_T represents the energy wasted by transmission, and n is o the number of transmissions performed by the sensor S in the interval $(0, t)$. Other energy wastes, such as consumption during battery measurement, are being neglected. One can easily add them by subtracting a factor like γt from battery charge calculations, where γ is the average consumption per second.

5. Case Study

Assumption 2. *For simplicity, we assume that the coverage area of a UAV, which is given by a circle of radius ρ whose center is the position of the UAV, is proportional to the energy spent on transmission. In mathematical terms, we assume that*

$$\rho = \rho_c E_T, \quad (12)$$

where ρ_c is the constant of proportionality.

We are now going to evaluate the performance of the system through simulations. For all cases studied in this work, we have $\eta = 1$, $C_{\max} = 100$, a carrier frequency of 400 MHz, $\sigma_{\text{Ab}} = 0.01$, $H_m = 1$, $\sigma_{\text{Ab}} = 0.1$, Combined antenna gain of 5 dBi, $T_E = 0.01$, and the sensors are uniformly spread over the area of interest, unless we specify the other values. The interval between the occurrences of the same event follows an Exponential distribution with mean $1/\lambda$. The other parameters and their respective impacts on the system are discussed below. It is important to state that the scenario analyzed here provides a benchmark case where the aspects indicated in Figure 1 are covered. Our rationale is to build a simple example that can numerically demonstrate the trade-offs involved in the system performance.

As we assumed earlier, the communication between the UAV and the decision-maker takes place without errors. Furthermore, a UAV within a range of the sensor nodes always correctly receives the event signaling, and then replies with the appropriate ACK. Through these assumptions, we can characterize the event signaling success rate R_s as a ratio proportional to the number of UAVs, the coverage area of a UAV, and the number of sensors. On the other hand, R_s is inversely proportional to the dimension of the area of interest. Thus, by assessing R_s, we are indirectly comprehending how the N_s and N_u parameters impact the system's performance. We estimate the success rate of sensor transmissions R_s by the total number of transmissions made by the sensors divided by the total number of transmissions actually received by the UAVs.

Other metrics adopted in this work are presented as follows:

- The event signaling rate R_{es} is defined as the average probability that a sensor node transmits when an event is identified—a sensor node may fail to signal an event due to low battery power or waiting for an ACK;

- The event detection rate R_{ed} is defined as the average probability that an event is detected by the decision-maker—a relevant parameter for the controller (decision maker); in a nutshell, it represents how many events are handled by the system;
- Age of Information (AoI) is a relatively new metric for measuring update data, such as status or control updates. We define AoI as the elapsed time between the physical event and the time the decision-maker becomes aware of the event. It should be noted that only events received by the decision-maker are taken into account in the calculations, i.e., a possible event that was not transmitted is disregarded;
- The empty battery rate R_{eb} is defined as the probability that a sensor's battery charge is below E_T—that is, that the sensor is unable to transmit.

5.1. Varying the Energy E_T

Consider a system with $N_s = 64$, $N_u = 5$, an area of interest measuring 40 m by 40 m, $\rho_c = 2$, the transmission power of UAVs of $P_{UAV} = 100\ C_{max}$—note that we do not deal directly with the transmitted power; instead, we represent it in a scale of the capacity of the sensors' batteries—$W = 10$ s and $\lambda = 1/200$. The following cases are assessed by varying E_T as $\{2.5, 5, 10, 15\}$. For $E_T = 2.5$, we have a consumption of 2.5% of the battery per transmission and a reduced coverage area of the UAVs ($\rho = 5$ m). On the other hand, a large coverage area ($\rho = 30$ m) can be achieved by setting $E_T = 15$ at the cost of high energy consumption. The intermediate values of E_T, $\{5.10\}$ enable analysis regarding the trade-off between energy cost and coverage area.

Figure 6 depicts a possible configuration of sensors and UAVs. In Figure 6, we have an arrangement of sensors and UAV at a given time within the area of interest for $E_T = 2.5$. The small blue circles are the sensors, the black × are the UAVs, and the big black circles represent the area that the UAV is covering. This hypothetical scenario highlights inadequate coverage of the areas of interest; due to an E_T being relatively low, the coverage area of each UAV is small, leading to a significant amount of sensors being out of reach of any UAV. In this scenario, we have $R_{es} = 0.16$; therefore, 83% of transmission attempts are unsuccessful. A high failure rate is expected due to poor coverage quality. On the other hand, we have the event detection rate $R_{ed} = 0.51$. Thus, we can say that at least this first case study points out that our proposed system may be able to handle event signaling in a very effective way.

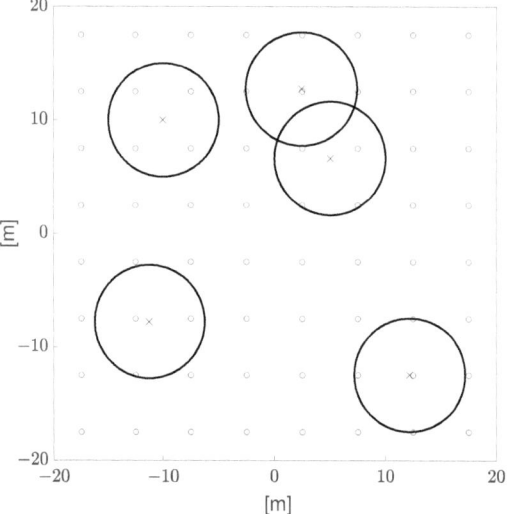

Figure 6. Arrangement of sensors and UAV at a given time within the area of interest for $E_T = 2.5$. The small blue circles are the sensors, the black × are the UAVs, and the big black circles represent the area that the UAV is covering.

We assess the performance of the system against three other conditions: $E_T = 5$, $E_T = 10$, and $E_T = 15$. Figure 7 presents the histogram of AoI for the four E_T conditions, and the main statistics of the AoI for the evaluated cases are summarized in Table 2. By observing the behavior of the histograms in Figure 7, one can say that AoI has a distribution similar to an exponential one. The positive impact of E_T on AoI is clear, i.e., AoI has its average, variance, and maximum value drastically reduced with the increase of E_T. Regardless of how strict the decision-maker requirement of AoI is, the proposed system can be expected to meet; as seen in Table 1, the system quickly converges to a minimum AoI—note that AoI $= 1$, in our case, means that the delay is the smallest physically possible—with the increase of E_T. Furthermore, AoI can be further reduced by decreasing W due to AoI values being scaled by W. We will see this effect in more detail later.

Table 2. Mean, variance, and maximum AoI for the evaluated cases.

	$E_T = 2.5$	$E_T = 5$	$E_T = 10$	$E_T = 15$
Mean	46.6615	14.7353	1.7323	1.0046
Variance	4476.7869	822.9471	22.7098	0.0703
Maximum	733	253	101	21

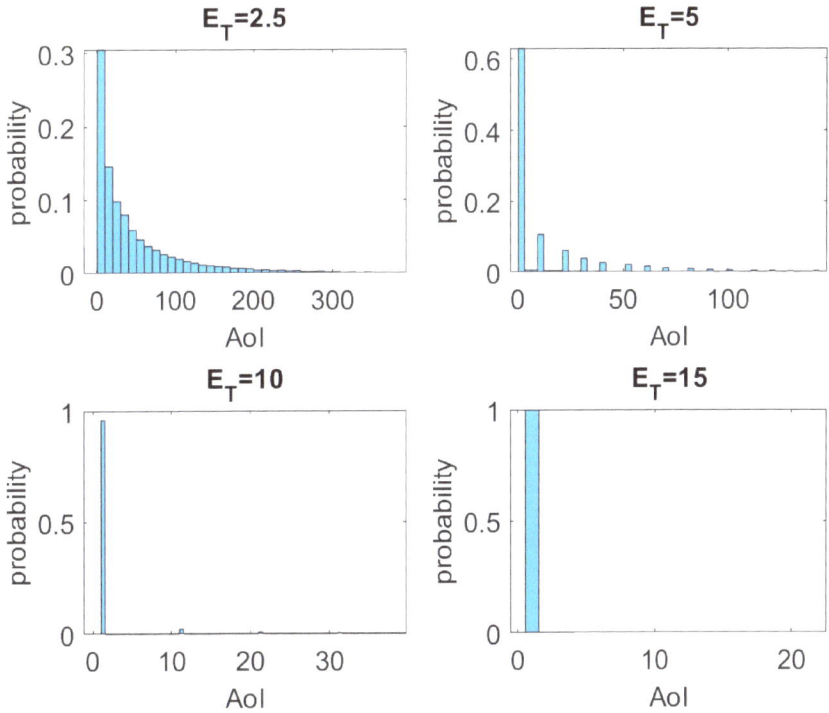

Figure 7. Histogram of AoI for $E_T = \{2.5, 5, 10, 15\}$.

Figure 8 shows the values of R_s, R_{es}, R_{ed}, and R_{eb} for the E_T varying as $\{2.5, 5, 10, 15\}$. R_s, R_{es}, and R_{ed} have an asymptotic growth with increasing E_T. On the other hand, although we are increasing the power used in transmission by the sensor nodes, R_{eb} is extremely low and even negligible for $E_T = \{10, 15\}$. A peculiar behavior of the system arises when $E_T = 15$; we have that R_s exceeds the others with $E_T = 15$. In addition, R_s reaches the value of 1 for $E_T = 15$. One can expect that, with increasing R_s, that is, increasing the efficiency with which messages reach a UAV, R_{ed} also increases. This effect is visible in Figure 8, besides being clear that $R_{es} = R_{ed}$ for $R_s = 1$. Such behavior can be explained by noting that every signaled event is received by the UAVs for $R_s = 1$.

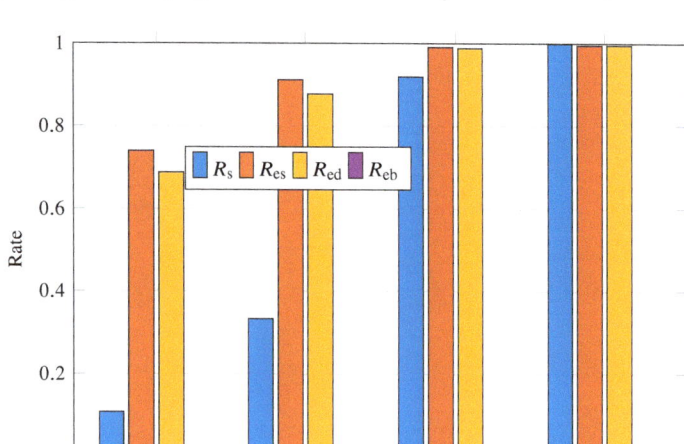

Figure 8. R_s, R_{es}, R_{ed}, and R_{eb} versus $E_T = \{2.5, 5, 10, 15\}$.

5.2. Varying W

Consider a system with $N_s = 64$, $N_u = 5$, an area of interest measuring 40 m by 40 m, $\rho_c = 0.8$, the transmission power of UAVs of $P_{UAV} = 100 C_{max}$, $E_T = 10$, $W = \{1, 100, 200, 500\}$ and $\lambda = 1/200$. The selected delays W somewhat cover four distinct behaviors of the system. $W = 1$ corresponds to a no-wait system, in which sensor nodes start retransmitting as soon as possible. As briefly discussed here, SFC enables extremely effective shared access of the medium, and thus our system in the no-wait condition ($W = 1$) does not deteriorate in performance due to collisions. On the other hand, it must be supported that the battery consumption is high for $W = 1$. An energetically more economical behavior than $W = 1$ is the case of $W = 100$. In this approach, the system waits half of the mean time between events to retransmit. $W = 200$ leads the sensor nodes to wait exactly the average time of the events, thus increasing the chance that a new event occurs before the retransmission. Finally, $W = 500$ is certainly the condition that offers the greatest energy savings at the cost of increasing non-signaling of events. This is an extreme condition, in which the probability of retransmitting with a new event occurring is high.

Figure 9 presents the histogram of AoI for $W = \{1, 100, 200, 500\}$, and the main statistics of the AoI for the evaluated cases are summarized in Table 3. As previously stated, AoI is indeed highly dependent on W. Note in Table 3 that AoI is capped at 10 for $W = 10$ and reaches 1386 for $W = 500$. On the other hand, the histograms in Figure 9 demonstrate that in general (around 60%) the AoI maintains its minimum value (AoI = 1). Furthermore, a certain behavior of periodic peaks can be noticed in AoI for $W = 100$ and $W = 200$. A possible explanation for such behavior is as follows. Let the event $\mathcal{E}$ be monitored by a sensor. Suppose $\mathcal{E}_1$ occurs at time t_1 and the sensor node transmits unsuccessfully. Then, a new transmission is performed successfully at $t_1 + W$ in order to signal $\mathcal{E}_1$. However, assume that a new event $\mathcal{E}_2$ (an identical event to $\mathcal{E}_1$ but at a different time) occurs in $t_1 < t_2 < t_1 + W$. In this hypothetical case, the AoI is calculated by AoI $= t_1 + W - t_2$. In contrast, AoI $= W$ if $t_2 > t_1 + W$. Therefore, we can say that, as long as the probability of $t_2 - t_1 > W$ is high enough, we have a predominance of AoI being a multiple of W. This characteristic can be noticed both in Figures 7 and 9. This explanation justifies the reason for obtaining AoI of non-multiple values of W.

Table 3. Mean, variance, and maximum AoI for the evaluated cases.

	$W = 1$	$W = 100$	$W = 200$	$W = 500$
Mean	1.7618	57.4874	68.7423	71.9253
Variance	3.2745	9122	14340	19514
Maximum	10	1334	1131	1386

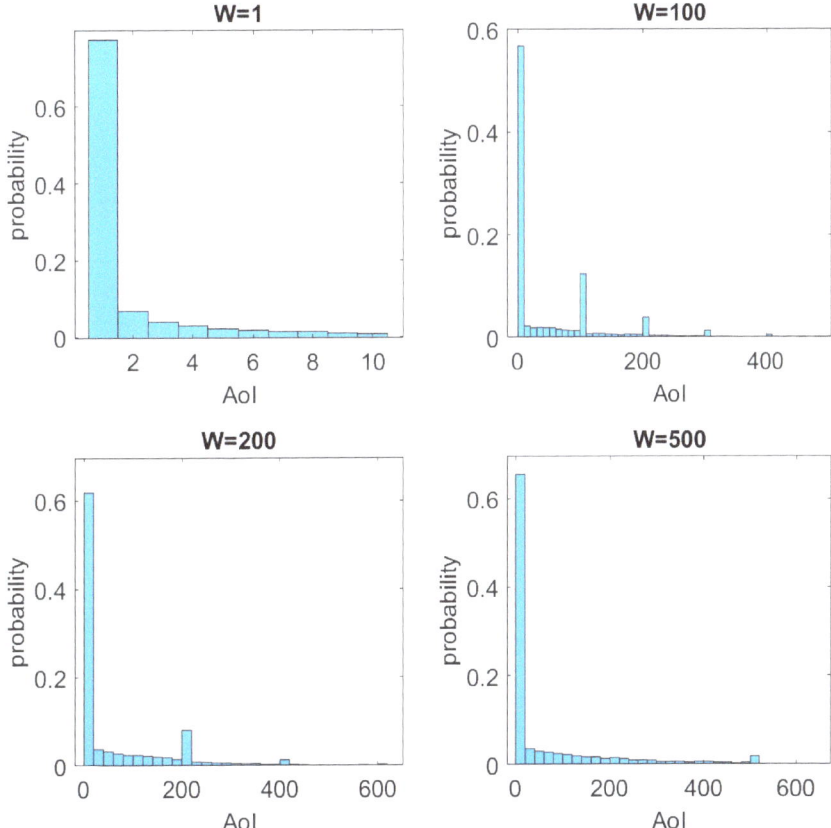

Figure 9. Histogram of AoI for $W = \{1, 100, 200, 500\}$.

Figure 10 shows the values of R_s, R_{es}, R_{ed}, and R_{eb} for the W varying as $\{1, 100, 200, 500\}$. The first point to note is the high energy consumption for retransmissions in short intervals, i.e., $W = 1$. Because battery recharge is insufficient for $W = 1$, the probability of battery depletion increases drastically. We can see a drop in R_{es} with increasing W; as W increases, the probability of a new event occurring during the wait increases. The effect of W on R_{ed} must be understood in conjunction with R_s as follows. Let $R_s(t)$ be the chance of successful transmission in time t. Thus, since the UAVs move randomly and continuously, one can expect the correlation between $R_s(t_1)$ and $R_s(t_2)$ to be inversely proportional to $|t_1 - t_2|$. The increasing curve that R_s presents for an increase in W corroborates this statement. In other words, as the waiting for retransmission increases, the chances of the sensor being out of coverage again are smaller. In addition, R_{ed} benefits from this R_s increase, but R_{ed} simultaneously deteriorates due to unsignaled events.

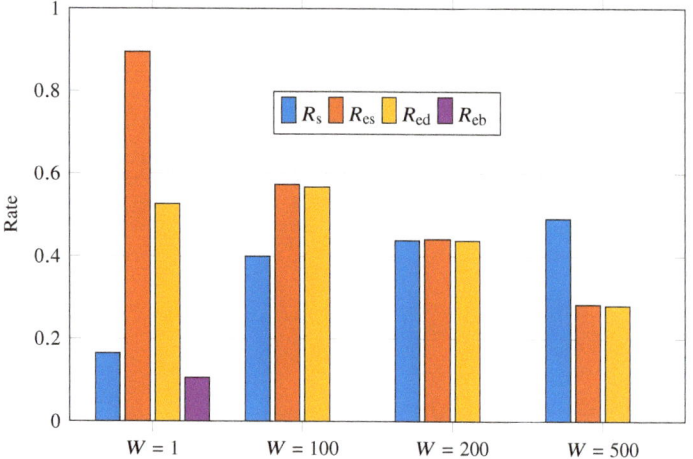

Figure 10. R_s, R_{es}, R_{ed}, and R_{eb} for $W = \{1, 100, 200, 500\}$.

5.3. Varying P_{UAV}

Consider a system with $N_s = 64$, $N_u = 5$, an area of interest measuring 40 m by 40 m, $\rho_c = 3$, transmission power of UAVs of $P_{UAV}/C_{max} = \{0.5, 2, 5, 10\}$, $E_T = 10$, $W = 20$ and $\lambda = 1/200$. Our objective is now to evaluate the behavior of the battery. In this way, we vary the power of the unmodulated carrier transmitted by the UAVs for the ACK signal. We started from a relatively weak signal, $P_{UAV}/C_{max} = 0.5$, to a significantly strong signal, $P_{UAV}/C_{max} = 10$.

Figure 11 shows the values of R_s, R_{es}, R_{ed}, and R_{eb} for the P_{UAV}/C_{max} varying as $\{0.5, 2, 5, 10\}$. In order to highlight the effect of R_{eb} on R_{es} and R_{ed}, we conveniently adjust the system parameters in such a way as to obtain $R_s = 1$. It is evident that, as R_{eb} decreases, both R_{es} and R_{ed} increase. This behavior is explained by the fact that, as the battery remains charged enough for transmission for a longer time, the greater the probability of signaling an event. Another notable feature in Figure 11 is the equality in the values of R_{es} and R_{ed}; it can be understood by keeping $R_s = 1$.

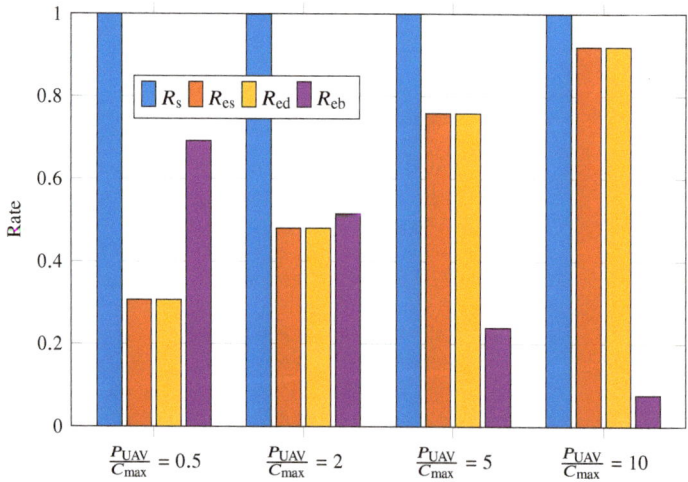

Figure 11. R_s, R_{es}, R_{ed}, and R_{eb} for $P_{UAV}/C_{max} = \{0.5, 2, 5, 10\}$.

Figure 12 presents the histogram of AoI for $P_{UAV}/C_{max} = \{0.5, 2, 5, 10\}$, and the main statistics of the battery charge for the evaluated cases are summarized in Table 4. The evaluated scenarios demonstrate that the battery charge pdf does not have a clear pattern. The pdf has, in a way, one characteristic for $P_{UAV}/C_{max} = 0.5$ and $P_{UAV}/C_{max} = 2$, and another distinct one for $P_{UAV}/C_{max} = 5$ and $P_{UAV}/C_{max} = 10$.

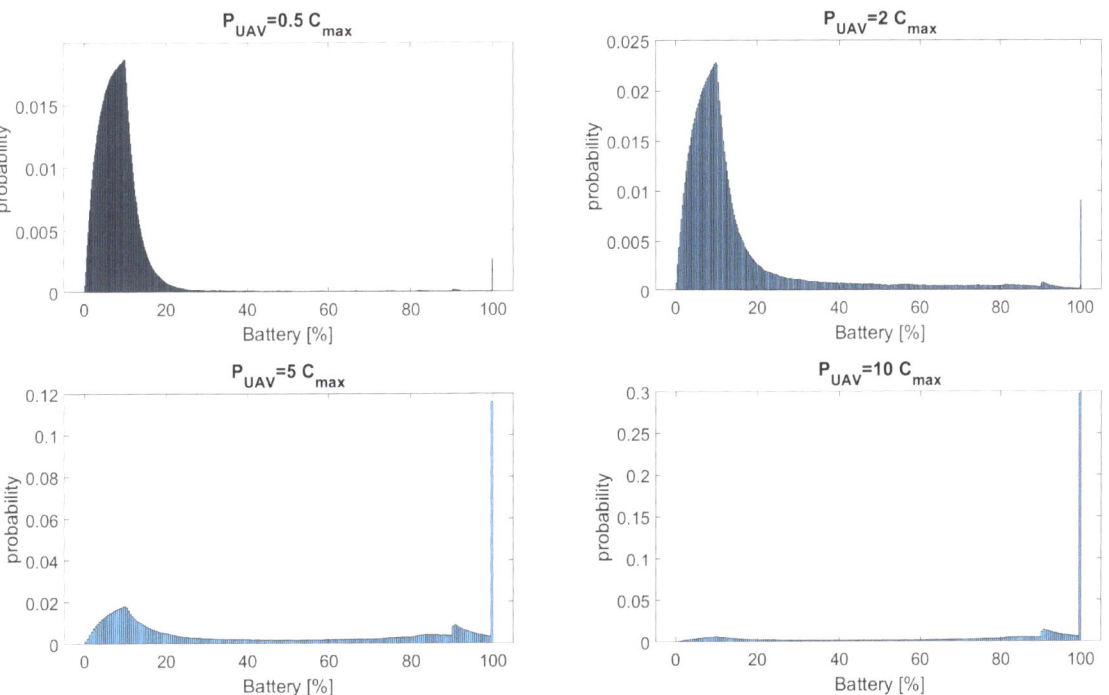

Figure 12. Histogram of battery charge for $P_{UAV}/C_{max} = \{0.5, 2, 5, 10\}$.

Table 4. Mean, variance, and R_{eb} of battery charge for the evaluated cases.

	$P_{UAV}/C_{max} = 0.5$	$P_{UAV}/C_{max} = 2$	$P_{UAV}/C_{max} = 5$	$P_{UAV}/C_{max} = 10$
Mean	10.0205	16.5881	44.2877	70.1149
Variance	140.7149	411	1340	1160
R_{eb}	0.68	0.52	0.24	0.08

6. Conclusions

This work addressed a remote wireless monitoring system composed of sensors and UAVs. The purpose of the system is to inform the decision-maker about events identified by the sensors. Furthermore, the sensors employ the RF-EH technique to recharge the batteries. We adopt an event transmission method called SFC that makes it possible to recharge the battery and transport information with the same energy packet. In this way, our proposal is disruptive because it can effectively and efficiently use the energy and data resources available. Simulations were performed to corroborate the advantages of our proposal in numerical terms. Through a representative case study, we discussed different aspects of the system and assessed the effect of the parameters in its performance including different metrics like error rates and AoI. Our numerical results indicated that the proposed system has the potential to be adjusted in order to meet the requirement measured in terms of AoI, which is determined by the specifications given by the end application in consideration. As a future research path, we plan to assess this framework in different realistic scenarios such as robots in an industrial plant and remote monitoring of power grids.

Author Contributions: Conceptualization, all; methodology, all; software, P.E.G.S.; validation, P.E.G.S.; formal analysis, all; investigation, all; writing—original draft preparation, all; writing—review and editing, all; visualization, P.E.G.S.; supervision, N.M., P.H.J.N., R.A.A.d.S.; funding acquisition, N.M., P.H.J.N., R.A.A.d.S. All authors have read and agreed to the published version of the manuscript.

Funding: This paper is partly supported by (1) Academy of Finland via: (a) FIREMAN consortium n.326270 as part of CHIST-ERA grant CHIST-ERA-17-BDSI-003, (b) EnergyNet Fellowship n.321265/n.328869/n.352654, and (c) X-SDEN project n.349965; (2) Jane and Aatos Erkko Foundation via STREAM project; (3) by CNPq (Grant 311470/2021-1); (4) by São Paulo Research Foundation (FAPESP) (Grant No. 2021/06946-0); and (5) by RNP, with resources from MCTIC, Grant No. 01245.010604/2020-14, under the Brazil 6G project of the Radiocommunication Reference Center (*Centro de Referência em Radiocomunicações*—CRR) of the National Institute of Telecommunications (*Instituto Nacional de Telecomunicações*—Inatel), Brazil; (5) It was also supported by the Science Foundation Ireland under grant 13/RC/2077_P2 (CONNECT).

Institutional Review Board Statement: Not applicable.

Informed Consent Statement: Not applicable.

Data Availability Statement: The codes used to generate the results presented in this paper are available upon request.

Conflicts of Interest: The authors declare no conflict of interest.

References

1. Ponnimbaduge Perera, T.D.; Jayakody, D.N.K.; Sharma, S.K.; Chatzinotas, S.; Li, J. Simultaneous Wireless Information and Power Transfer (SWIPT): Recent Advances and Future Challenges. *IEEE Commun. Surv. Tutorials* **2018**, *20*, 264–302. [CrossRef]
2. Sharma, P.; Singh, A.K. A survey on RF energy harvesting techniques for lifetime enhancement of wireless sensor networks. *Sustain. Comput. Inform. Syst.* **2023**, *37*, 100836. [CrossRef]
3. Ibrahim, H.H.; Singh, M.J.; Al-Bawri, S.S.; Ibrahim, S.K.; Islam, M.T.; Alzamil, A.; Islam, M.S. Radio Frequency Energy Harvesting Technologies: A Comprehensive Review on Designing, Methodologies, and Potential Applications. *Sensors* **2022**, *22*, 4144. [CrossRef]
4. Mou, X.; Sun, H. Wireless Power Transfer: Survey and Roadmap. In Proceedings of the 2015 IEEE 81st Vehicular Technology Conference (VTC Spring), Glasgow, Scotland, 11–14 May 2015; pp. 1–5. [CrossRef]
5. Guo, J.; Zhang, H.; Zhu, X. Theoretical Analysis of RF-DC Conversion Efficiency for Class-F Rectifiers. *IEEE Trans. Microw. Theory Techn.* **2014**, *62*, 977–985. [CrossRef]
6. Kim, J.; Kwon, I. Design of a High-Efficiency DC-DC Boost Converter for RF Energy Harvesting IoT Sensors. *Sensors* **2022**, *22*, 10007. [CrossRef]
7. de Castro Tomé, M.; Gutierrez-Rojas, D.; Nardelli, P.H.; Kalalas, C.; Da Silva, L.C.P.; Pouttu, A. Event-driven Data Acquisition for Electricity Metering: A Tutorial. *IEEE Sensors J.* **2022**, *22*, 1–9. [CrossRef]
8. Luo, K.; Jiao, Y. Automatic fault detection of sensors in leather cutting control system under GWO-SVM algorithm. *PLoS ONE* **2021**, *16*, e0248515. [CrossRef] [PubMed]
9. Silva, P.E.G.; de Souza, R.A.A.; da Costa, D.B.; Moualeu, J.M.; Yacoub, M.D. Error Probability of M-Phase Signaling With Phase Noise Over Fading Channels. *IEEE Trans. Veh. Technol.* **2020**, *69*, 6766–6770. [CrossRef]
10. Silva, P.E.G.; de Souza, R.A.A.; Nardelli, P.J.; Moualeu, J.M.M.; da Costa, D.B. Performance of MPSK modulation with imperfect phase-recovery under severe fading conditions. *Electron. Lett.* **2022**, *58*, 333–335. [CrossRef]
11. Gautam, S.; Solanki, S.; Sharma, S.K.; Chatzinotas, S.; Ottersten, B. Boosting Quantum Battery-Based IoT Gadgets via RF-Enabled Energy Harvesting. *Sensors* **2022**, *22*, 5385. [CrossRef]
12. Gória Silva, P.E.; Dester, P.S.; Siljak, H.; Marchetti, N.; Nardelli, P.H.; de Souza, R.A. Semantic-functional Communications for Multiuser Event Transmissions via Random Maps. *arXiv* **2022**, arXiv:2204.
13. Chinipardaz, M.; Amraee, S. Study on IoT networks with the combined use of wireless power transmission and solar energy harvesting. *Sadhana* **2022**, *47*, 86. [CrossRef]
14. Alves, H.; Lopez, O.A. *Wireless RF Energy Transfer in the Massive IoT Era: Towards Sustainable Zero-Energy Networks*; Wiley: Hoboken, NJ, USA, 2021.
15. Nardelli, P.H. *Cyber-Physical Systems: Theory, Methodology, and Applications*; John Wiley & Sons: Hoboken, NJ, USA, 2022.
16. Knecht, O.; Bosshard, R.; Kolar, J.W. High-Efficiency Transcutaneous Energy Transfer for Implantable Mechanical Heart Support Systems. *IEEE Trans. Power Electron.* **2015**, *30*, 6221–6236. [CrossRef]

17. Mou, X.; Gladwin, D.; Jiang, J.; Li, K.; Yang, Z. Near-Field Wireless Power Transfer Technology for Unmanned Aerial Vehicles: A Systematical Review. *IEEE J. Emerg. Sel. Top. Ind. Electron.* **2023**, *4*, 147–158. [CrossRef]
18. Hamam, R.E.; Karalis, A.; Joannopoulos, J.; Soljačić, M. Efficient weakly-radiative wireless energy transfer: An EIT-like approach. *Ann. Phys.* **2009**, *324*, 1783–1795. [CrossRef]
19. Sample, A.P.; Meyer, D.T.; Smith, J.R. Analysis, Experimental Results, and Range Adaptation of Magnetically Coupled Resonators for Wireless Power Transfer. *IEEE Trans. Ind. Electron.* **2011**, *58*, 544–554. [CrossRef]
20. Tamura, M.; Murai, K.; Naka, Y. Capacitive Coupler Utilizing Electric Double Layer for Wireless Power Transfer Under Seawater. In Proceedings of the 2019 IEEE MTT-S International Microwave Symposium (IMS), Boston, MA, USA, 2–7 June 2019; pp. 1415–1418. [CrossRef]
21. Hui, S.Y.R.; Zhong, W.; Lee, C.K. A Critical Review of Recent Progress in Mid-Range Wireless Power Transfer. *IEEE Trans. Power Electron.* **2014**, *29*, 4500–4511. [CrossRef]
22. Wang, Y.; Wei, G.; Dong, S.; Dong, Y.; Li, X. Design of a microwave power transmission demonstration system for space solar power station. *Int. J. Microw.-Comput.-Aided Eng.* **2022**, *32*, e23523. [CrossRef]
23. Zhang, R.; Ho, C.K. MIMO Broadcasting for Simultaneous Wireless Information and Power Transfer. *IEEE Trans. Wirel. Commun.* **2013**, *12*, 1989–2001. [CrossRef]
24. Zhou, X.; Zhang, R.; Ho, C.K. Wireless Information and Power Transfer in Multiuser OFDM Systems. *IEEE Trans. Wirel. Commun.* **2014**, *13*, 2282–2294. [CrossRef]
25. Ng, D.W.K.; Lo, E.S.; Schober, R. Wireless Information and Power Transfer: Energy Efficiency Optimization in OFDMA Systems. *IEEE Trans. Wirel. Commun.* **2013**, *12*, 6352–6370. [CrossRef]
26. Zhou, X. Training-Based SWIPT: Optimal Power Splitting at the Receiver. *IEEE Trans. Veh. Technol.* **2015**, *64*, 4377–4382. [CrossRef]
27. Bai, Y.; Liu, Q.; Chen, R.; Zhang, Q.; Wang, W. Long-Range Optical Wireless Information and Power Transfer. *IEEE Internet Things J.* **2023**, *10*, 1617–1627. [CrossRef]
28. Valenta, C.R.; Durgin, G.D. Harvesting Wireless Power: Survey of Energy-Harvester Conversion Efficiency in Far-Field, Wireless Power Transfer Systems. *IEEE Microw. Mag.* **2014**, *15*, 108–120. [CrossRef]
29. Liang, Z.; Yuan, J. Modelling and Prediction of Mobile Service Channel Power Density for RF Energy Harvesting. *IEEE Wirel. Commun. Lett.* **2020**, *9*, 741–744. [CrossRef]
30. Sait, S.M.; Ahmed, S.F.; Rafiq, M.R. Experimental study on broadband radiofrequency electromagnetic radiations near cellular base stations: A novel perspective of public health. *J. Therm. Anal. Calorim.* **2020**, *143*, 1935–1942. [CrossRef]
31. Goldsmith, A. *Wireless Communications*; Cambridge University Press: Cambridge, UK, 2005. [CrossRef]
32. Yu, H.; Chen, T.; Hao, F. A New Event-Triggered Control Scheme for Stochastic Systems. *IEEE Trans. Autom. Control.* **2022**. [CrossRef]
33. García-Orellana, C.J.; Macías-Macías, M.; González-Velasco, H.M.; García-Manso, A.; Gallardo-Caballero, R. Low-Power and Low-Cost Environmental IoT Electronic Nose Using Initial Action Period Measurements. *Sensors* **2019**, *19*, 3183. [CrossRef]

Disclaimer/Publisher's Note: The statements, opinions and data contained in all publications are solely those of the individual author(s) and contributor(s) and not of MDPI and/or the editor(s). MDPI and/or the editor(s) disclaim responsibility for any injury to people or property resulting from any ideas, methods, instructions or products referred to in the content.

Article

Outage Analysis of Unmanned-Aerial-Vehicle-Assisted Simultaneous Wireless Information and Power Transfer System for Industrial Emergency Applications

Aleksandra Cvetković [1,*], Vesna Blagojević [2], Jelena Anastasov [3], Nenad T. Pavlović [1] and Miloš Milošević [1]

1. Faculty of Mechanical Engineering, University of Niš, 18000 Niš, Serbia; nenad.t.pavlovic@masfak.ni.ac.rs (N.T.P.); milos.milosevic@masfak.ni.ac.rs (M.M.)
2. School of Electrical Engineering, University of Belgrade, 11000 Belgrade, Serbia; vesna.golubovic@etf.rs
3. Faculty of Electronic Engineering, University of Niš, 18000 Niš, Serbia; jelena.anastasov@elfak.ni.ac.rs
* Correspondence: aleksandra.cvetkovic@masfak.ni.ac.rs

Abstract: In the scenario of a natural or human-induced disaster, traditional communication infrastructure is often disrupted or even completely unavailable, making the employment of emergency wireless networks highly important. In this paper, we consider an industrial Supervisory Control and Data Acquisition (SCADA) system assisted by an unmanned aerial vehicle (UAV) that restores connectivity from the master terminal unit (MTU) to the remote terminal unit (RTU). The UAV also provides power supply to the ground RTU, which transmits the signal to the end-user terminal (UT) using the harvested RF energy. The MTU-UAV and UAV-RTU channels are modeled through Nakagami-m fading, while the channel between the RTU and the UT is subject to Fisher–Snedecor composite fading. According to the channels' characterization, the expression for evaluating the overall probability of outage events is derived. The impact of the UAV's relative position to other terminals and the amount of harvested energy on the outage performance is investigated. In addition, the results obtained based on an independent simulation method are also provided to confirm the validity of the derived analytical results. The provided analysis shows that the position of the UAV that leads to the optimal outage system performance is highly dependent on the MTU's output power.

Keywords: energy harvesting; industrial emergency applications; outage performance; simultaneous wireless information and power transfer; unmanned aerial vehicle

1. Introduction

The advantages and benefits of utilizing unmanned aerial vehicles (UAVs) have attracted significant attention in both the academic and industrial spheres. UAVs have been extensively utilized across various sectors, including surveillance, military operations, health services, etc. [1]. Furthermore, it has been shown that UAVs are applicable for infrastructure monitoring without imposing danger to humans. Namely, in [2], a UAV-assisted setup equipped with cameras and sensors was utilized for railway inspection and monitoring. Also, UAVs have become indispensable in precision agriculture for essential functions such as crop monitoring, plant health assessment, pest and disease detection, and optimization of irrigation and fertilizer applications [3]. The integration of UAVs in industrial systems has great potential due to their possible applications in maintenance, process monitoring and management, as well as manufacture automatization [4]. Due to their rapid response and flexibility, UAVs can provide stable communication or establish temporary communication links in disaster-stricken areas, especially when existing infrastructure has been damaged or destroyed [5].

The performance of various UAV-assisted systems has been analyzed in the scientific literature [5–8], with the aim of providing satisfactory solutions for communication needs in emergency-saving scenarios. The issues related to optimization of the UAV position and coping with network destruction in a natural disaster, with and without ground base stations, were examined in [5]. The novel cluster-based mechanism for sensor networks aided by UAVs, which enable data collection over shorter propagation paths and thus improve system performance, was proposed in [6]. Authors in [7] focused on a unified framework for a UAV-assisted emergency multihop device-to-device (D2D) network in disaster areas. The presented results showed a performance improvement in terms of the throughput and outage probability achieved by implementing the UAV-assisted wireless coverage approach. The boundaries of UAV technology applications in industrial disasters and other important directives for further research in this field were identified in [8]. Recently, UAV assistance has also been promoted for general Supervisory Control and Data Acquisition (SCADA) architecture in [9,10].

In recent years, the Internet of Things (IoT) systems and applications have rapidly become ubiquitous and the growth in the number of connected devices has brought higher reliability requests, increased data rates, and energy efficiency. Furthermore, the large number of devices in current networks has a significant influence on the changes in the traditional approaches regarding the powering of transmission nodes. Namely, although battery usage is appealing in some scenarios, due to the large number of nodes, this way of providing energy turns out to be impractical. Wireless power supply represents the alternative approach, where the energy in the environment can be harvested from existing sources such as solar, wind energy, and radio-frequency (RF) energy [11,12]. Within the energy harvesting (EH) approach, the harvesting of RF energy has a special feature, i.e., it can be used for simultaneous information and power transfer (SWIPT) [13,14]. Additionally, RF energy can be purposely transferred to the desired node [15,16]. As SWIPT exploits the principle of energy harvesting from RF signals, that carry both information and energy components, it is suitable for implementation in networks with energy-constrained devices, such as low-power communication devices and sensors, allowing them to recharge their batteries while simultaneously performing the communication tasks. In the case when power is intentionally sent to one of the nodes in the system, additional relaying nodes can be incorporated by applying the time-switching (TS) or power-splitting (PS) protocol [17]. In the TS-based concept, the energy harvester timely switches between the EH and the information transmission phases. The device uses the harvested energy for data transmission. In the PS-based SWIPT concept, the received energy is divided into two portions, where the first one is used for energy harvesting, while the remaining is dedicated for information transmission. This approach allows continuous energy harvesting while still maintaining the system ability to transmit data. When the power reduction in the PS protocol is applied, the device can allocate a larger portion of harvested energy for data transmission, therefore mitigating the impact on the data rate. The decrease in the information transmission time in the TS-based protocol leads to the conservation of energy and the decrease in data rate, while in the PS-based protocol, the transmission time remains constant as the power allocated for transmission is subject to adjustment. The optimal power splitting relay-based cooperative selection scheme was analyzed in [18] for the communication in IoT systems.

In the recently published literature, it is considered that UAVs are capable of fulfilling multiple purposes in the areas of IoT, industry, and other wireless communications. Although they can primarily be employed for information gathering [19,20] and information transfer [21], a very important class of functionalities also encompasses enabling the wireless power supply using RF energy [22,23], as UAVs can provide the power supply in areas that are not accessible when conventional approaches or unmovable nodes are used. Consequently, there are a number of published scientific papers that investigate their system performance in the case when a UAV acts as a harvester or supplier [24–28]. In the case when UAV devices have limited energy and therefore constrained duration of the operating time in the air, they can harvest RF energy for battery charging [24]. Moreover, a signal from a UAV can be used as an RF energy source for supplying energy-constrained devices on the ground [25–28]. The optimization of a UAV's trajectory or the minimization of the overall energy consumption in various IoT system scenarios was investigated in [25,28]. The outage and error performance of an IoT system with multiple UAV relays using TS and PS energy harvesting relaying protocols were defined in [24]. In addition, the EH method has been extensively investigated in industrial IoT or in networks for industry automation. In [29], the EH was utilized for information and power transfer to the server machine, which forwards information from the data center to multiple destination machines using a reconfigurable intelligent surface (RIS). In addition, in [30], the RIS was also used to maintain communication among satellite and multiple users in the proposed relay network. The extension of conventional communication to the non-terrestrial systems with/without UAV including communication with satellites in order to fulfil the power constraints of certain nodes was provided in [31–33].

Motivation and Contribution

In modern industry concepts, it is necessary to provide automatized control functions for the end users such as machines, IoT devices, actuators, and sensors. In most scenarios, information is transferred among communication nodes over wireless channels. The outage performance analysis of dual-hop relaying systems with ground relays used for both information signal retransmission and RF energy harvesting can be found in [34–38], for Nakagami-*m* and Fisher–Snedecor fading environments. In our work, we extend the number of communication nodes in accordance with the specific application for industrial purposes. The scenario under consideration corresponds to the industrial SCADA architecture, which consists of the master terminal unit (MTU), the remote terminal unit (RTU), and the user terminal (UT). In a natural or man-made disaster situation, the direct communication between the MTU and the RTU can be corrupted and the powering of the RTU can also be compromised due to the loss of conventional power sources or limited battery life. Since the requirements for the transmission of critical information to the UT node should be met, in this paper, we incorporate a UAV as an additional relay and supplier for the RTU, for such an emergency scenario. In the survey paper about UAV applications [1], a similar system setup was proposed, omitting performance metrics analysis. In [26,27], unified frameworks for charging strategies of a UAV were proposed, which enables SWIPT for IoT nodes or cluster heads with an aim to enhance their further functioning in the networks. With the motivation based on the previously mentioned works, we relate our research to the performance analysis of a SCADA system aided by a UAV, which relays information from the MTU to RTU and enables RTU retransmission to the UT, by supplying the RTU with energy. In the considered analysis, the UAV and the RTU utilize the DF protocol.

In this paper, we provide the system performance analysis and derive the expressions of the probability of outage and the system throughput. We assume that the first link and the second link are Nakagami-*m* fading channels, which represent channel modeling widely utilized in the literature for communication links between the ground nodes and UAV. The third link, RTU-UT, relates to D2D communication over short distances and thus is modeled as a Fisher–Snedecor fading channel. To the best of the authors' knowledge, the outage performance analysis of Nakagami-*m*/Nakagami-*m*/Fisher–Snedecor relaying systems with the UAV that is employed both as a relay and power supplier for the RTU node has not been previously reported.

This paper's objectives and main contributions are as follows:

(1) We investigate an industrial SCADA system, assisted by a UAV in a hazardous disaster scenario when the RTU is disconnected from the MTU and also left without power supply.

(2) The novel outage probability and throughput expressions are derived for the considered multihop Nakagami-*m*/Nakagami-*m*/Fisher–Snedecor relaying system, with a UAV employed as a DF relay and energy supplier for RTU.

(3) The outage performance and end-to-end system throughput analysis is carried out, in detail, aiming to adjust the UAV's relative position above the MTU-RTU link and the amount of supplied power in relation to other essential parameters.

(4) The important insights in the interplay of environmental parameters are provided, such as the amount of MTU output power or harvesting power as well as the UAV position in relation to other communication units, with the aim of enhancing system reliability.

(5) The numerical results based on analytical expressions are provided and compared with simulation results based on developed Monte Carlo simulation model, in order to demonstrate the validity of the derived analytical expressions. Based on the obtained results, additional conclusions concerning the impact of the system and channel model on outage performances are derived.

In brief, this paper is structured as follows. In Section 2, we present the statistical characterization of communication links and describe the RF energy harvesting protocol employed at the RTU. In Section 3, we derive the analytical expression for the outage system performance, whereby part of the mathematical derivation of the performance is provided in detail in the Appendix A. Numerical and simulation results are presented and discussed in Section 4. The conclusion is given in Section 5.

2. System and Channel Model

We consider an industrial-service communication system illustrated in Figure 1, as a solution for the natural disaster scenarios. The traditional communication infrastructure is often disrupted or even completely unavailable due to natural or human disasters (such as earthquakes, floods, bushfires, and tornadoes), demanding the employment of emergency wireless networks. We consider a general system model that can be applied as a SCADA system, in the scenario when it is disabled due to destruction or inefficiency of the existing infrastructure. SCADA, as an important part of industrial systems, consists of an MTU, an RTU, and an end user terminal [9,10], where the RTUs are used at remote destinations and are usually placed in outdoor inaccessible environments. SCADA architecture is utilized for monitoring and management of industrial processes. Therefore, the MTU can send information to the RTU to provide an emergency shutdown of the process, to prevent hazardous situations by starting or stopping pumps or adjusting the speed of pumps, and to regulate the flow of fluids or gases by opening/closing valves [9]. Namely, the industrial control functions are performed by using the communication link, which consists of the MTU-RTU and the RTU-UT hops, but according to the predicted scenario, the communication between the master and the remote unit is disrupted. In addition, the RTU is an energy-limited device left without conventional power supply due to disaster conditions. In such a scenario, the UAV is employed as a relay for information transfer between the MTU and the RTU. It also serves as an energy supplier for the RTU, thus enabling data transmission from the MTU to the end user.

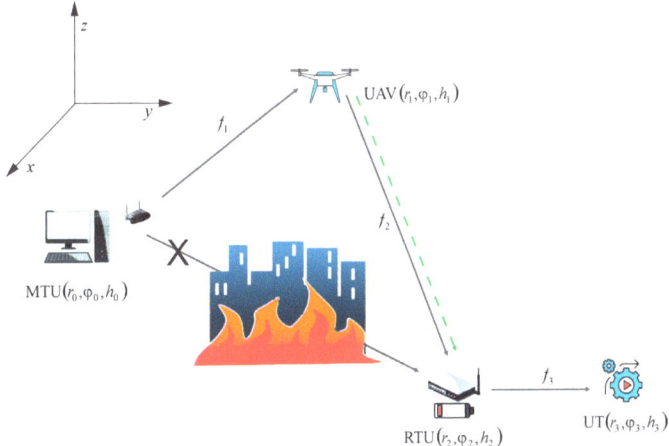

Figure 1. UAV-assisted industrial system for emergency applications.

2.1. Channel Model

According to a wireless model of propagation [39], the received signal is a sum of multipath delayed components mostly caused by reflection, diffraction, and scattering propagation mechanisms. Consequently, the signal level at the receiver is a random variable and should be statistically determined by the corresponding probability density function (PDF) in order to evaluate the important system performance metrics.

We adopt the assumption that the multipath propagation over communication links between the UAV and the ground nodes is modeled using the Nakagami-m distribution, regarding that it is convenient for describing both line-of-sight and non-line-of-sight channel conditions. The received signal envelopes in the MTU-UAV, the UAV-RTU, and the RTU-UT links are denoted by f_1, f_2, and f_3, respectively. Therefore, the channel power gains of the MTU-UAV and the UAV-RTU links are denoted by $g_1 = |f_1|^2$ and $g_2 = |f_2|^2$, respectively, and can be statistically characterized by the following Gamma PDFs [39]:

$$p_{g_i}(\gamma) = \frac{1}{\Gamma(m_i)} \left(\frac{m_i}{\overline{\gamma}_i}\right)^{m_i} \gamma^{m_i - 1} e^{-\frac{m_i}{\overline{\gamma}_i}\gamma}, \; i = 1, 2, \quad (1)$$

where $\Gamma(\cdot)$ denotes the Gamma function ([40], (8.310)), m_i denotes the multipath fading parameter, which depends on the signal propagation environment; and $\overline{\gamma}_i = E[\gamma_i], i = 1, 2$.

The communication between the RTU and the UT is typical D2D communication, characterized by relatively small link distances. Accordingly, the RTU-UT link can be described by the Fisher–Snedecor F composite channel, which is proposed as the most appropriate fit model to empirical data of the D2D wireless communication [41,42]. The channel power gain of the RTU-UT link can be denoted by $g_3 = |f_3|^2$ and the corresponding PDF is then formulated as [41]

$$p_{g_3}(x) = \frac{\left(\frac{m_3}{m_{s3}\overline{\gamma}_3}\right)^{m_3} x^{m_3 - 1}}{B(m_3, m_{s3}) \left(\frac{m_3 x}{m_{s3}\overline{\gamma}_3} + 1\right)^{m_3 + m_{s3}}}, \quad (2)$$

where $B(\cdot, \cdot)$ denotes the Beta function ([40], (8.380)), $\overline{\gamma}_3 = E[\gamma_3]$ is the average power gain over the RTU-UT link; while m_3 and m_{s3} denote the multipath fading and shadowing shaping parameters, respectively.

2.2. System Model

We assume that the UAV is utilized to establish the communication from the MTU to the RTU in the case when the direct MTU-RTU link connection is not achievable, and to enable the powering of the RTU. For the considered industrial application of the UAV, its communication requirements must comply with the existing regulative framework provided in [43]. The UAV and the RTU employ the DF relaying scheme to provide transmission of the signal to the end user terminal. Also, the UAV enables energy for the RTU following the PS protocol. We consider the SWIPT protocol based on PS, as it leads to a smaller data rate loss when the same parts of power and time are applied for harvesting in PS and TS protocols, respectively [44].

In the intended communication, the entire time frame period, T, is divided into three equal time slots, as shown in Figure 2. The first time slot is used for the MTU-UAV information transmission, the third slot is used for the RTU-UT information transmission, while the second time slot is determined for both the UAV-RTU information transmission and the energy harvesting. During the second time slot, the part of the total power P of the received signal, $\theta \times P$, is utilized for the harvesting, while the rest of the power $(1 - \theta) \times P$ is dedicated to the UAV-RTU information transmission. The parameter θ ($0 \leq \theta \leq 1$) denotes the power-splitting factor.

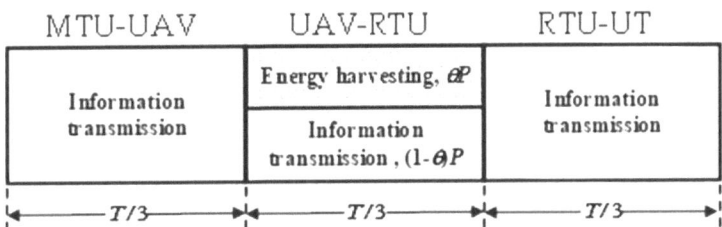

Figure 2. The frame structure during the period T.

Let us assume that the MTU transmits signal x_1 with the power P_S. The received signal at the UAV is then given by

$$y_1 = \sqrt{\frac{P_S}{d_1^{\delta_1}}} f_1 x_1 + n_1, \tag{3}$$

where δ_1 is the path loss exponent of the first link, MTU-UAV; n_1 denotes the additive white Gaussian noise (AWGN) at the UAV; and d_1 is the distance between the MTU and the UAV. Further, the received signal-to-noise ratio (SNR) at the UAV has the following form:

$$\gamma_1 = \frac{P_S |f_1|^2}{d_1^{\delta_1} \sigma_1^2} = \frac{P_S}{d_1^{\delta_1} \sigma_1^2} g_1 \tag{4}$$

where σ_1^2 denotes the variance of the AWGN.

The UAV decodes and re-encodes the received signal and transmits signal x_2 with power P_{UAV}. Thus, the received signal at the RTU, based on the PS protocol, can be expressed as

$$y_2 = \sqrt{\frac{(1-\theta) P_{UAV}}{d_2^{\delta_2}}} f_2 x_2 + n_2, \tag{5}$$

where d_2 is the distance between the UAV and the RTU, δ_2 is the corresponding path loss exponent, and n_2 is the AWGN at the RTU. Accordingly, the received SNR at the RTU can be defined as

$$\gamma_2 = \frac{(1-\theta) P_{UAV}}{d_2^{\delta_2} \sigma_2^2} g_2, \tag{6}$$

where σ_2^2 is the variance of the AWGN.

Relying on the PS protocol, the remaining part of the available power is intended for the energy supply of the RTU battery. The total harvested energy by RTU (at the end of the second time slot) can be calculated as

$$E_H = \eta \frac{\theta P_{UAV} g_2}{d_2^{\delta_2}} \frac{T}{3}, \tag{7}$$

where $0 < \eta < 1$ is the energy conversion efficiency.

As the total amount of harvested energy is used for the further transmission, the corresponding transmit power of the RTU can be determined as

$$P_R = \frac{E_H}{T/3} = \frac{\eta \theta P_{UAV}}{d_2^{\delta_2}} g_2, \tag{8}$$

and the received signal at the end UT can be formulated as

$$y_3 = \sqrt{\frac{P_R}{d_3^{\delta_3}}} f_3 x_3 + n_3. \tag{9}$$

In addition, based on (8) and (9), the received SNR at the end UT is given as

$$\gamma_3 = \frac{\eta\theta P_{UAV}}{d_2^{\delta_2} d_3^{\delta_3} \sigma_3^2} g_2 g_3. \quad (10)$$

In Equations (9) and (10), d_3 denotes the distance between the RTU and the end UT, δ_3 is the corresponding path loss exponent, while n_3 is AWGN with the variance σ_3^2 at the end user terminal.

3. Outage Performance

Statistically, the outage probability is defined as the probability that the instantaneous SNR falls below predefined threshold, γ_{th}. The outage threshold, γ_{th}, represents the SNR value that is a boundary between correct system functioning and the system outage. It depends on the specific application and system parameters, such as the modulation format, implementation of the receivers, and bit rates.

The system under consideration will be in outage (or will not function correctly) if any of the three communication links is in outage. Thus, the outage performance of the overall system can be calculated as

$$P_{out} = \Pr\{\gamma_1 \leq \gamma_{th}\} + \Pr\{\gamma_2 \leq \gamma_{th}\}\Pr\{\gamma_1 > \gamma_{th}\} \\ + \Pr\{\gamma_1 > \gamma_{th}\}\Pr\{\gamma_3 \leq \gamma_{th}, \gamma_2 > \gamma_{th}\}, \quad (11)$$

where $\Pr\{\cdot\}$ denotes the probability.

As the variable γ_1 is independent of the random variables γ_2 and γ_3, the probabilities $\Pr\{\gamma_1 \leq \gamma_{th}\}$ and $\Pr\{\gamma_1 > \gamma_{th}\}$ in (11) can be determined as

$$\Pr\{\gamma_1 \leq \gamma_{th}\} = \Pr\left\{g_1 \leq \frac{d_1^{\delta_1}\sigma_1^2}{P_S}\gamma_{th}\right\} = F_{g_1}\left(\frac{d_1^{\delta_1}\sigma_1^2}{P_S}\gamma_{th}\right), \quad (12)$$

and

$$\Pr\{\gamma_1 > \gamma_{th}\} = 1 - \Pr\{\gamma_1 \leq \gamma_{th}\} = 1 - F_{g_1}\left(\frac{d_1^{\delta_1}\sigma_1^2}{P_S}\gamma_{th}\right), \quad (13)$$

where $F_{g_1}(.)$ is the cumulative distribution function (CDF) of the Gamma variable. Relying on the PDF expression in (1), the CDF can be expressed as [39]

$$F_{g_i}(\gamma) = 1 - \frac{\Gamma\left(m_i, \frac{m_i}{\overline{\gamma}_i}\gamma\right)}{\Gamma(m_i)}, \; i = 1, 2. \quad (14)$$

Further, recalling (6), the probability $\Pr\{\gamma_2 \leq \gamma_{th}\}$, can be defined as

$$\Pr\{\gamma_2 \leq \gamma_{th}\} = \Pr\left\{g_2 \leq \frac{d_2^{\delta_2}\sigma_2^2\gamma_{th}}{(1-\theta)P_{UAV}}\right\} = F_{g_2}\left(\frac{d_2^{\delta_2}\sigma_2^2\gamma_{th}}{(1-\theta)P_{UAV}}\right). \quad (15)$$

By introducing $a = \frac{(1-\theta)P_{UAV}}{d_2^{\delta_2}\sigma_2^2}$, the instantaneous SNR in (6) becomes $\gamma_2 = ag_2$, and the PDF of γ_2 is defined, following the relation $p_{\gamma_2}(\gamma) = p_{g_2}(\gamma/a)/a$ [45], as

$$p_{\gamma_2}(\gamma) = \frac{1}{\Gamma(m_2)}\left(\frac{m_2}{a\overline{\gamma}_2}\right)^{m_2}\gamma^{m_2-1}e^{-\frac{m_2}{a\overline{\gamma}_2}\gamma}. \quad (16)$$

Finally, by making change $b = \frac{\eta\theta\sigma_2^2}{(1-\theta)d_3^{\delta_3}\sigma_3^2}$ in (10), we obtain $\gamma_3 = b\gamma_2 g_3$, and the probability $\Pr\{\gamma_3 \leq \gamma_{th}, \gamma_2 > \gamma_{th}\}$ can be rewritten as

$$\Pr\{\gamma_3 \leq \gamma_{th}, \gamma_2 > \gamma_{th}\} = \Pr\{b\gamma_2 g_3 \leq \gamma_{th}, \gamma_2 > \gamma_{th}\} = \Pr\left\{g_3 \leq \frac{\gamma_{th}}{b\gamma_2}, \gamma_2 > \gamma_{th}\right\}. \quad (17)$$

The analytical derivation of Equation (17) is provided in detail in Appendix A, and the solution can be expressed as its approximate closed form as

$$\Pr\{\gamma_3 \leq \gamma_{th}, \gamma_2 > \gamma_{th}\} \cong \frac{1}{\Gamma(m_2)\Gamma(m_3+m_{s3})B(m_3,m_{s3})}\left(\frac{m_2\gamma_{th}}{a\overline{\gamma}_2}\right)^{m_2}$$
$$\cdot\left[\left(\frac{m_3}{m_{s3}\overline{\gamma}_3 b}\right)^{m_2} G_{2,3}^{2,2}\left(\frac{m_2 m_3 \gamma_{th}}{a\overline{\gamma}_2 m_{s3}\overline{\gamma}_3 b}\middle|\begin{array}{c} 1-m_2, 1-m_2-m_{s3} \\ 0, m_3-m_2, -m_2 \end{array}\right)\right. \quad (18)$$
$$\left. -\left(\frac{m_3}{m_{s3}\overline{\gamma}_3 b}\right)^{m_3} G_{3,3}^{2,2}\left(\frac{m_{s3}\overline{\gamma}_3 b}{m_3}\middle|\begin{array}{ccc} 1-m_2+m_3, & 1, & 1+m_3 \\ m_3, & m_2+m_{s3}, & m_3-m_2 \end{array}\right)\right],$$

where $G_{p,q}^{m,n}(\cdot)$ denotes Meijer's G function ([40], (9.301)).

Hence, by substituting the derived equations (Equations (12), (13), (15) and (18)) into (11), we obtain the exact closed-form approximate result for the probability of outage, i.e., the probability of a system being in a failure.

Moreover, we define the maximum data rate, achievable in the channel as a consequence of a deep fading, so-called the outage capacity. The DF system is said to be in outage in the case when the SNR at any of the receive nodes in all three hops is lower than the predetermined threshold, γ_{th}. For the probability of an outage equal to $P_{out}(\gamma_{th})$, the normalized capacity is given by the following expression [17]:

$$C_{out} = \frac{1}{3\ln 2}(1 - P_{out}(\gamma_{th}))\ln(1+\gamma_{th}). \quad (19)$$

Thus, the achievable throughput T_{out} is determined as

$$T_{out} = C_{out}. \quad (20)$$

4. Numerical Results

In this section, we present numerical results based on the analysis formulated in the previous section and develop an independent Monte Carlo simulation method, with an aim to investigate the impact of various system and channel parameters on the outage performance. Numerical results for the outage probability and achievable throughput are obtained based on derived analytical expressions and are compared with the simulation results obtained using the independent Monte Carlo simulation model. From the obtained results, it can be concluded that the results based on the simulation method and analytical expression are in excellent agreement, showing the accuracy of the developed analysis.

The location of each network node is determined in the cylindrical coordinate system as MTU(r_0, φ_0, h_0), UAV(r_1, φ_1, h_1), RTU(r_2, φ_2, h_2), and UT(r_3, φ_3, h_3). For the sake of simplicity, the coordinates of the nodes' positions, for the presented numerical results, are MTU(0 m, 0 rad, 0 m), UAV(r_1, 0 rad, H), RTU(400 m, 0 rad, 0 m), and UT(450 m, 0 rad, 0 m). Consequently, the system model can be redrawn as in Figure 3. The distances between nodes d_i, $i = 1, 2, 3$, are defined by the Euclidean norm. Further, the following system parameters are set as $\theta = 0.8$, $\delta_1 = \delta_2 = \delta_3 = 2.05$, $\sigma_1^2 = 10^{-5}$ mW, and $\sigma_2^2 = \sigma_3^2 = 10^{-7}$ mW. In addition, the fading parameters that describe conditions of channels are set as $m_1 = 5$ for the MTU-UAV link, $m_2 = 2$ for the UAV-RTU link, and $m_3 = 3.5$ and $m_{s3} = 5$ for the RTU-UT link.

According to Figure 3, the specified distances in the MTU-UAV, the UAV-RTU, and the RTU-UT links are $d_1 = \sqrt{r_1^2 + H^2}$, $d_2 = \sqrt{(r_2-r_1)^2 + H^2}$, and $d_3 = |r_3 - r_2| = 50$ m, respectively.

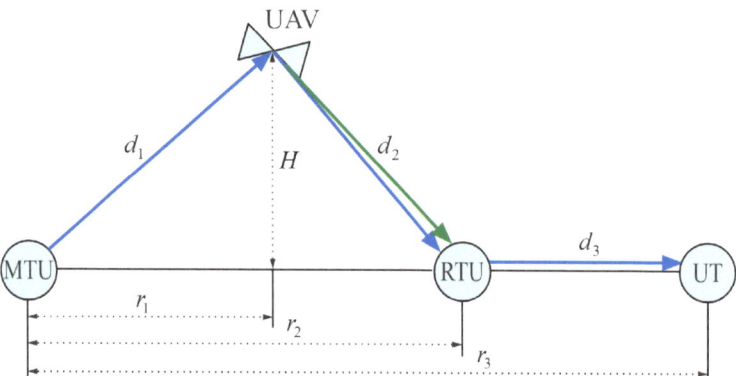

Figure 3. Nodes' positions in UAV-assisted industrial system for emergency applications.

The outage probability versus (vs.) the MTU transmit power P_S is presented in Figure 4, for different values of power-splitting factor θ. The results are obtained for two different values of horizontal distances in the MTU-UAV link, i.e., for $r_1 = 200$ m and $r_1 = 400$ m. It can be observed that with the increase in the MTU transmit power, the probability of outage decreases, up to a certain value after which the probability tends to a constant value, i.e., it enters saturation. The further power increase does not improve the outage performance, which can be explained by the fact that the system performance is dominantly determined by the link with the worst conditions and, for high values of P_S, indicates that the rest of the system is interrupted, regardless of the transmit power of the MTU-UAV link. The outage probability is also dependent on the power-splitting factor since higher values of θ indicate a greater value of the harvested power at the RTU, consequently leading to the higher RTU transmit power and enhanced system performance. The increase in the horizontal distance component in the MTU-UAV link results in the increase in the total MTU-UAV distance d_1, to higher path losses of the observed link, and, thus, to worse system performance for smaller values of P_S (up to 10 dBm). For output power values above $P_S = 10$ dBm, better system performance is achieved when the MTU-UAV distance is larger because the UAV-RTU distance has a lower value, and subsystem UAV-RTU-UT dominantly determines the system performance. For $r_1 = r_2 = 400$ m, the distance between the UAV and RTU is smallest and equals H (as the UAV is directly above the RTU), the path-loss is reduced, and for all analyzed scenarios, the probability of system failure is lower.

In Figure 5, the outage performance dependence on the UAV altitude is shown, for the case when the UAV is above RTU ($r_1 = r_2 = 400$ m). Numerical and simulation results are obtained for different values of MTU transmit power P_S and power-splitting factor θ. For lower values of P_S, the MTU-UAV link represents the critical one for the outage, resulting in a higher probability of system failure. In this case, the influence of the amount of harvested energy and the influence of UAV altitude is not significant. By increasing the MTU output power, the probability of an outage event dominantly depends on the failure of the rest of system, and thus the power-splitting ratio has a significant impact on the system outage. The performance improves when the collected energy on the RTU is higher, i.e., when θ is larger. In addition, with increasing UAV position height, the outage probability increases due to higher path-loss, and the influence of UAV altitude on the performance is significant.

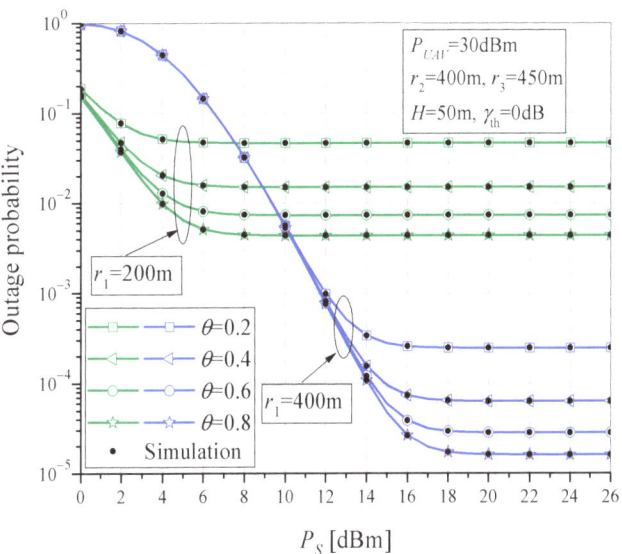

Figure 4. The outage probability vs. MTU output power for various power-splitting factors.

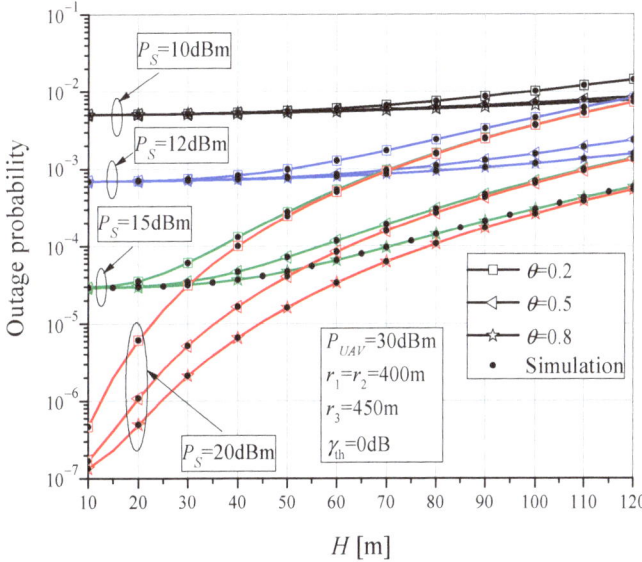

Figure 5. The outage probability vs. UAV altitude for various power-splitting factors.

The dependence of the outage probability on horizontal MTU-UAV distance for various values of P_S is presented in Figure 6. When the UAV is located closer to the MTU ($r_1 < 200$ m), the impact of the transmitted MTU power on the outage probability is negligible, and by increasing the distance r_1, the outage probability decreases. For the certain horizontal MTU-UAV distance value, the minimum probability of outage occurs. The outage probability increases with the further increase in r_1. This effect can be intuitively explained by the fact that if the wireless power transfer is applied for the RTU power supply, the best performance is obtained when the UAV is directly above the RTU. However, this fact is valid only for higher values of P_S; thus, the outage of the MTU-UAV link does not affect the overall outage performance. In general, the optimal location of the UAV that contributes to the minimum of the outage probability is located between the MTU and the RTU. For smaller values of P_S, the optimal performance is obtained when the UAV is positioned closer to the MTU, and vice versa. The best outage performance is obtained in the case of high MTU output power, when the UAV is positioned directly above the RTU, which harvests the energy from the UAV.

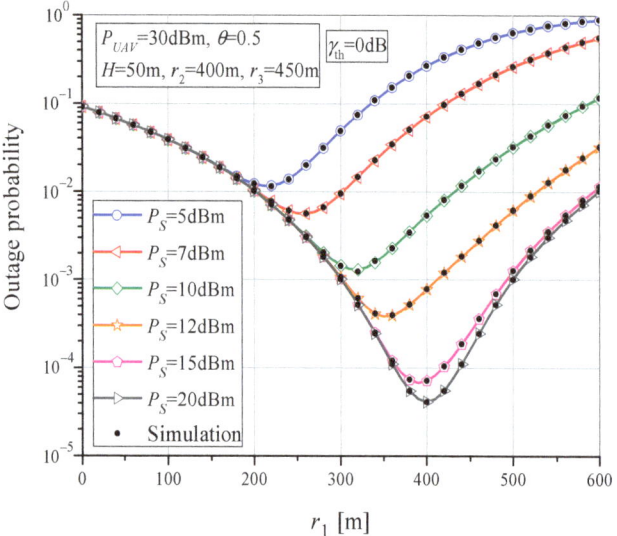

Figure 6. The outage probability vs. MTU-UAV horizontal distance for various values of MTU output power.

In Table 1, the values of UAV position r_1 that lead to the minimum outage probability as well as the corresponding outage probability values are listed for various P_S values and the following set of parameters: $\eta = 0.5$, $\gamma_{th} = 0$ dB, $P_{UAV} = 30$ dBm, and $H = 50$ m. It can be noticed that for the given set of parameters, increasing the MTU power beyond 20 dBm does not lead to a further improvement in system performance. For higher values of MTU output power, the optimal UAV distance r_1 is equal to r_2.

Table 1. Optimal values of r_1 to achieve minimum of outage probability for different P_S values.

P_S [dBm]	r_{1opt} [m]	P_{out}
5	215.10	1.16×10^{-2}
6	235.81	8.28×10^{-3}
7	256.70	5.64×10^{-3}
8	277.42	3.65×10^{-3}
9	297.60	2.24×10^{-3}
10	316.90	1.30×10^{-3}
11	335.10	7.22×10^{-4}
12	352.02	3.87×10^{-4}
13	367.50	2.04×10^{-4}
14	380.99	1.12×10^{-4}
15	390.99	6.88×10^{-5}
16	396.47	5.09×10^{-5}
17	398.75	4.42×10^{-5}
18	399.58	4.19×10^{-5}
19	399.86	4.12×10^{-5}
20	399.95	4.09×10^{-5}
21	399.98	4.08×10^{-5}
22	399.99	4.08×10^{-5}

The contour plot of the outage probability dependence on the UAV position defined by (r_1, H) is presented in Figure 7, for P_S = 15 dBm, P_{UAV} = 30 dBm, and θ = 0.5. The results present a set of values of the UAV height H and the MTU-UAV horizontal distance r_1, which lead to the predefined outage probability. When the UAV is positioned at a certain height, changing the MTU-UAV horizontal distance could lead to the predefined outage probability and vice versa. For instance, to obtain an outage probability smaller than 10^{-3}, the distance r_1 should be in the range $r_1 \in (300 \text{ m}, 500 \text{ m})$ whereby the height of the UAV can be up to 110 m. To achieve an outage probability smaller than 10^{-4}, the r_1 distance should be between 350 m and 450 m, while the maximal UAV's height H can be around 50 m.

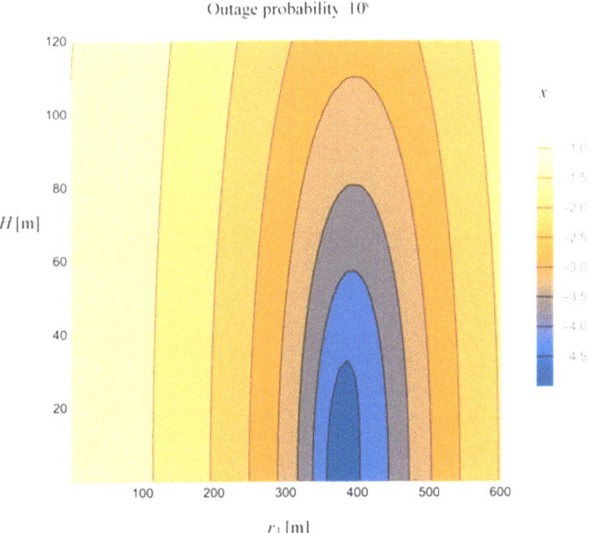

Figure 7. The contour plot of outage probability vs. UAV position r_1 and H for P_S = 15 dBm.

In Figure 8, the throughput T_{out} is shown as a function of MTU-UAV horizontal distance, r_1, for parameter values θ = 0.7, H = 50 m, and γ_{th} = 5 dB. The results are presented for different values of MTU output power, P_S. It can be noticed that for each value of P_S, there is an optimal UAV position at which maximum throughput is achieved. When the UAV is closer to the MTU, the transmitted MTU power does not affect the throughput and it is determined by the UAV-RTU-UT subsystem. However, as the UAV-MTU distance increases, the throughput also increases due to smaller UAV-RTU distance. At a certain distance r_1, the maximum throughput can be reached. With the increase in MTU transmit power P_S, the value of maximum throughput also increases and it is achieved for higher values of distance r_1. With a further increase in r_1 (beyond the one that maximizes throughput), the throughput decreases and the overall system performance deteriorates.

The throughput dependence on the horizontal MTU-UAV distance for various values of power-splitting factors θ is shown in Figure 9. When the UAV position is closer to the MTU, the throughput does not depend on Ps (as in Figure 8), but only on the amount of harvested energy at the RTU. With an increase in the power-splitting factor θ, the amount of harvested energy at the RTU is larger, which allows the maximum throughput value to be achieved at larger UAV-RTU distances, i.e., at a smaller r_1. Also, when the power-splitting value is smaller, the maximum value of the throughput is smaller. Overall, it is noticeable that the maximum throughput is highly dependent on the position of the UAV, the transmit power of MTU, and the amount of harvested energy.

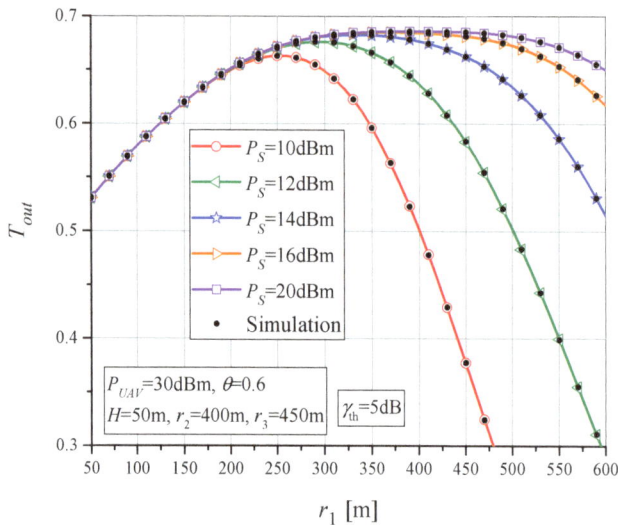

Figure 8. The throughput vs. MTU-UAV horizontal distance for various values of MTU output power.

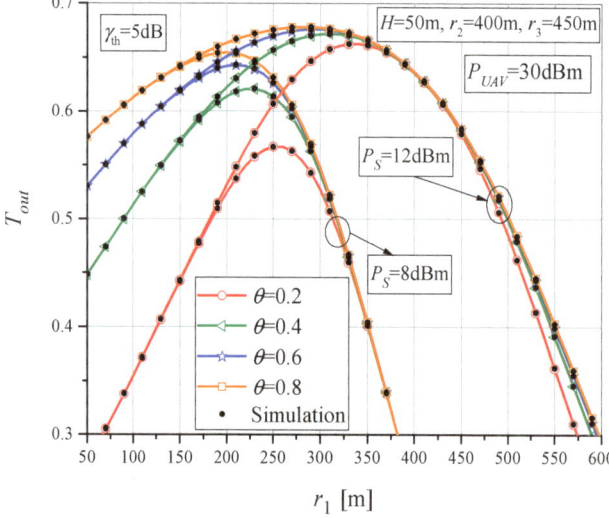

Figure 9. The throughput vs. MTU-UAV horizontal distance for various power-splitting factors.

Figure 9 shows the dependence of the optimal value r_1 on the power-splitting factor θ (and therefore on the amount of collected energy at the RTU). In order to investigate the dependence of the optimal value of the horizontal position between the UAV and the MTU on the power-splitting factor and the MTU transmit power, we present the results in Figure 10. For $P_S = 10$ dBm and $\theta = 0.8$, the optimal horizontal UAV-MTU distance is $r_{1opt} = 300$ m, while for $\theta = 0.2$, the optimal value is $r_{1opt} = 350$ m, due to the smaller value of the collected power at the RTU. Therefore, in the case of smaller power-splitting factor values, the optimal position r_{1opt} is higher in order to reduce the UAV-RTU distance and corresponding path loss. When $P_S > 18$ dBm, the optimal MTU-UAV horizontal distance is independent of the power-splitting factor and corresponds to the position when the UAV is above the RTU.

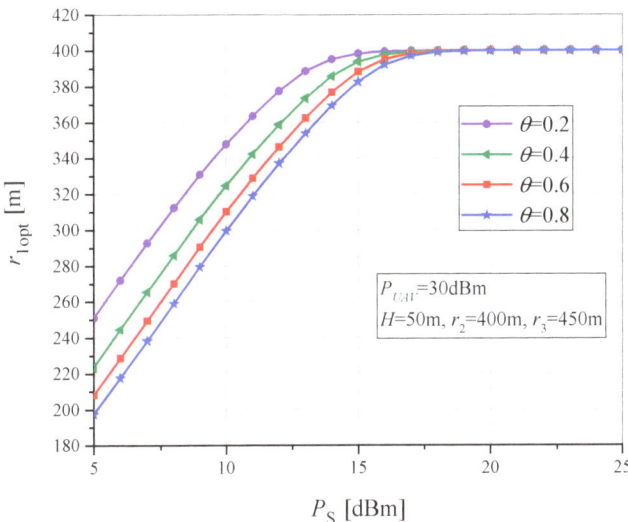

Figure 10. The optimal values of MTU-UAV horizontal distance vs. MTU output power for various power-splitting factor.

5. Conclusions

In this paper, we propose and analyze the outage performance of an industrial system assisted by an unmanned aerial vehicle, which is resilient to the emergency scenario when direct communication between the master terminal unit and the remote terminal unit is disabled or the RTU is left without power supply due to an unpredictable disaster. In the proposed system, the UAV is utilized as both a relay for information communication and as a power supplier for the RTU, which forwards decoded information further to the intended end node. The analytical expressions of the outage event probability and system throughput are derived and the impact of system parameters on the system performance is examined.

The obtained results show that the probability of an outage and the achievable throughput are highly dependent on the position of the UAV relative to the MTU and the RTU, the MTU output power, and the power splitting factor values, i.e., the amount of harvested energy. For lower values of the MTU output power, the UAV should be positioned closer to the MTU. Then, the outage probability depends strongly on the amount of harvested energy and D2D link characteristics. Otherwise, if the MTU power is larger, to ensure that the first link is not in failure, the position of the UAV should be close to the RTU in order to provide the RTU with sufficient output power. It has been shown that the optimal values of the UAV position highly depend on the MTU output power and the amount of harvested energy at the RTU, and these values are calculated for the considered system and channel parameters.

The presented results are useful in the design of an industrial system resilient to the emergency outage scenario, in terms of making an efficient tradeoff between system parameters and the level of outage events, to assure reliable signal transmission.

Author Contributions: Conceptualization, A.C., V.B. and J.A.; methodology, A.C., V.B. and M.M.; software, A.C. and N.T.P.; validation, J.A. and N.T.P.; formal analysis, A.C. and V.B.; investigation, A.C. and V.B.; resources, V.B., J.A. and M.M.; data curation, A.C. and V.B.; writing—original draft preparation, A.C., V.B. and J.A.; writing—review and editing, A.C., V.B., J.A., N.T.P. and M.M.; visualization, A.C. and J.A.; supervision, N.T.P. and M.M.; project administration, A.C., V.B. and J.A. All authors have read and agreed to the published version of the manuscript.

Funding: This research received no external funding.

Institutional Review Board Statement: Not applicable.

Informed Consent Statement: Not applicable.

Data Availability Statement: Not applicable.

Acknowledgments: The work of A. Cvetković, N.T. Pavlović, and M. Milošević was financially supported by the Ministry of Science, Technological Development and Innovation of the Republic of Serbia (Contract No. 451-03-47/2023-01/200109). V. Blagojević acknowledges the support of the Science Fund of the Republic of Serbia, grant No 7750284 (Hybrid Integrated Satellite and Terrestrial Access Network—hi-STAR).

Conflicts of Interest: The authors declare no conflict of interest.

Appendix A

In order to obtain the analytical form of the overall outage probability, the probability in Equation (17), $\Pr\{\gamma_3 \leq \gamma_{th}, \gamma_2 > \gamma_{th}\}$, is expressed as

$$\Pr\{\gamma_3 \leq \gamma_{th}, \gamma_2 > \gamma_{th}\} = \int_{\gamma_{th}}^{\infty} F_{g_3}\left(\frac{\gamma_{th}}{b\gamma_2}\right) p_{\gamma_2}(\gamma_2) d\gamma_2 =$$
$$= \int_0^{\infty} F_{g_3}\left(\frac{\gamma_{th}}{b\gamma_2}\right) p_{\gamma_2}(\gamma_2) d\gamma_2 - \int_0^{\gamma_{th}} F_{g_3}\left(\frac{\gamma_{th}}{b\gamma_2}\right) p_{\gamma_2}(\gamma_2) d\gamma_2 = I_1 - I_2, \quad (A1)$$

with

$$I_1 = \int_0^{\infty} F_{g_3}\left(\frac{\gamma_{th}}{b\gamma_2}\right) p_{\gamma_2}(\gamma_2) d\gamma_2, \quad (A2)$$

and

$$I_2 = \int_0^{\gamma_{th}} F_{g_3}\left(\frac{\gamma_{th}}{b\gamma_2}\right) p_{\gamma_2}(\gamma_2) d\gamma_2, \quad (A3)$$

where $F_{g_3}(.)$ is the CDF of a Fisher–Snedecor F variable. The CDF is defined as [42]

$$F_{g_3}(\gamma) = \left(\frac{m_3 \gamma}{m_{s3} \overline{\gamma}_3}\right)^{m_i} \frac{1}{\Gamma(m_3 + m_{s3}) B(m_3, m_{s3})} G_{2,2}^{1,2}\left(\frac{m_3 \gamma}{m_{s3} \overline{\gamma}_3} \middle| \begin{array}{c} 1 - m_3, 1 - m_3 - m_{s3} \\ 0, -m_3 \end{array}\right). \quad (A4)$$

By substituting Equations (A4) and (1) into Equation (A2), the integral I_1 can be rewritten as

$$I_1 = \frac{1}{\Gamma(m_3 + m_{s3}) B(m_3, m_{s3})} \left(\frac{m_3 \gamma_{th}}{m_{s3} \overline{\gamma}_3 b}\right)^{m_3} \frac{1}{\Gamma(m_2)} \left(\frac{m_2}{a\overline{\gamma}_2}\right)^{m_{m2}}$$
$$\cdot \int_0^{\infty} \gamma_2^{-m_3+m_2-1} G_{2,2}^{1,2}\left(\frac{m_3 \gamma_{th}}{m_{s3} \overline{\gamma}_3 b \gamma_2} \middle| \begin{array}{c} 1 - m_3, 1 - m_3 - m_{s3} \\ 0, -m_3 \end{array}\right) e^{-\frac{m_2}{a\overline{\gamma}_2} \gamma_2} d\gamma_2. \quad (A5)$$

In order to solve integral in I_1, the exponential function is expressed in the form of Meijer's G function ([46], (01.03.26.0004.01)) as

$$e^{-\frac{m_2}{a\overline{\gamma}_2} \gamma_2} = G_{0,1}^{1,0}\left(\frac{m_2}{a\overline{\gamma}_2} \gamma_2 \middle| \begin{array}{c} - \\ 0 \end{array}\right), \quad (A6)$$

and relying on the argument simplification of Meijer's G function ([46], (07.34.16.0002.01))

$$G_{2,2}^{1,2}\left(\frac{m_3 \gamma_{th}}{m_{s3} \overline{\gamma}_3 b \gamma_2} \middle| \begin{array}{c} 1 - m_3, 1 - m_3 - m_{s3} \\ 0, -m_3 \end{array}\right) = G_{2,2}^{2,1}\left(\frac{m_{s3} \overline{\gamma}_3 b \gamma_2}{m_3 \gamma_{th}} \middle| \begin{array}{c} 1, 1 + m_3 \\ m_3, m_3 + m_{s3} \end{array}\right), \quad (A7)$$

the integral in Equation (A5) becomes

$$I_1 = \frac{1}{\Gamma(m_3+m_{s3})B(m_3,m_{s3})}\left(\frac{m_3\gamma_{th}}{m_{s3}\bar{\gamma}_3 b}\right)^{m_3}\frac{1}{\Gamma(m_2)}\left(\frac{m_2}{a\bar{\gamma}_2}\right)^{m_2}$$
$$\cdot \int_0^{\infty}\gamma_2^{-m_3+m_2-1}G_{2,2}^{2,1}\left(\frac{m_{s3}\bar{\gamma}_3 b\gamma_2}{m_3\gamma_{th}}\left|\begin{array}{c}1,1+m_3\\ m_3, m_3+m_{s3}\end{array}\right.\right)G_{0,1}^{1,0}\left(\frac{m_2}{a\bar{\gamma}_2}\gamma_2\left|\begin{array}{c}-\\0\end{array}\right.\right)\mathrm{d}\gamma_2. \quad (A8)$$

Further, with the help of ([46], (07.34.21.0011.01)), the previous integral is solved in the exact closed-form, and I_1 is represented as the following form:

$$I_1 = \frac{1}{\Gamma(m_2)\Gamma(m_3+m_{s3})B(m_3,m_{s3})}$$
$$\cdot\left(\frac{m_2 m_3\gamma_{th}}{a\bar{\gamma}_2 m_{s3}\bar{\gamma}_3 b}\right)^{m_2} G_{2,3}^{2,2}\left(\frac{m_2 m_3\gamma_{th}}{a\bar{\gamma}_2 m_{s3}\bar{\gamma}_1 b}\left|\begin{array}{c}1-m_2, 1-m_2-m_{s3}\\ 0, m_3-m_2, -m_2\end{array}\right.\right). \quad (A9)$$

Furthermore, by substituting Equations (A4) and (1) into Equation (A3), with appropriate indexed parameters, the expression of I_2 is rewritten as

$$I_2 = \frac{1}{\Gamma(m_2)\Gamma(m_3+m_{s3})B(m_3,m_{s3})}\left(\frac{m_3\gamma_{th}}{m_{s3}\bar{\gamma}_3 b}\right)^{m_3}\left(\frac{m_2}{a\bar{\gamma}_2}\right)^{m_2}$$
$$\cdot \int_0^{\gamma_{th}}\gamma_2^{-m_3+m_2-1}G_{2,2}^{1,2}\left(\frac{m_3\gamma_{th}}{m_{s3}\bar{\gamma}_3 b\gamma_2}\left|\begin{array}{c}1-m_3, 1-m_3-m_{s3}\\ 0, -m_3\end{array}\right.\right)e^{-\frac{m_2}{a\bar{\gamma}_2}\gamma_2}\mathrm{d}\gamma_2. \quad (A10)$$

The integral in I_2 cannot be obtained in the closed form, following the approach used for solving the integral in I_1. Since the argument of the exponential function is a small value (i.e., $a \gg 1$), by expanding the exponential function into a series ([46], (01.03.06.0002.01)) and taking into account only the first term, without loss of generality, we obtain the following approximate form:

$$I_2 \cong \frac{1}{\Gamma(m_2)\Gamma(m_3+m_{s3})B(m_3,m_{s3})}\left(\frac{m_3\gamma_{th}}{m_{s3}\bar{\gamma}_3 b}\right)^{m_3}\left(\frac{m_2}{a\bar{\gamma}_2}\right)^{m_2}$$
$$\cdot \int_0^{\gamma_{th}}\gamma_2^{-m_3+m_2-1}G_{2,2}^{1,2}\left(\frac{m_3\gamma_{th}}{m_{s3}\bar{\gamma}_3 b\gamma_2}\left|\begin{array}{c}1-m_3, 1-m_3-m_{s3}\\ 0, -m_3\end{array}\right.\right)\mathrm{d}\gamma_2. \quad (A11)$$

Now, the previous integral can be solved utilizing ([46], (07.34.21.0084.01)), and the integral I_2 becomes

$$I_2 \cong \frac{\gamma_{th}^{m_2-m_3}}{\Gamma(m_2)\Gamma(m_3+m_{s3})B(m_3,m_{s3})}\left(\frac{m_3\gamma_{th}}{m_{s3}\bar{\gamma}_3 b}\right)^{m_3}\left(\frac{m_2}{a\bar{\gamma}_2}\right)^{m_2}$$
$$\cdot G_{3,3}^{2,2}\left(\frac{m_{s3}\bar{\gamma}_3 b}{m_3}\left|\begin{array}{ccc}1-m_2+m_3, & 1, & 1+m_3\\ m_3, & m_2+m_{s3}, & m_3-m_2\end{array}\right.\right). \quad (A12)$$

Finally, the probability $\Pr\{\gamma_3 \leq \gamma_{th}, \gamma_2 > \gamma_{th}\}$ is obtained by substituting Equations (A9) and (A12) into Equation (A1) in the form of Equation (19).

References

1. Alzahrani, B.; Oubbati, O.S.; Barnawi, A.; Atiquzzaman, M.; Alghazzawi, D. UAV assistance paradigm: State-of-the-art in applications and challenges. *J. Netw. Comput. Appl.* **2020**, *166*, 102706. [CrossRef]
2. Banić, M.; Milentijević, A.; Pavlović, M.; Ćirić, I. Intelligent machine vision based railway infrastructure inspection and monitoring using UAV. *FU Mech. Eng.* **2019**, *17*, 357–364. [CrossRef]
3. Basiri, A.; Mariani, V.; Silano, G.; Aatif, M.; Iannelli, L.; Glielmo, L. A survey on the application of path-planning algorithms for multi-rotor UAVs in precision agriculture. *J. Navig.* **2022**, *75*, 364–383. [CrossRef]
4. Mourtzis, D.; Angelopoulos, J.; Panopoulos, N. UAVs for industrial applications: Identifying challenges and opportunities from the implementation point of view. *Procedia Manuf.* **2021**, *55*, 183–190. [CrossRef]
5. Zhao, N.; Lu, W.; Sheng, M.; Chen, Y.; Tang, J.; Yu, F.R.; Wong, K.-K. UAV-assisted emergency networks in disasters. *IEEE Wirel. Commun.* **2019**, *26*, 45–51. [CrossRef]
6. Shah, A.F.M.S. Architecture of Emergency Communication Systems in Disasters through UAVs in 5G and Beyond. *Drones* **2023**, *7*, 25. [CrossRef]
7. Liu, X.; Li, Z.; Zhao, N.; Meng, W.; Gui, G.; Chen, Y.; Adachi, F. Transceiver design and multihop D2D for UAV IoT coverage in disasters. *IEEE Internet Things J.* **2019**, *6*, 1803–1815. [CrossRef]

8. Aiello, G.; Hopps, F.; Santisi, D.; Venticinque, M. The employment of unmanned aerial vehicles for analyzing and mitigating disaster risks in industrial sites. *IEEE Trans. Eng. Manag.* **2020**, *67*, 519–530. [CrossRef]
9. Pliatsios, D.; Sarigiannidis, P.; Lagkas, T.; Sarigiannidis, A.G. A Survey on SCADA Systems: Secure Protocols, Incidents, Threats and Tactics. *IEEE Commun. Surv. Tutor.* **2020**, *22*, 1942–1976. [CrossRef]
10. Xiang, X.; Gui, J.; Xiong, N.N. An Integral Data Gathering Framework for Supervisory Control and Data Acquisition Systems in Green IoT. *IEEE Trans. Green Commun.* **2021**, *5*, 714–726. [CrossRef]
11. Ibrahim, H.H.; Singh, M.J.; Al-Bawri, S.S.; Ibrahim, S.K.; Islam, M.T.; Alzamil, A.; Islam, M.S. Radio Frequency Energy Harvesting Technologies: A Comprehensive Review on Designing, Methodologies, and Potential Applications. *Sensors* **2022**, *22*, 4144. [CrossRef] [PubMed]
12. Sherazi, H.H.R.; Zorbas, D.; O'Flynn, B. A Comprehensive Survey on RF Energy Harvesting: Applications and Performance Determinants. *Sensors* **2022**, *22*, 2990. [CrossRef] [PubMed]
13. Ashraf, N.; Sheikh, S.A.; Khan, S.A.; Shayea, I.; Jalal, M. Simultaneous Wireless Information and Power Transfer with Cooperative Relaying for Next-Generation Wireless Networks: A Review. *IEEE Access* **2021**, *9*, 71482–71504. [CrossRef]
14. Masotti, D.; Shanawani, M.; Murtaza, G.; Paolini, G.; Costanzo, A. RF Systems Design for Simultaneous Wireless Information and Power Transfer (SWIPT) in Automation and Transportation. *IEEE J. Microw.* **2021**, *1*, 164–175. [CrossRef]
15. Kozić, N.; Blagojević, V.; Cvetković, A.; Ivaniš, P. Performance Analysis of Wirelessly Powered Cognitive Radio Network with Statistical CSI and Random Mobility. *Sensors* **2023**, *23*, 4518. [CrossRef] [PubMed]
16. Tin, P.T.; Dinh, B.H.; Nguyen, T.N.; Ha, D.H.; Trang, T.T. Power Beacon-Assisted Energy Harvesting Wireless Physical Layer Cooperative Relaying Networks: Performance Analysis. *Symmetry* **2020**, *12*, 106. [CrossRef]
17. Blagojevic, V.M.; Cvetkovic, A.M.; Ivanis, P. Performance analysis of energy harvesting DF relay system in generalized-K fading environment. *Phys. Commun.* **2018**, *28*, 190–200. [CrossRef]
18. Zou, Y.; Zhu, J.; Jiang, X. Joint Power Splitting and Relay Selection in Energy-Harvesting Communications for IoT Networks. *IEEE Internet Things J.* **2020**, *7*, 584–597. [CrossRef]
19. Cvetković, A.; Blagojević, V.; Manojlović, J. Capacity Analysis of Power Beacon-Assisted Industrial IoT System with UAV Data Collector. *Drones* **2023**, *7*, 146. [CrossRef]
20. Nguyen, M.T.; Nguyen, C.V.; Do, H.T.; Hua, H.T.; Tran, T.A.; Nguyen, A.D.; Ala, G.; Viola, F. UAV-assisted Data Collection in Wireless Sensor Networks: A Comprehensive Survey. *Electronics* **2021**, *10*, 2603. [CrossRef]
21. Chen, Y.; Zhao, N.; Ding, Z.; Alouini, M.-S. Multiple UAVs as relays: Multi-hop single link versus multiple dual-hop links. *IEEE Trans. Wirel. Commun.* **2018**, *17*, 6348–6359. [CrossRef]
22. Liu, Y.; Dai, H.N.; Wang, H.; Imran, M.; Wang, X.; Shoaib, M. UAV-enabled data acquisition scheme with directional wireless energy transfer for Internet of Things. *Comput. Commun.* **2020**, *155*, 184–196. [CrossRef]
23. Hu, Y.; Yuan, X.; Zhang, G.; Schmeink, A. Sustainable wireless sensor networks with UAV-enabled wireless power transfer. *IEEE Trans. Veh. Technol.* **2021**, *70*, 8050–8064. [CrossRef]
24. Ji, B.; Li, Y.; Zhou, B.; Li, C.; Song, K.; Wen, H. Performance Analysis of UAV Relay Assisted IoT Communication Network Enhanced with Energy Harvesting. *IEEE Access* **2019**, *7*, 38738–38747. [CrossRef]
25. Hu, H.; Xiong, K.; Qu, G.; Ni, Q.; Fan, P.; Letaief, K.B. AoI-minimal trajectory planning and data collection in UAV-assisted wireless powered IoT networks. *IEEE Internet Things J.* **2021**, *8*, 1211–1223. [CrossRef]
26. Hassan, A.; Ahmad, R.; Ahmed, W.; Magarini, M.; Alam, M.M. UAV and SWIPT Assisted Disaster Aware Clustering and Association. *IEEE Access* **2020**, *8*, 204791–204803. [CrossRef]
27. Feng, W.; Tang, J.; Yu, Y.; Song, J.; Zhao, N.; Chen, G.; Wong, K.K. UAV-Enabled SWIPT in IoT Networks for Emergency Communications. *IEEE Wirel. Commun.* **2020**, *27*, 140–147. [CrossRef]
28. Hu, Y.; Yuan, X.; Xu, J.; Schmeink, A. Optimal 1D trajectory design for UAV-enabled multiuser wireless power transfer. *IEEE Trans. Commun.* **2019**, *67*, 5674–5688. [CrossRef]
29. Dhok, S.; Raut, P.; Sharma, P.K.; Singh, K.; Li, C.-P. Non-linear energy harvesting in RIS-assisted URLLC networks for industry automation. *IEEE Trans. Commun.* **2021**, *69*, 7761–7774. [CrossRef]
30. Lin, Z.; Niu, H.; An, K.; Wang, Y.; Zheng, G.; Chatzinotas, S.; Hu, Y. Refracting RIS-Aided Hybrid Satellite-Terrestrial Relay Networks: Joint Beamforming Design and Optimization. *IEEE Trans. Aerosp. Electron. Syst.* **2022**, *58*, 3717–3724. [CrossRef]
31. Lin, Z.; Lin, M.; Zhu, W.-P.; Wang, J.-B.; Cheng, J. Robust Secure Beamforming for Wireless Powered Cognitive Satellite-Terrestrial Networks. *IEEE Trans. Cogn. Commun. Netw.* **2021**, *7*, 567–580. [CrossRef]
32. Lin, Z.; Lin, M.; Champagne, B.; Zhu, W.-P.; Al-Dhahir, N. Secrecy-Energy Efficient Hybrid Beamforming for Satellite-Terrestrial Integrated Networks. *IEEE Trans. Commun.* **2021**, *69*, 6345–6360. [CrossRef]
33. Lin, Z.; Lin, M.; Cola, T.; Wang, J.-B.; Zhu, W.-P.; Cheng, J. Supporting IoT with Rate-Splitting Multiple Access in Satellite and Aerial-Integrated Networks. *IEEE Internet Things J.* **2021**, *8*, 11123–11134. [CrossRef]
34. Zhong, S.; Huang, H.; Li, R. Outage probability of power splitting SWIPT two-way relay networks in Nakagami-m fading. *EURASIP J. Wirel. Commun. Netw.* **2018**, *11*, 1–8. [CrossRef]
35. Hoang, T.M.; Nguyen, B.C.; Tran, P.T.; Dung, L.T. Outage Analysis of RF Energy Harvesting Cooperative Communication Systems Over Nakagami-m Fading Channels with Integer and Non-Integer m. *IEEE Trans. Veh. Technol.* **2020**, *69*, 2785–2801. [CrossRef]
36. Nawaz, F.; Hassan, S.; Aissa, S.; Saleem, S. Outage Probability for a Decode-and-Forward SWIPT Relaying System in Nakagami Fading. *Internet Technol. Lett.* **2017**, *1*, e13. [CrossRef]

37. Makarfi, A.U.; Kharel, R.; Rabie, K.M.; Li, X.; Badarneh, O.S.; Nauryzbayev, G.; Arzykulov, S. Performance Analysis of SWIPT Networks over Composite Fading Channels. In Proceedings of the IEEE Eighth International Conference on Communications and Networking (ComNet), Hammamet, Tunisia, 27–30 October 2020. [CrossRef]
38. Simonović, M.; Cvetković, A.; Manojlović, J.; Nikolić, V. Outage performance evaluation of device-to-device system with energy harvesting relay. *Therm. Sci.* **2020**, *25*, 1771–1780. [CrossRef]
39. Goldsmith, A. *Wireless Communications*; Cambridge University Press: Cambridge, UK, 2005.
40. Gradshteyn, I.S.; Ryzhik, I.M. *Table of Integrals, Series, and Products*, 7th ed.; Elsevier/Academic Press: London, UK, 2007.
41. Yoo, S.K.; Cotton, S.L.; Sofotasios, P.C.; Matthaiou, M.; Valkama, M.; Karagiannidis, G.K. The Fisher–Snedecor F distribution: A simple and accurate composite fading model. *IEEE Commun. Lett.* **2017**, *21*, 1661–1664. [CrossRef]
42. Kong, L.; Kaddoum, G. On physical layer security over the Fisher-Snedecor F wiretap fading channels. *IEEE Access* **2018**, *6*, 39466–39472. [CrossRef]
43. Fotouhi, A.; Qiang, H.; Ding, M.; Hassan, M.; Giordano, L.G.; Garcia-Rodriguez, A.; Yuan, J. Survey on UAV Cellular Communications: Practical Aspects, Standardization Advancements, Regulation, and Security Challenges. *IEEE Commun. Surv. Tutor.* **2019**, *21*, 3417–3442. [CrossRef]
44. Kim, Y.H.; Chowdhury, I.A.; Song, I. Design and Analysis of UAV-Assisted Relaying with Simultaneous Wireless Information and Power Transfer. *IEEE Access* **2020**, *8*, 27874–27886. [CrossRef]
45. Papoulis, A. *Probability, Random Variables, and Stochastic Processes*; McGraw-Hill: New York, USA, 1991.
46. Wolfram Research. Available online: https://functions.wolfram.com/ (accessed on 1 May 2023).

Disclaimer/Publisher's Note: The statements, opinions and data contained in all publications are solely those of the individual author(s) and contributor(s) and not of MDPI and/or the editor(s). MDPI and/or the editor(s) disclaim responsibility for any injury to people or property resulting from any ideas, methods, instructions or products referred to in the content.

Article

UAV-Based Servicing of IoT Nodes: Assessment of Ecological Impact

Jarne Van Mulders *,†, Jona Cappelle †, Sarah Goossens, Lieven De Strycker and Liesbet Van der Perre

ESAT-DRAMCO, Ghent Technology Campus, KU Leuven, 9000 Ghent, Belgium
* Correspondence: jarne.vanmulders@kuleuven.be
† These authors contributed equally to this work.

Abstract: Internet of Things (IoT) nodes get deployed for a variety of applications and often need to operate on batteries. This restricts their autonomy and/or can have a major ecological impact. The core idea of this paper is to use a unmanned aerial vehicle (UAV) to provide energy to IoT nodes, and hence prolong their autonomy. In particular, the objective is to perform a comparison of the total energy consumption resulting from UAV-based recharging or battery replacement versus full provisioning at install time or remote RF-based wireless power transfer. To that end, an energy consumption model for a small license-free UAV is derived, and expressions for system efficiencies are formulated. An exploration of design and deployment parameters is performed. Our assessment shows that UAV-based servicing of IoT nodes is by far more beneficial in terms of energy efficiency when nodes at distances further than a few meters are serviced, with the gap increasing to orders of magnitude with the distance. Our numerical results also show that battery swapping from an energy perspective outperforms recharging in the field, as the latter increases hovering time and the energy consumption related to that considerably. The ecological aspects of the proposed methods are further evaluated, e.g., considering toxic materials and e-waste.

Keywords: Internet of Things; energy efficiency; wireless power transfer; sustainability

Citation: Van Mulders, J.; Cappelle, J.; Goossens, S.; De Strycker, L.; Van der Perre, L. UAV-Based Servicing of IoT Nodes: Assessment of Ecological Impact. *Sensors* **2023**, *23*, 2291. https://doi.org/10.3390/s23042291

Academic Editors: Onel Luis Alcaraz López and Katsuya Suto

Received: 12 December 2022
Revised: 13 February 2023
Accepted: 16 February 2023
Published: 18 February 2023

Copyright: © 2023 by the authors. Licensee MDPI, Basel, Switzerland. This article is an open access article distributed under the terms and conditions of the Creative Commons Attribution (CC BY) license (https://creativecommons.org/licenses/by/4.0/).

1. Introduction

The introduction of the IoT has created the opportunity to develop a myriad of applications. The number of IoT edge devices that will be deployed is predicted to grow fast, from 5–10 billion in 2020 up to 200+ billion in 2030 [1]. However, installing and maintaining these countless devices can be cumbersome, if not impossible, if done manually. It is evident that, for an efficient deployment of these sky-high numbers, any dedicated installation or maintenance effort should be minimized. The 'fire and forget' credo, towards which recent IoT devices are aimed, does not properly consider the End of Life (EoL), especially since toxic materials (e.g., in batteries) can remain in nature and cause pollution. From an ecological point of view, it would be problematic if even a small percentage of these battery-powered devices were not recovered from the environment. Therefore, systems for automatic battery charging/replacement and commissioning are indispensable.

Solid technological advances in the IoT have focused on low-power communication and design [2]. This has led to the development of energy-efficient long-range wide-area networks (LoRaWANs), strict energy management, load switching to reduce standby/sleep currents, and the use of low-power Microcontroller Units (MCUs). The lifetime of battery-powered devices can be extended by applying low-power strategies such as 'think before you talk', going to sleep as much as possible, etc. [2]. Applying these strategies has its limits, especially for time-critical applications, which require more frequent measurements and communication. As a result, there is a need for human intervention to recharge or replace the battery, or even replace the complete IoT device.

Automation is required for an efficient maintenance of massively deployed IoT devices to ensure their longevity. Lessons learned from early IoT adopters [3] include the existence of a 'skills gap' between what is currently required to deploy and maintain IoT systems versus the abilities of maintenance people. Automation would eliminate the need for technical knowledge and save time and money. Moreover, an IoT network can be applied in a variety of environments that can be difficult to reach for humans, e.g., environmental monitoring of remote field sites. Locations such as these bring practical challenges, including physical installation and maintenance difficulties.

The challenges regarding the autonomy of IoT nodes and the overhead involved in manual interventions to recharge or replace batteries motivate the main idea and objective of this paper: to study the opportunities in utilizing unmanned vehicles (UVs) to increase the lifespan of IoT devices. We present three scenarios, for each of which the energy budget is analyzed. These include noncoupled wireless power transfer (WPT) technologies and UV-based solutions. We assess the energy budget for UVs that serve as mobile power banks and show that this approach can lead to a more sustainable integration. All these scenarios create, in theory, an infinite autonomy and, moreover, enable a reduction in manual labor for use cases that cannot rely on their own source of harvested green energy.

The remainder of this paper is structured as follows. Section 2 discusses the energy delivery scenarios and associated design and deployment assumptions. A model for the energy consumption of a UV has been adapted based on actual measurements. This model is then used in the quantitative study. The overall energy budget calculations for each energy delivery scenario are explained in Section 3. Section 4 compares the three solutions by calculating their respective performances. Sustainability gains are demonstrated in Section 5, and Section 6 concludes this manuscript.

2. System and IoT Node Architecture and Related Work

Higher autonomy of battery-powered embedded devices can be achieved by making better use of the available energy. However, adopting storage technologies with higher energy densities can also result in lifespan gains due to the availability of more energy. Lithium thionyl chloride (LTC) batteries with extremely high energy densities of around ≈ 1000 Wh/l are often selected.

If longevity proves insufficient, harvesting energy from environmental sources is often explored as an interesting solution. Solar, thermal, and mechanical sources with power densities up to 10^5, 10^3, and 10^3 µW/m^3, respectively [2], are often sufficient to cover the full energy consumption of the applications. Where these solutions cannot be implemented or when the autonomy is still too limited, the final option is a human intervention to recharge or replace the drained battery or the whole device. This is labor-intensive and leads to high maintenance costs and extra traffic on the road. To avoid intensive maintenance, this article evaluates novel techniques for extending the lifetime of IoT devices, comparing the following three options:

1. Using uncoupled wireless power transfer approaches such as laser power transfer or radio frequency (RF) power transfer to recharge the IoT node's batteries. These techniques feature wireless power over longer distances combined with lower efficiencies compared with the coupled WPT techniques [4].
2. Engaging UVs deployed as a mobile power bank to charge batteries on-site.
3. Employing UVs acting as a technician to replace batteries autonomously.

Figure 1 shows the scenarios discussed in this article. To assess and compare system performance and clarify the benefits of each approach, design and deployment assumptions need to be made. Different types of UVs are available, e.g., UAV, unmanned ground vehicles (UGVs), or autonomous underwater vehicles (AUVs). This work focuses on a UAV. The UAV is expected to fly to the IoT applications and deliver the energy. We compare the radio frequency power transfer (RFPT) and UAV approaches departing or delivering from the same initial location, as illustrated in Figure 1. The distance between IoT node and both UAV charge station and the RFPT transmitter is assumed to be equal to the line-of-sight (LoS) distance. This work focuses on the remote charging of a single IoT node.

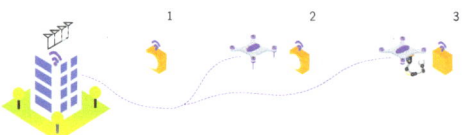

Figure 1. Overview of different solutions: (1) uncoupled wireless power transfer, (2) wireless charging by UAV equipped with WPT transmitter in proximity of the IoT node, and (3) battery replacement by UAV.

2.1. Uncoupled Wireless Power Transfer

We present a concise study of the effectiveness of lifetime enhancements of IoT nodes through uncoupled WPT techniques to serve as a reference for the UAV-based servicing options. In uncoupled systems, no magnetic or electrical coupling is present between the transmit and receive sides. The wireless power delivery happens over distances where the receiver is located in the far field of the transmitter. Moreover, it is feasible to power multiple devices at the same time, in contrast to coupled technologies. Uncoupled approaches fully rely on EM waves captured by the receiver device. In the literature, two technologies are distinguished to deliver power in the far field: RFPT and laser power transfer (LPT). In this comparative study, RFPT technology is considered, as it currently is the most convenient of the two in terms of requirements for an unobstructed path between transmitter and receiver.

The achievable distance using RFPT is restricted to some meters, due to the high path loss and limited sensitivity of the harvester circuits. This is further elaborated in Section 3.1. In most cases, such as with radio frequency identification (RFID) systems, this technology is selected with the aim of powering very low-energy-consuming nodes and rarely for substantially charging devices. The energy storage will rather consist of a storage buffer capacitor instead of a battery.

The main objective of our current study is to evaluate efficiency as a function of distance. Necessary assumptions are made in order to obtain quantitative estimations. The power output of the RFPT, specified as effective isotropic radiated power (EIRP), is limited to comply with regulatory European Telecommunications Standards Institute (ETSI) constraints [5]. Low-complexity models are used in a single-input single-output (SISO) configuration with a single-tone waveform, and a single frequency is considered. The power amplifier (PA) of the RFPT is assumed to have a fixed efficiency.

Related work. Recently, a lot of research has been conducted regarding waveform design to increase RF/DC efficiency [6] and multiband harvesting solutions [7] to increase harvested power. Distributed beamforming to increase the received power and coverage is investigated, a.o., in the European project REINDEER [8]. By coherent operation of the individual devices, a power spot at a certain location could be created. This RadioWeaves approach has already been introduced in [9].

2.2. UV with Charging Facility

Quite recently, approaches to energize nodes based on a UAV capable of transferring energy to an IoT node have been introduced. As such, this study aims to provide a proper estimate of UAV energy requirements in relation to the practical parameters, design choices, and various assumptions.

On site, the UAV should refill the application storage buffer with several hundred to thousands of joules of energy. Two scenarios can be assumed. The UAV first lands, then starts transferring charge (1), or hovers while delivering energy simultaneously (2). Both wireless and cable solutions would be available in these scenarios. Figure 2 summarizes the four procedures for transferring energy.

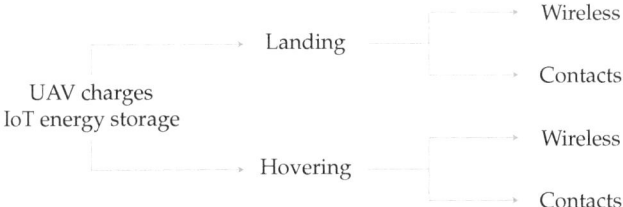

Figure 2. Overview of different subscenarios for IoT charging with UAV.

Providing a landing spot at every node is often not practical, hence this option is not further examined in this research. Charging during hovering offers more flexibility because no landing spot is required. Moreover, a wireless power transmission solution is opted for because of less stringent alignment constraints that do exist when using contacts. Article [10] already summarized the wireless power transfer approaches that could be implemented on a UAV. The fine-grained position system to align the node and UAV is out of the scope of this discussion. The underlined procedure in Figure 2 is elaborated in Section 3.2.

We further take into account the constraint of a small <250 g UAV, as for these no registration or training is required according to the European Union Aviation Safety Agency (EASA) [11]. The impact of wind and gusts in the UAV flight are neglected, and flights are considered to be performed at sea level.

Related work. Tiurlikova et al. [12] propose a UAV equipped with an inductive WPT link to charge LoRaWAN-enabled devices. Alternatively, Tiurlikova et al. [13] present a multi-UAV energy harvesting network with multiple UAVs transmitting energy to several nodes through RF links.

This article primarily focuses on the feasibility of UAV-based systems, with an emphasis on energy and achievable efficiency. Control of the UAV and routing (navigation and maintenance scheduling) algorithms, such as the ones elaborated in [14,15], are considered out of the scope of this paper.

2.3. IoT Energy Storage Replacement

In this scenario, we propose the replacement of the IoT battery by engaging a UAV. Here, the UAV will rather get the function of a packet delivery system, as described by, e.g., Pugliese et al. in [16]. The main difference is the payload consisting of an energy storage buffer that is mechanically and electronically compatible with the IoT battery slot, instead of a packet. The constraint of a 250 g UAV still applies, thus limiting carried weight. Similar to the option described in the previous section, a distinction between landing and hovering can be made. We further assume the worst-case scenario in terms of energy consumption, i.e., the UAV hovers during the swapping process. So, the UAV flies to the IoT location autonomously and swaps a battery of similar weight. The extra UAV power consumption related to the additional weight is taken into account in the model parameters in Section 2.4. The analysis is elaborated in Section 3.3.

2.4. UAV Energy Consumption Model

Scenarios 2 and 3 from Figure 1 require a UAV to transfer the energy. To properly estimate the overall energy in Section 3, the energy consumption of the self-constructed UAV, depicted in Figure 3, is needed. To obtain representative estimates of the power consumption and enable analysis and optimized system design and operation, a mathematical model is derived. This model for the power consumption should at least incorporate the impact of the weight (including payload) and speed of the UAV.

Related work. Previous research has shown that UAVs are difficult to model because of the large number of parameters and variables involved. A comprehensive analysis by Zhang et al. [17] present multiple solutions to model the UAV power consumption. In their work, the authors define four categories of parameters that influence the power drawn from the UAV battery: UAV design, the environment (flight conditions), UAV dynamics, and delivery operation. Furthermore, Zhang et al. [17] describe five fundamental models for steady-level flight. Some energy consumption models, such as the one by Dorling et al. [18], ignore the impact of a UAV's speed on its energy consumption. Other models, such as the work of Ferrandez et al. [19], apply to much heavier UAVs, making them less representative for the use-case considered in this manuscript. Regression based on field trials may also be used to estimate UAV power consumption; however, this eliminates the flexibility of later changing, e.g., the payload weight. The model of Kirschstein [20] is the most comprehensive model and considers component-based parameters of the UAV. The parameters in a UAV component model are obtained by decomposing several individual forces of the UAV and describing them by separated models. The extended model from Kirschstein [20], based on previous research described in [21], estimates multiple constants from a set of equations, provided further in this paper, by conducting three experiments. The latter model considers vertical speed and airspeed combined with the total weight (incl. payload). The survey of Zhang et al. [17] shows that avionics power constitutes a significant share of the total consumption, especially in smaller UAVs. Our proposed model for the tiny, lightweight (170 g) UAV of Figure 3 should hence take into account avionics power, which is essentially the current draw of the onboard components without active motors. A more comprehensive analysis of UAV models is beyond the scope of this paper.

Figure 3. Picture of the drone considered in this study.

We hence based the UAV power consumption analysis on the model of Liu et al. [21], supplemented by the missing avionics power. This study assumes the UAV will be in steady-state flight most of the time, yielding zero net acceleration. The flight path of the UAV generally consists of a steady-state ascent, followed by a forward flight, and finally a descent. The total power P_{uav} drawn from the battery is decomposed into 4 components (Equation (1)) and is impacted mainly by the total mass m, the vertical and horizontal speed V_{ver} and V_{hor}, and the thrust T, representing the amount of upward force that the UAV is able to generate: the induced power P_i (Equation (2)), profile power P_p (Equation (3)), parasitic power P_{par} (Equation (4)), and avionic power P_{avio}. The induced power provides thrust by pushing air downwards, the profile power overcomes the rotational drag encountered by the rotating propeller blades, and the parasitic power accounts for the air drag generated in forward motion. In this model, the impact of wind is neglected, so that the airspeed equals the ground speed, denoted as V_{hor}. The total trust T is represented by Equation (5), with α the angle of attack, m the total mass, and g the gravitational acceleration.

$$P_{\text{uav}} = P_{\text{avio}} + P_{\text{i}} + P_{\text{p}} + P_{\text{par}} \qquad (1)$$

$$P_{\text{i}}(T, V_{\text{vert}}) = k_1 T \left[\frac{V_{\text{vert}}}{2} + \sqrt{\left(\frac{V_{\text{vert}}}{2}\right)^2 + \frac{T}{k_2^2}} \right] \qquad (2)$$

$$P_{\text{p}}(T, V_{\text{hor}}) = c_2 T^{3/2} \qquad (3)$$

$$P_{\text{par}}(V_{\text{hor}}) = c_3 V_{\text{hor}}^3 \qquad (4)$$

$$T = \sqrt{\left(mg - c_4 (V_{\text{hor}} \cos \alpha)^2\right)^2 + \left(c_3 V_{\text{hor}}^2\right)^2} \qquad (5)$$

The UAV of Figure 3 consists of four brushless DC (BLDC) motors controlled by the onboard electronic speed controllers (ESCs) of the Toothpick F722 flight controller. The STM32F722 MCU on the flight controller board contains modified firmware to send out relevant data through the telemetry output via a Bluetooth point-to-point connection. A Python script receives the Bluetooth messages and stores them to estimate constants in postprocessing. The simplex communication from UAV to a computer includes information on battery voltage, power consumption, flight altitudes, vertical and horizontal speed, accelerations in XYZ directions, the angle of attack, yaw–pitch–roll angles, etc.

Depending on the environment, flight times can differ significantly. For example, less lift is produced for the same amount of energy in hot air because it is less dense than cold air. Similarly, lift decreases when increasing in altitude, or if humidity increases. The type of flying will also impact the flight time. Flight control algorithms which let the UAV fly as smoothly as possible are recommended, as they will yield the longest flight time.

2.4.1. Completing the Proposed Model

To derive the model constants $\{k_1, k_2\}$ and $\{c_1, c_2, c_3, c_4\}$ from Equations (2) to (5), a total of three experiments in steady-state are required (the constant c_1 is defined as $\frac{k_1}{k_2}$).

Experimental Measurement 1: Hovering UAV with Payloads

First, a hovering experiment with different calibrated payloads m_{payload} is conducted, resulting in different average power consumption. Both the horizontal (V_{hor}) and vertical speeds (V_{vert}) are zero, thus the relation between power and weight during hovering can be written as Equation (6).

$$P_{\text{uav,hover}} = P_{\text{avio}} + (c_1 + c_2)(mg)^{3/2} \qquad (6)$$

The measured avionic power (P_{avio}) (i.e., power when UAV is active but not flying) equals 10.6 W. The data points are scattered, as shown in Figure 4a, and the relations for the constants are estimated by the least squares (LS) fit via Equation (6) and gives $c_1 + c_2 = 12.88$.

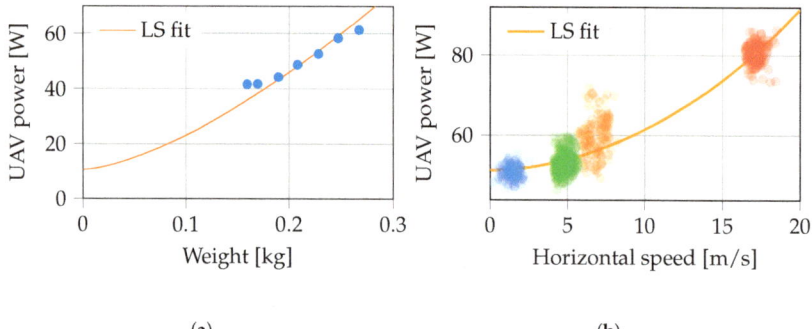

(a)

(b)

Figure 4. Experiment 1 gauged the hovering power related to UAV weight and payload. (**a**) LS fit on hovering data; Experiment 3 measured the UAV power and angle related to the horizontal speed. (**b**) LS fit on horizontal flight data. ($m_{payload}$ = 50 g).

Experimental Measurement 2: Vertical Movements

Second, the power consumption was measured and averaged during ascent and descent at a constant vertical speed to heights of up to 50 m. Multiple repeated descents and ascents yielded four data points, where the vertical velocity V_{vert} is constant. $Cos\ \alpha$ and accelerations in the horizontal UAV axes were not allowed in these measurements, meaning that the horizontal speed (V_{hor}) equals zero, and so the thrust T equals mg. Equation (7) shows the corresponding formula for the vertical power. Here, c_2 can be replaced by $12.88 - c_1$ and k_1 by $k_2 \cdot c_1$. Table 1 contains the parameters extracted from the LS fitting. The fitting was performed with five data points, two data points for ascending and two for descending, along with the energy consumption during hovering.

$$P_{uav,vertical} = k_1 mg \left[\frac{V_{vert}}{2} + \sqrt{\left(\frac{V_{vert}}{2}\right)^2 + \frac{mg}{k_2^2}} \right] + c_2 (mg)^{3/2} \qquad (7)$$

Table 1. Custom UAV parameters extracted from real-life experiments.

Parameter	Value [-]	Parameter	Value [-]
k_1	1.560	c_2	9.451
k_2	0.467	c_3	−0.0037
c_1	3.428	c_4	−0.0044

Experimental Measurement 3: Steady-Level Flight

Last, to calculate flight forward power, the UAV is flown forward at a number of constant speeds. Data points taken during high accelerations (i.e., at speed transitions) are neglected. Figure 4b shows the measurement data and corresponding least squares fitting. Equation (8) depicts the resulting formula. Since no useful angle of attack data could be extracted from the UAV measurements, $cos\ \alpha$ is assumed equal to 1. By fitting our model parameters to the resulting data, the UAV power consumption model is considered accurate enough for our purposes. The fitting here includes the cosine dependency in the model parameters, more specifically c_3 and c_4 of Equation (8).

To improve model parameters, outliers were removed. The final model parameters derived for the tiny UAV focused on are summarized in Table 1.

$$P_{uav,horizontal} = (c_1 + c_2) \left[\sqrt{\left(mg - c_4(V_{hor} \cos \alpha)^2\right)^2 + \left(c_3 V_{hor}^2\right)^2} \right]^{3/2} + c_3 V_{hor}^3 \qquad (8)$$

2.4.2. Optimal Energy Per Meter

In the proposed UAV model, an optimal propulsion speed exists to minimize the amount of energy per meter (E_{pm}). This phenomenon occurs as a large amount of hovering energy is supplemented by only a fraction of this hovering energy related to speed, and it is illustrated in Figure 5. In normal conditions, the mass is 170 g, and an optimal speed of 19.5 m/s will result in an E_{pm} of 3.7 J/m. The most efficient flight forward speed will shift to higher values as the weight of the UAV increases. In the following Sections 3.2 and 3.3, the optimal speed for the corresponding UAV weight is taken as actual value in the calculations.

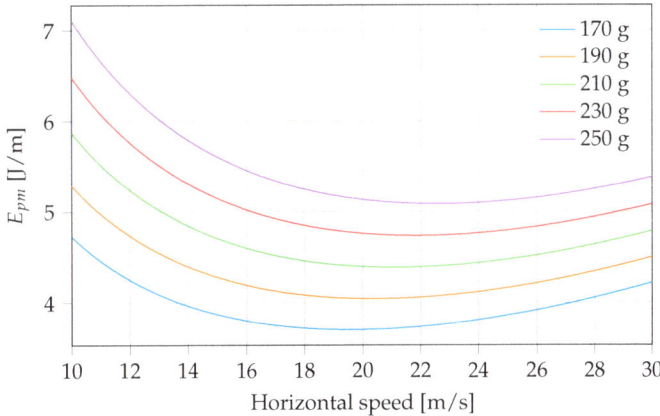

Figure 5. Energy per meter related to flight forward speed.

2.4.3. Travel Energy

The UAV flies to the location via a given path with a total energy consumption of $E_{uav,travel}$. The path can be divided into ascent, forward flight, and descent, corresponding to energy quantities $E_{uav,ascent}$, $E_{uav,ff}$, and $E_{uav,descent}$, respectively. Before the horizontal movement is initiated, the UAV rises at a velocity V_{vert} to a predefined altitude h_1, it then flies horizontally to the location to subsequently descend to an altitude h_2. During horizontal displacements, the equations assume the shortest distance to move from the initial location to the IoT node at a horizontal velocity V_{hor} over a distance d. The second vertical displacement to the nodes location at height h_2 amounts to $h_1 - h_2$. Moreover, the total mass of the UAV m is included in the travel energy. The formula accounting for the different parts of the path is shown in Equation (9) and uses the formulas of the UAV model from Section 2.4.

$$E_{uav,travel}(h_1, h_2, V_{vert}, d, V_{hor}, m) = \underbrace{P_{uav}(V_{vert}, m) \cdot \frac{h_1}{V_{vert}}}_{E_{uav,ascent}} + \underbrace{P_{uav}(V_{hor}, m) \cdot \frac{d}{V_{hor}}}_{E_{uav,ff}} \\ + \underbrace{P_{uav}(-V_{vert}, m) \cdot \frac{h_1 - h_2}{V_{vert}}}_{E_{uav,descent}} \quad (9)$$

3. Models for Energy Provisioning

The scenarios explained above are elaborated here with the aim of indicating how much energy is required under different conditions, such as energy needed on the IoT side, distance to the base station, etc.

3.1. Powering IoT Devices with RFPT

An RF transmitter sends electromagnetic energy as radio waves to the receiver antenna. The received power is converted to electrical energy. The system successively contains, from transmitter to receiver (from left to right in Figure 6), a waveform generator, a PA, an impedance matching circuit, two antennas, a secondary tuning network, a rectifier, a charge pump, and most often a maximum power point tracking (MPPT) DC/DC converter. Each element causes losses, which are included in the analysis and depicted in Figure 6.

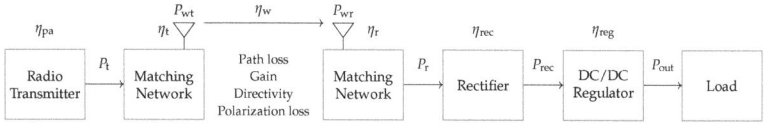

Figure 6. Transmission and efficiency model for RF power transfer [4].

Equation (10) combines all efficiency contributions to the total RFPT efficiency η_{rfpt}, with η_{pa} the PA efficiency, η_t the transmit antenna efficiency, η_w the wireless link efficiency, η_r the receive antenna efficiency, η_{rec} the rectifier efficiency, and η_{reg} the DC/DC conversion efficiency. The wireless link efficiency includes path loss, antenna gains or directivity, and potential polarization mismatch losses. Typically, the path losses are responsible for the highest loss contribution in a radio frequency power transfer system.

$$\eta_{\text{rfpt}} = \eta_{\text{pa}} \cdot \eta_t \cdot \eta_w \cdot \eta_r \cdot \eta_{\text{rec}} \cdot \eta_{\text{reg}} \tag{10}$$

3.1.1. Required Effective Isotropic Radiated Power

An expression is needed for the required EIRP for a given distance. Since EIRP represents the power just after the transmit antenna, the η_{pa} and η_t from Equation (10) can be temporarily omitted and briefly addressed in Section 3.1.3. Expressions for the remaining efficiencies n_w, n_r, η_{rec}, and η_{reg} are needed in this elaboration. The wireless link efficiency, η_w, mainly depends on the distance and the carrier frequency. The reflection coefficient of the receiver antenna is given by η_r. This antenna delivers RF energy to the harvester circuit. The harvester is responsible for converting the RF energy to direct current (DC) energy, where η_r and η_{rec} depend to some extent on the amount of incoming RF power. To find a relation between transmitted RF power and received DC power, some assumptions are made. In this analysis, we consider a SISO system without beamsteering or beamforming capabilities.

The transmission formula of Friis, depicted in Equation (11), gives the expressions for efficiency parameters η_t, η_w, and η_r, with $|\Gamma_t|$ and $|\Gamma_r|$ the matching loss at transmitter and receiver, respectively, d the distance between them, G_t and G_r the antenna gains, and λ the wavelength. Thus, $P_t \cdot G_t \cdot (1 - |\Gamma_t|^2) \cdot |\rho_t|^2$ is equal to the EIRP. Consequently, the received power P_r [dBm] depends on the distance, wavelength, EIRP, and the receiver antenna gain, as formulated in Equation (12), expressed as -values. Furthermore, this assessment assumes no polarization losses $\rho_t = \rho_r = 1$ and no matching losses $|\Gamma_t| = |\Gamma_r| = 0$.

$$\frac{P_r}{P_t} = \underbrace{(1 - |\Gamma_t|^2)}_{\text{TX matching } \eta_t} \underbrace{(1 - |\Gamma_r|^2)}_{\text{RX matching } \eta_r} \underbrace{\left(\frac{\lambda}{4\pi d}\right)^2 G_t G_r}_{\text{Path loss} \eta_w} \overbrace{|\rho_t \cdot \rho_r|^2}^{\text{Polarization loss}} \tag{11}$$

$$P_r = EIRP + G_r - 20 \log\left(\frac{4\pi d}{\lambda}\right) \tag{12}$$

To obtain a realistic estimate of the remaining conversion losses, a state-of-the-art energy harvester circuit is selected. The AEM40940 evaluation module can provide efficient DC power based on fluctuating input power using an MPPT algorithm. In summary, it can charge batteries and capacitors with RF energy and integrates a matching circuit, rectifier, and DC/DC converter [22].

The datasheet of the AEM40940 RF energy harvester provides a comprehensive discussion on the features of this IC. In addition, it includes performance curves relating the overall efficiency to the incoming RF energy. The overall efficiency is defined as the ratio of the DC power available at the output of the internal boost converter (set to 4.5V) and the RF power at the input of the matching circuit (P_{in}). The data are presented for multiple operating frequencies (867 MHz, 921 MHz, and 2.4 GHz), extracted from the datasheet [22] and transformed to the relation between the input RF energy and the DC output power ($P_{\text{dc,out}}$). The raw data points and the fitted relation are shown in Figure 7.

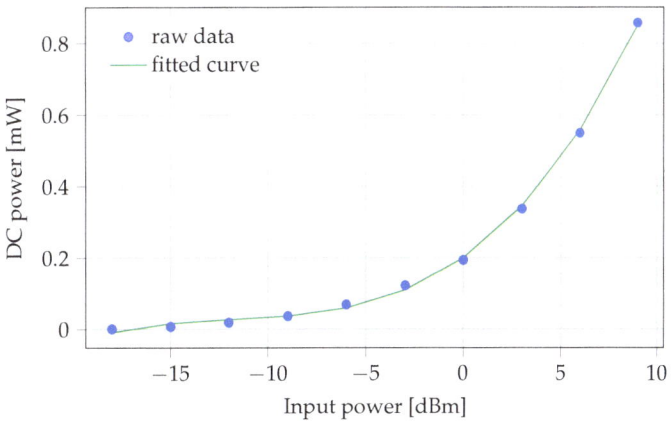

Figure 7. Fitted curve of the raw data of the AEM40940 energy harvester [22].

A third-degree polynomial fitting, as formulated in Equation (13), results in values $a = 8.45 \times 10^{-5}$, $b = 3 \times 10^{-3}$, $c = 3.84 \times 10^{-2}$, and $d = 2 \times 10^{-1}$.

$$P_{\text{dc,out}}[\text{mW}] = a \cdot P_{\text{in}}^3 + b \cdot P_{\text{in}}^2 + c \cdot P_{\text{in}} + d \quad (13)$$

We calculated the required EIRP as a function of distance and delivered DC power by combining Equations (12) and (13) with $P_{in} = P_r$. Figure 8 shows the relationship for output powers 10, 50, 100, 500 uW, and 1 mW. A working frequency of 868 MHz and half-wave dipole receiver antenna with a gain of 2.15 dBi are assumed. The former frequency was chosen due to its lower path loss compared with 2.4 GHz and because of fewer regulatory limitations on emitted power.

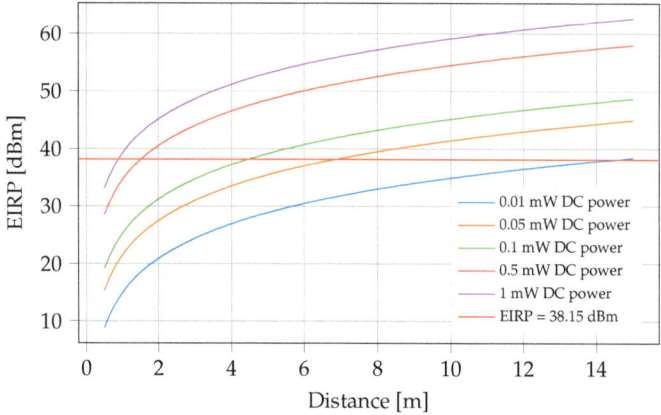

Figure 8. Required transmit power for an RF SISO system, expressed in EIRP.

3.1.2. Regulatory Restrictions

The harmonized ETSI standard *ETSI EN 302 208* discusses the radio frequency identification equipment restrictions. The equipment can operate in the 865 MHz to 868 MHz band with power levels up to 2 W and in the 915 MHz to 921 MHz band with power levels up to 4 W [5]. The levels, detailed in the standard, are expressed in effective radiated power (ERP). When using a half-wave dipole antenna, the EIRP is 2.15 dB higher than the corresponding ERP value. In conclusion, 38.15 dBm EIRP is the maximum power level allowed that can be lawfully emitted in this Industrial, Scientific and Medical (ISM) band. This limits the maximum distance when a certain DC power is required at the receiver. Figure 8 depicts the required EIRP power as a function of the distance and Table 2 shows the maximum distances for varying receive powers.

Table 2. Maximum harvesting distance 38.15 dBm EIRP related to varying DC output power levels.

DC Power (mW) (Required)	RF Input Power (dBm) (Harvester)	Max Distance (m) ($EIRP = 38.15$ dBm)
0.01	−14.15	14.5
0.05	−7.62	6.9
0.1	−3.88	4.7
0.5	5.43	1.6
1	10.06	0.9

Techniques such as beamforming give a more directive beam but will not directly increase the operating distance. The reason is that the EIRP limit still applies. However, it is important to note that the antenna beamwidth can affect the EIRP values. As stated in [5], keeping the beamwidth within certain limits, results in higher permissible transmit power. The antenna beamwidth can be shaped by selecting patch antennas or implementing beamforming with multiple antennas.

3.1.3. Power Amplifier

The preceding analysis omitted the PA losses. This component has a significant contribution to the total RFPT losses and should be included in efficiency analyses. Power amplifiers ensure amplified modulated signals. These circuits are connected to, and matched with, the transmit antenna. RFPT requires only an amplified single-tone signal, without modulated signals, connected to the transmit antenna. An amplified high-frequency sine wave can be generated via different circuits, each coming with their corresponding maximum achievable efficiencies.

Three types of efficiencies for power amplifiers are used in the literature [23]: the drain efficiency (DE), power-added efficiency (PAE) and η_{overall}. In this paper, the drain efficiency DE suffices, meaning the RF output power $P_{\text{RF,out}}$ divided by the input DC power $P_{\text{DC,in}}$, represented in Equation (14), also denoted as PA efficiency (η_{pa}).

$$\eta_{\text{pa}} = DE = \frac{P_{\text{RF,out}}}{P_{\text{DC,in}}} \tag{14}$$

The overall RFPT efficiency is further formulated in Section 4.1.1.

3.2. Energy Delivery to Battery-Powered Devices

A UAV can be used to provide power to IoT nodes. The total UAV power consumption is primarily determined by three factors:

1. UAV flight and avionics consumption is a combination of hovering, ascending, descending, and steady-level flight energy. The model for the UAV is detailed in Section 2.4. The total energy consumption to fly to the node's location and vice versa is called the total travel energy and is given in Section 2.4.3.
2. The selected energy storage at the IoT node determines the charge rate and the amount of power transmitted to the node. A brief overview is given in Section 3.2.1.
3. Power transfer technology between the UAV and IoT node imposes additional constraints on the system, such as limited distance between WPT-transmitter and receiver, efficiency, losses, and maximum transferable power, as discussed in Section 3.2.2.

3.2.1. Energy Storage Selection

In applications, an IoT node is typically designed with a certain autonomy in mind. Even if a UAV-based approach is pursued, it is not practically feasible that batteries need to be recharged too frequently. Therefore, the storage technology must have a certain capacity to last several weeks/months on a single charge, and the battery charge time should be as short as possible, since the UAV is hovering in the meantime. In this article, three different types of rechargeable battery compositions are suggested for use in embedded devices. The three storage solutions, i.e., electrostatic double-layer capacitors (EDLC), lithium titanate (LTO), and lithium-ion capacitor (LIC), are compared in the spin diagram in Figure 9. These EDLC, LIC, and LTO cells can be charged much faster than conventional lithium cells, for which lithium cobalt oxide (LCO) is the most common composition [24,25]. This is a critical aspect in the proposed approach to avoid long hovering times.

One of the most crucial parameters during the selection of the storage solution is the self-discharge. We are carrying out a long-term experiment tracking self-discharge, as depicted in Figure 10. The results clearly indicate that EDLC cells have an excessive self-discharge rate. They are therefore not considered further.

The remaining options are further compared. Figure 9 shows that LTO or LIC chemistries come with reasonable volumetric energy densities while having very low internal resistance and high charge rates. Both technologies were charged at the maximum speed as advised in the manual and represented in Figure 11. An LTO battery with a measured capacity (KORAD KEL103) of 66.3 mAh is charged according to a constant current (CC) constant voltage (CV) method. With a 663 mA (10 C) charge current and a 2.8 V maximal voltage, the battery state of charge (SoC) rises to 90% after 6 min, corresponding with 581 J stored energy. Similarly, the charging process of the LIC cell with a capacity of 50 F was measured. With a charging current of 2.8 A and a maximum voltage of 4.2 V [26], it took approximately 2.6 min to achieve a SoC of 90%, corresponding with 472 J stored energy. The charge power of the LTO cell is more constant in time compared with the LIC cell. LIC cells initially charge faster, resulting in the need for high current peaks. Charging components have to be dimensioned according to the higher peak current, as possible, with the fast-charging LIC cells. This can result in a higher bill of materials (BOM).

We further define the SoC at the start of the charging process as $SoC_{initial}$ and the SoC at which charging is terminated as SoC_{target}. A cut-off current, as defined in the manual, is used to determine the end of the charging process, i.e., at a 100% SoC. The time to achieve a fully recharged cell is rather high compared with the time to achieve a 90% SoC, since the amount of stored energy per unit time decreases when approaching a 100% SoC. Figure 11 also depicts this charging rate decrease.

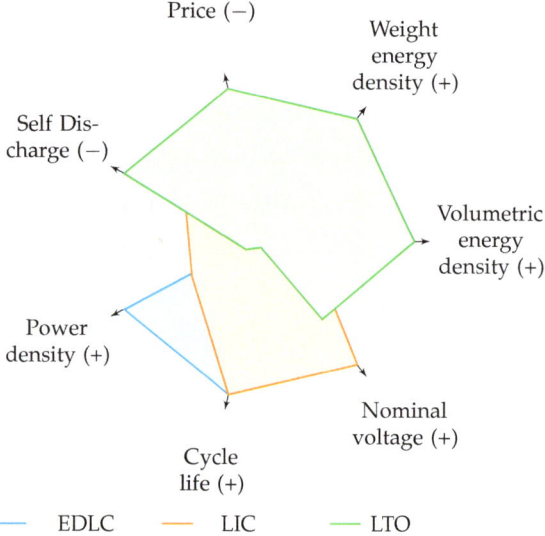

Figure 9. Spin chart comparing properties of energy storage compounds EDLC (orange) LIC (blue), and LTO (green). (+) higher is better, (−) lower is better.

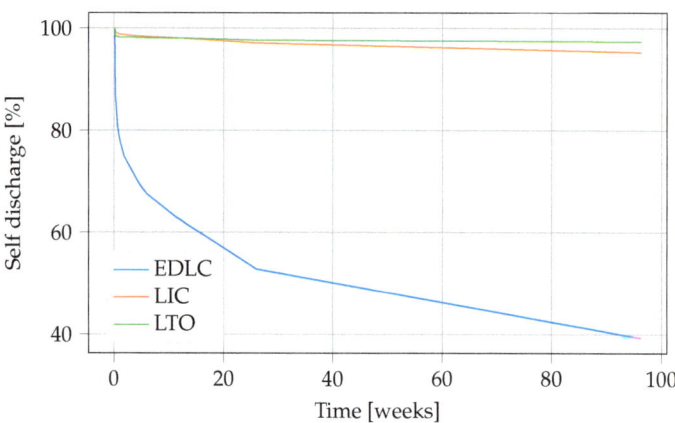

Figure 10. Self-discharge of EDLC, LIC, and LTO cells measured over approximately two years. The self-discharge is calculated as the percentage of remaining cell voltage compared with the initial voltage as a function of time.

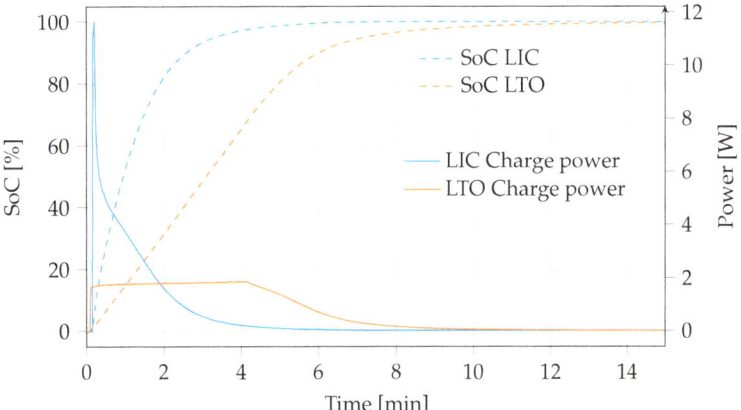

Figure 11. Measured charge curve of a 66.3 mA h LTO battery and a 50 F LIC capacitor.

Furthermore, the initial or remaining SoC has a positive impact on charge time but is not advantageous to the amount of transferred energy. For example, for the LTO cell, the time in CC-mode will decrease if $SoC_{initial} > 0\%$. The average amount of stored energy per unit time will become lower. In short, it is better that the IoT cell is sufficiently discharged at the time the UAV comes to deliver energy.

LTO technology is further assumed to be implemented in the IoT devices for the following key reasons: the fast charge time in combination with the low self-discharge and the relatively high energy density.

3.2.2. Energy Transfer Link

The survey in [4] summarizes the wireless power transmission technologies, subdivided in coupled and uncoupled approaches. The latter allows higher operating distances compared with the coupled technologies. Due to restrictions in radiation levels of both light and RF sources, LPT and RFPT systems are not suitable to quickly recharge the IoT energy storage. In addition to uncoupled technologies, the coupled approaches such as capacitive power transfer (CPT) or inductive power transfer (IPT) need a very strict alignment method. In the scenarios considered in this article, strong alignment is difficult to achieve, and thus the CPT and strongly coupled IPT systems are not good choices to transfer the energy. An inductive system with both receiver and transmitter in resonance, i.e., Magnetic Resonance Coupling (MRC), provides the most appropriate solution. IPT systems operate with highly coupled and aligned coils (coupling factor $k > 0.3$ [27]). MRC systems, on the other hand, can operate on coupling factors below 0.1, giving these resonant systems more spatial freedom and better handling of fluctuations in k values. MRC is based on two coils tuned to the same resonance frequency, in which, even at low coupling factors, high link efficiencies can be achieved. The amount of power required through the link will range from a few 100 mW to several watts, which is transferable with MRC systems.

To accomplish further analysis in Section 3.2.3, an assessment of the efficiency of the WPT system is performed. The transmitter PCB coil fits with the proposed UAV frame of Figure 3 with dimensions 145 by 165 mm and a total of two windings. The receiver PCB coil has four windings and dimensions of 100 by 100 mm. Via Ansys Maxwell finite element analysis, self-inductances ($L_1 L_2$), series resistances ($R_1 R_2$), and mutual inductance (M) between the coils are obtained. The configured frequency is set to 6.78 MHz, as also used by the Airfuel Alliance Resonance [28]. The coupling factor (k_{mrc}) is calculated with Equation (15).

$$k_{mrc} = \frac{M}{\sqrt{L_1 + L_2}} \quad (15)$$

Based on the coupling factor, the maximum link efficiency (η_{mrc}) can be calculated using Equation (16) [29], where Q_T and Q_R are the quality factors of the transmit and receive coils, respectively. This assumes the most optimal load on the receiver side. If the receiver circuit is not loaded with the optimal resistance, the link efficiency decreases. Additional analysis assumes an optimal load.

$$\eta_{\text{mrc}} = \frac{k_{\text{mrc}}^2 Q_T Q_R}{1 + k_{\text{mrc}}^2 Q_T Q_R} \quad (16)$$

We conducted finite element simulations for two vertical misalignments of 100 mm and 150 mm. Lateral misalignment is being neglected, and angular misalignment is minimal, because the UAV coil is located parallel to the receiver coil during hovering at the node's location. The results are shown in Table 3.

Table 3. Calculated link efficiency using simulation data obtained via the finite element analysis for two vertical misalignments.

(x, y, z) [mm]	M [µH]	L_1 [µH]	R_1 [mΩ]	L_2 [µΩ]	R_2 [mΩ]	k_{mrc} [-]	η_{mrc} [%]
(0, 0, 100)	0.083	2.165	489	3.488	612	0.035	76.5
(0, 0, 150)	0.032	2.166	488	3.483	610	0.014	51.0

The simulations show that the maximum link efficiency will achieve percentages of 50% or more for the selected coils at distances of up to 150 mm. Additionally, the total efficiency of the WPT system (η_{wpt}) is composed of several other individual component contributions, such as the preamplifier ($\eta_{\text{pre-amp}}$), power inverter (η_{inv}), rectifier (η_{rect}), and battery charger (η_{smps}). The η_{wpt} efficiency is stated in Equation (17).

$$\eta_{\text{wpt}} = \eta_{\text{pre-amp}} \cdot \eta_{\text{inv}} \cdot \eta_{\text{mrc}} \cdot \eta_{\text{rect}} \cdot \eta_{\text{smps}} \quad (17)$$

It would take the reader too long to elaborate on all components of an MRC system individually. We assume low losses in the remaining components. Due to new technologies such as GaN FETs, high efficiencies (>0.9) in DC/AC inverters and DC/DC converters are feasible. In the further analyses, we assume a total efficiency η_{wpt} of 50%. While we so far only discussed the efficiency of a WPT system, evidently, limitations in energy transfer due to component selection also exist.

3.2.3. Limitations of IoT Storage Capacity

A first analysis investigates the maximum size of the IoT storage capacity. The capacity cannot be selected as infinitely large, since the UAV would have to hover for a very long time to transfer energy to the node. Furthermore, the available energy of the UAV is limited to the selected UAV battery capacity, which consequently translates to limitations in the size of the IoT storage capacity. For the IoT storage buffer, we assume a 90% charge after each intervention. The nodes are reloaded from an initial SoC of 0% to the target SoC of 90%.

The analysis is performed for the UAV in Figure 3 weighing 170 g with a battery capacity of 9.62 W h. The distance between departure and arrival location is given by d and can be gradually increased to test the limits of this setup. The optimal speed from Section 2.4.2, here named V_{hor}, is 19.5 m/s. It is further assumed that the UAV flies to 50 m (h_1) altitude at the departure location and hovers to 10 m (h_2) altitude at arrival. In both cases, the vertical speed is set at 10 m/s (V_{vert}). These proposed parameters allow the UAV travel energy, explained in Section 2.4.3, to be calculated.

In this example, an LTO cell is charged at the node's location. We assume, the charge process follows a simplified linear model. Mei et al. [30] already measured the charge time for LTO cells related to different C-rates as being the rate at which a battery can be charged relative to its capacity, specifically in CC-mode. The charging time (in CC-mode) decreases from 27.5 min to approximately 6 min when changing the charge rate from 2 C to 8 C, respectively. As shown in Figure 11, the charging curve proceeds linearly to about 90% SoC. In our measurement, the maximum C-rate is set to 10 C. The maximum charge rate may vary depending on the brand of LTO cell selected [31,32].

We assume a UAV equipped with an inductive transmitter system with an efficiency of 50%, as explained in Section 3.2.2. The WPT link is limited to a certain transfer power $P_{ch,max}$. This analysis assumes a received power of 1 W. Based on the received power, the C-rate can be determined via Equation (18) together with the battery capacity E_{node}. The analysis additionally limits the maximal C-rate to 10 C.

$$C_{rate} = \frac{P_{ch,max}}{E_{node}} \cdot 3600 \leq 10\,C \qquad (18)$$

Based on the parameters such as the output power sourced from the WPT receiver, the target SoC, and lastly the size of the IoT storage, an estimate of the charge time t_{ch} can be found according to Equation (19).

$$t_{ch} = \frac{3600}{C_{rate}} \cdot (SoC_{target} - SoC_{initial}) \qquad (19)$$

The total energy $E_{uav,overall}$ required for an intervention can be calculated with Equation (20) and is subdivided into three parts:

1. The travel energy $E_{uav,travel}$ from Equation (9) for the outward and return flights. We multiply the traveling energy by two. Although, if h_1 and h_2 are different, there is a small deviation between the outward and return travel energy present. This small deviation is neglected here.
2. The hovering energy $E_{uav,hover}$, which depends on the charge time and hovering power from Equation (6).
3. The total amount of energy consumed by the WPT transmitter $E_{wpt,transmitter}$, with η_{wpt} the efficiency determined by the energy delivered by the UAV battery relative to the energy stored in the node.

$$E_{uav,overall} = 2 \cdot E_{uav,travel}(h_1, h_2, v_v, d, v_h, m) + \underbrace{t_{ch} \cdot P_{uav,hover}}_{E_{uav,hover}} + \underbrace{(\eta_{wpt})^{-1} \cdot E_{IoT,storage}}_{E_{wpt,transmitter}} \qquad (20)$$

Figure 12 shows the total energy requirement related to the size of the IoT battery for multiple distances. Taking into account the considered boundary conditions, the selected UAV will be able to recharge a storage of ≈900 J over a distance of 500 m.

The slope remains rather flat for smaller IoT batteries, due to the charge rate limitation of the chosen battery composition, in this case, 10 C for the LTO cell. A lower travel distance logically provides the opportunity to transfer larger amounts of energy and consequently store higher quantities of energy. In contrary, at increased distances, limitations in storage sizes will arise due to UAV battery capacity limits, indicated by the dotted line. Based on this graph and the daily consumption of the application, the autonomy of the IoT device can be determined. Subsequently, the number of yearly interventions can be calculated.

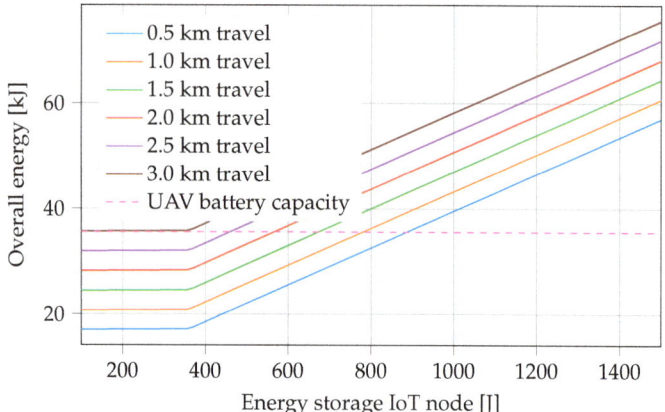

Figure 12. Maximum IoT storage capacity, limited by the battery capacity of the UAV, represented by a dotted line. The UAV battery capacity is 9.62 W h, and the maximum charge power $P_{ch,max}$ is set to 1 W. The energy storage IoT node involves an LTO cell with a charge rate of 10 C.

3.2.4. Duration of an Intervention

Figure 13 gives an estimate of the total intervention duration. The calculation accounts for the UAV battery capacity limitations, resulting in finite curves when battery capacity is insufficient. Remarkably, longer fly times can occur for shorter travel distances (e.g., $d = 0.5$ km) with the same UAV battery capacity. In that case, most of the intervention time is attributable to the node's charging process and not due to travel time, since this is constant for each distance. This analysis was performed under the same conditions used in Figure 12.

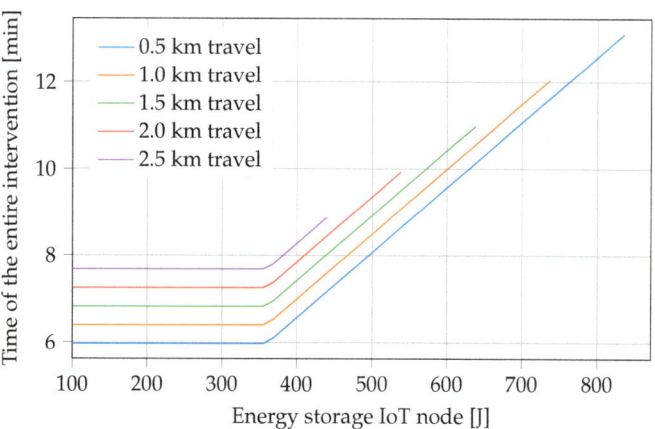

Figure 13. Overall operation time during one intervention. The same assumptions were adopted from Figure 12.

3.3. Swapping Embedded Batteries

In this third scenario, we consider a UAV capable of replacing the IoT node's battery. Rechargeable battery compositions make sense to use for environmental reasons, as elaborated in Section 5. non-rechargeable batteries can also be used for battery swapping, although their lifespan is much shorter than that of rechargeable batteries, which can be reused many times. For the actual energy calculations in this analysis, it is irrelevant whether the battery is rechargeable or not. The properties of the selected battery composition are more important.

Batteries with high weight energy densities are recommended here. These battery types are usually charged by means of lower C-rates than, e.g., the already discussed LTO, LIC or EDLC compositions. These higher storage capacities can, on the one hand, enable applications with higher energy needs or, on the other hand, increase autonomy. The latter consequently also reduces the number of UAV interventions. The drained batteries can be recharged at slow C-rates in the same place as the UAV battery, since there is no need for the recharging to happen on-site. Table 4 lists some commonly used batteries and shows the nickel cobalt aluminum (NCA) battery composition will provide the best weight energy density.

Table 4. Common rechargeable batteries with their weight energy density. The first four are lithium-based, while the latter two are nickel batteries.

Composition	Weight Energy Density [Wh/kg]	Reference
NMC	≈217	[33]
LFP	≈114	[34]
NCA	≈237	[35]
LTO	≈80	[31]
NiCd	≈36	[36]
NiMH	≈86	[37]

Other important parameters, such as self-discharge, sustainability aspects, life cycle assessment (LCA) of the batteries, shelf life, etc., are neglected here. These factors may impact the rudimentary choice to select for the highest weight energy density. Section 5 discusses briefly the environmental impacts of these different battery types.

This approach poses mainly practical implementation challenges, such as the implementation and design of the swapping mechanism. Existing work has already investigated battery-swapping mechanisms, more specifically, the replacement of a UAV battery [38,39]. Similar implementations can be explored to replace the IoT node's battery. Ultimately, this process must take place autonomously, whereby the efficiency highly depends on the replacement time ($t_{replacement}$), i.e., the time to complete the battery-swapping process. The battery should be easily detachable from the system or perhaps connected contactlessly to provide a smooth replacement. For the purpose of this paper, we neglect the mechanical difficulties of the battery-swapping system and focus on the energy reloading capabilities and efficiency.

The previous Equation (20) can be adapted to Equation (21) to estimate the overall energy consumption in the battery-swapping operation. Note that the added weight of the battery ($m_{payload}$) to the UAV weight (m_{uav}) will have a significant impact on the results. Several assumptions are made to estimate the UAV power consumption. The model from Section 2.4 is used, and the UAV is considered to travel at the weight-dependent optimal speed, as calculated in Section 2.4.3. A very cautious estimate of 60 s, i.e., $t_{replacement}$, for a battery swap is assumed. In terms of weight, both outbound and return flights carry the same payload.

$$E_{uav,overall} = 2 \cdot E_{uav,travel}(h_1, h_2, v_v, d, v_{h,optimal}(m_{uav} + m_{payload}), m_{uav}, m_{payload}) \\ + \underbrace{t_{replacement} \cdot P_{uav,hover}(m_{uav} + m_{payload})}_{E_{uav,hover}} \quad (21)$$

Figure 14 depicts the UAV energy needs for varying payloads and servicing distances. The UAV can carry a maximal payload of 80 g, and the selected battery contains a maximum usable energy of 34.63 kJ. These restrictions are represented by the green shading in the figure. For example, taking into account the discussed assumptions with a 20 g IoT battery and a travel distance of 2 km, a UAV energy of approximately 20 kJ is required to perform the task.

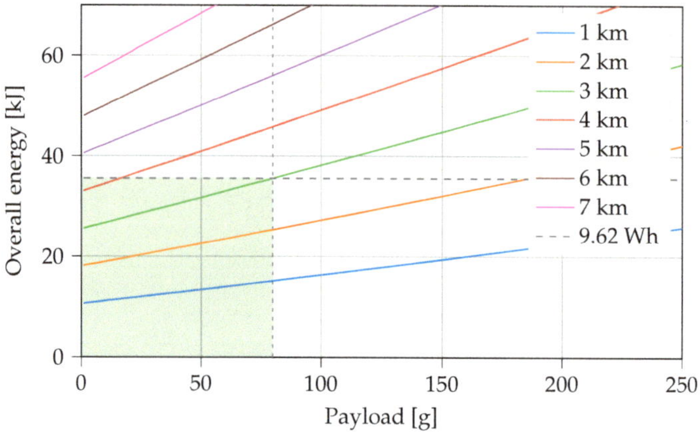

Figure 14. UAV energy needs for varying payloads and servicing distances. Delineated in green are the practically feasible cases, due to battery energy and payload limits. Assuming one UAV battery (9.62 W h), the payload can be as high as 80 g.

4. Comparative Survey and Results

This section evaluates the three scenarios based on figure of merit (FoM) parameters, complemented by a comparative discussion. It further investigates whether the results can be improved by selecting other types of UAVs.

4.1. Figure of Merit Analysis

To provide a means to assess the advantages of the scenarios introduced in Section 3, this section makes a comparison and defines a FoM. We first briefly summarize how the scenarios differ from each other.

(S1) Section 2.1 describes an RFPT SISO system. Since this can theoretically be switched on indefinitely, and hence time is not an issue, the power consumption is the main focus. The efficiency η_{rfpt} is calculated through the relation between DC output power and DC input power.

(S2) Section 3.2 rather concentrates on providing energy on site. The efficiency $\eta_{delivery}$ is defined by the stored energy in the IoT device related to consumed UAV energy.

(S3) The approach, explained in Section 3.3, requires another FoM calculation to evaluate the performance. The UAV with battery-swapping capabilities transports energy based on a separate energy storage medium, i.e., a reusable battery. The energy efficiency here no longer depends on how efficiently the UAV's battery is used but rather on the amount of energy delivered compared with the amount of initial energy. This initial energy is made up of the combined capacities of the carried and UAV batteries.

4.1.1. Radio Frequency Power Transfer Efficiency

The models from Section 2.1 defined the RF transmit power to get a certain amount of DC power on the receiver. Efficiency is quantified based on these estimates. In order to describe the RFPT overall efficiency, a realistic nonideal PA is assumed. Examples from Johansson and Fritzin [40] show that PAs can achieve drain efficiencies of up to 77% for power outputs above 31 dBm. The results depend on the distance and, to a lesser extent, the required receiving power. The efficiency is illustrated in Figure 15. It is important to notice that all efficiency values <0.1%, while only ranges lower than 5 m are considered for this scenario.

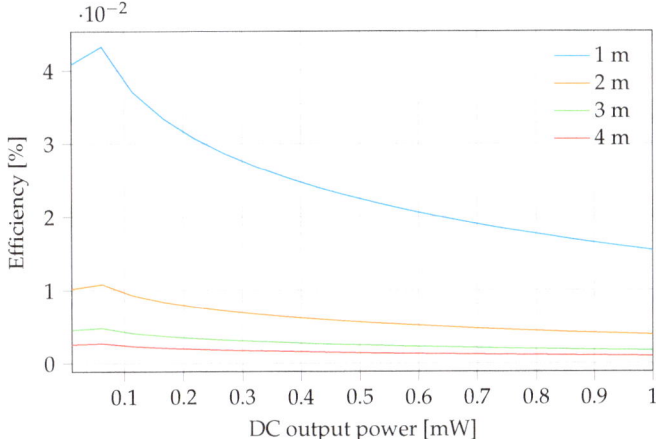

Figure 15. Estimate of the achievable efficiency levels for a SISO RFPT system with a PA with 77% efficiency. The maximum EIRP limit of 38.15 dbm is not taken into account.

4.1.2. UAV as Mobile Powerbank: Overall Efficiency

The overall efficiency $\eta_{\text{uav,charge}}$ is a straightforward extension of the previous analysis and can be determined by Equation (22). $E_{\text{uav,overall}}$ is elaborated in Section 3.2 and consists of the travel energy for outward and return flights, combined with the hovering energy while charging, as well as the energy consumed in the WPT transmitter. E_{node} is the energy received at the IoT node.

$$\eta_{\text{uav,charge}} = \frac{E_{\text{node}}}{E_{\text{uav,overall}}} \cdot 100\,\% \tag{22}$$

Figure 16 presents the overall efficiency related to the storage capacity of the IoT device. On the one hand, tiny IoT storage capacities logically result in lower efficiencies. On the other hand, oversized IoT storage capacities result in a situation in which no solution for the overall efficiency can be found. This is due to limitations in UAV battery capacity. The required energy (travel energy, hovering energy and UAV transmitter energy) exceeds the battery capacity, leading to an incalculable efficiency. Note that at the end of the curve, the UAV battery is fully utilized.

The dotted lines represent estimates of the UAV equipped with a dual battery. Obviously, this extra weight causes an increase in hovering energy. Therefore, the overall efficiency decreases for similar distances, despite the fact that the IoT storage capacities may increase. This is represented in the graphs of Figure 16.

A brief analysis of the parameter impact on overall efficiency is essential here. Consulting Equation (22) reveals that primarily $P_{\text{ch,max}}$ strongly influences the overall efficiency. More specifically, $\eta_{\text{uav,charge}}$ increases by a factor of ≈ 7, if the $P_{\text{ch,max}}$ increases from 1 W to 10 W. In addition, variations in WPT efficiency could also influence the overall energy efficiency and show to have a higher impact when $P_{\text{ch,max}}$ is large. For example, the variation from 10% to 50% WPT maximum achievable efficiency at $P_{\text{ch,max}}$ of 1 W and a travel distance of 1000 m indicate an increase from 1.7% ($E_{\text{node}} = 615$ J) to 2.1% ($E_{\text{node}} = 754$ J). A $P_{\text{ch,max}}$ of 10 W for similar travel distance provides an increase in η_{overall} from 4.2% ($E_{\text{node}} = 1487$ J) to 14.1% ($E_{\text{node}} = 5005$ J).

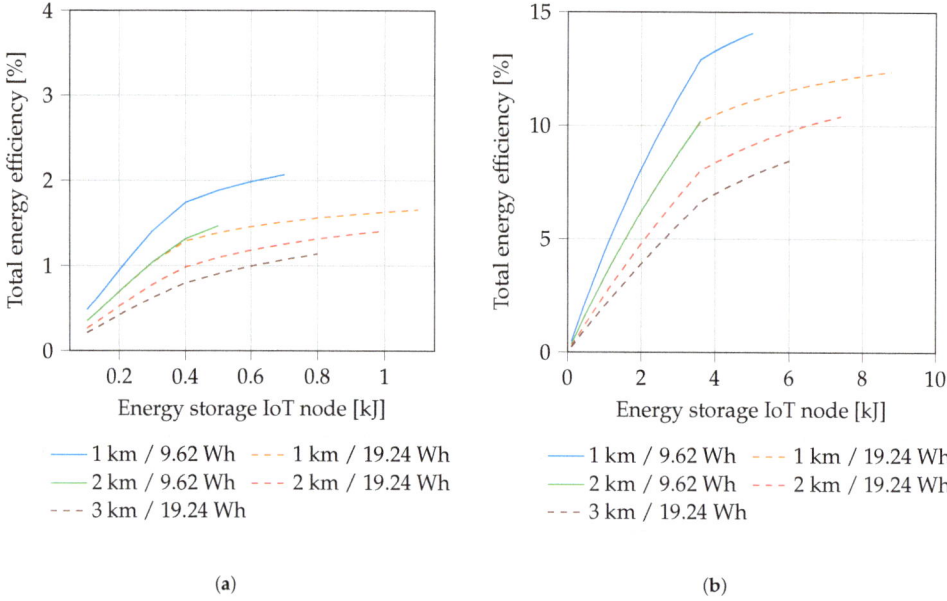

(a)   (b)

Figure 16. Overall energy efficiency including travel, WPT, and hover losses. A 3 km travel distance with a single UAV battery is not achievable, as also noted in previous analyses. (**a**) $P_{ch,max} = 1\,W$; (**b**) $P_{ch,max} = 10\,W$.

4.1.3. Swapping IoT Batteries: Overall Efficiency

To evaluate this third scenario, the efficiency ($\eta_{uav,swap}$) in Equation (23) is determined here, now in a slightly different way. Since the delivered energy is not extracted from the UAV battery, the carried energy must be included in the denominator. $E_{uav,overall}$ equals Equation (21), which amounts to two times the travel energy, along with the hover energy during the swapping process.

$$\eta_{uav,swap} = \frac{E_{node}}{E_{uav,overall} + E_{node}} \cdot 100\,\% \quad (23)$$

The efficiency is a function of the payload and depends heavily on the weight energy density of the battery technology. An NCA battery is assumed to be swapped in this graph. As shown in Figure 17, the curves increase when heavier batteries are selected. It is unrealistic to assume that the payload exceeds the weight of the UAV multiple times. As mentioned earlier, we consider a UAV of 250 g, implying a residual maximal payload of 80 g. Figure 17b shows the adapted curves, when the UAV's battery is doubled. In the latter case, the payload is limited to 21 g.

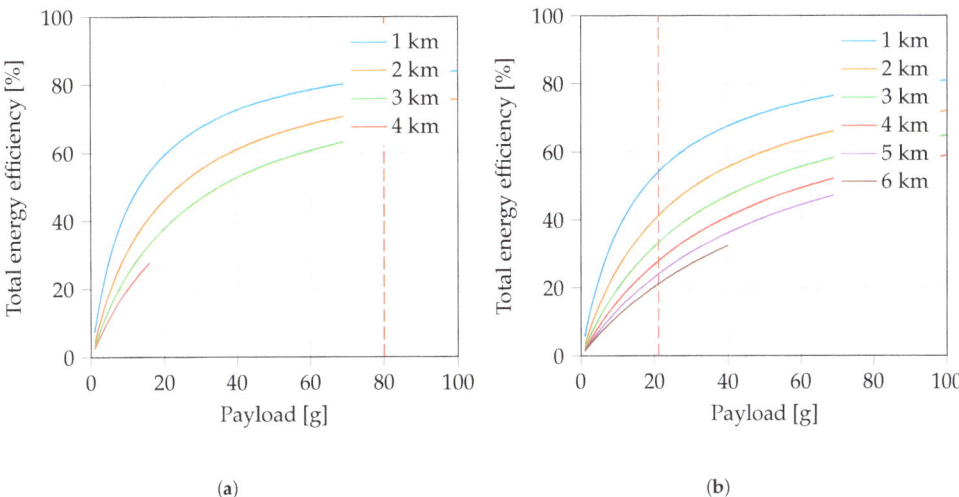

Figure 17. An NCA battery, with a weight energy density of 237 Wh/kg, assumed to be transported by the UAV. Assuming the total weight of the UAV should not exceed 250 g, the maximum payload of 80 g (one UAV battery) and 21 g (two UAV batteries) is indicated by the dotted vertical line. (**a**) One UAV battery; (**b**) two UAV batteries.

4.2. Figure of Merit Comparison and Overall Discussion

As earlier demonstrated in Section 4.1.1, the RFPT system reaches very low efficiency values, i.e., of a few hundredths of a percent and for distances in the range of meters. This tremendous disadvantage means that this technology has very limited use cases, i.e., only usable in very low-power applications. The overall advantage of RFPT is the nearly infinite amount of time that small amounts of energy can be delivered. Furthermore, this technology can serve multiple devices simultaneously, in contrast to the UAV approaches.

Unlike RFPT, UAV-based servicing has clear advantages due to the large serviceable area. The disadvantage is the more complex implementation, requiring flight planning algorithms and battery-swapping mechanisms, posing wireless charging challenges, etc. This section primarily aims to evaluate efficiencies in various situations. It does not make sense to compare the short wireless power range with a possible long-range UAV solution. It is more meaningful to compare the charge and swap process efficiencies from Sections 4.1.2 and 4.1.3 respectively and not further consider the RFPT technology here.

Table 5 compares several situations. Note that this comparison focuses on the side of the energy transfer to the node. The relation with parameters impacting the energy consumption of the node consumption, e.g., number and size of packets, is out of scope of this comparison. We refer to related work clarifying these aspects [2,10]. In all cases, the optimal speed criterion for the UAV, depicted in Section 2.4, is met. The battery technologies LTO and NCA at the IoT node are assumed for the charging and swapping process, respectively. The payload (including the optional second UAV battery) is still limited to 80 g. The maximum charge power is set to $P_{ch,max} = 10$ W. Based on these assumptions, along with the number of UAV batteries, travel distance, and IoT storage capacity, the FoM values are calculated. The UAV flying characteristics and weather conditions are assumed equal in the comparisons. The remaining SoC_{uav} was added to the table to make a fair comparison. Below, these four situations are briefly discussed.

Table 5. Comparison of FoM values for different scenarios.

Cf. nr.	Scenario	Payload	IoT Battery Composition	# UAV Batteries	Distance [km]	Storage Capacity [kJ]	Remaining SoC_{uav} [%]	FoM Value [%]
1	Charging	0 g	LTO	1	2.5	1.8	0.3	5.1
	Swapping	80 g	NCA	1	2.5	68.3	14.4	69.2
2	Charging	59 g	LTO	2	3.7	5	0.8	7.1
	Swapping	64.9 g	NCA	2	6.3	5	7.8	7.1
3	Charging	59 g	LTO	2	1	4	46.5	10.5
	Swapping	63.7 g	NCA	2	1	4	80.0	22.0
4	Charging	59 g	LTO	2	4	4	6.5	6
	Swapping	63.7 g	NCA	2	4	4	39.4	8.5

(Cf. 1) Battery swapping, even with a full payload, is more efficient than in situ wireless battery recharging. In this case, the UAV consumption is even smaller than the transported energy. Figure 17 already showed that higher payloads lead to higher FoM values, although an amount of 68.3 kJ of rechargeable IoT storage is rather unrealistic for low-power IoT integrations. They do not benefit from an oversized battery, since this reduces the environmental friendliness of the approach. A more energy-efficient solution is to use the full payload capability, which means transporting multiple IoT batteries at the same time to serve more devices in a given area. This analysis is not covered in this study.

(Cf. 2) An equally large IoT battery capacity of 5 kJ with a similar FoM demonstrates that the swapping process appears to be the most advantageous once again. In this case, the distance can reach 2.6 km further located nodes while maintaining the same FoM value.

(Cf. 3) If both numbers of UAV batteries, travel distance, and storage capacity are assumed equal, the swapping process is still advantageous. Furthermore, the UAV battery may preserve a significant amount of residual energy.

(Cf. 4) Continuing on (Cf. 3), the difference in FoM values becomes smaller with increasing distance. In this example, there is only a 2.5% difference. Obviously, the remaining SoC is higher when swapping, since the charge energy comes from the UAV battery.

Another way to compare wireless charging with battery swapping is to plot the FoM with respect to the distance. As shown in the previous analysis from Table 5, the UAV battery is not necessarily fully drained after an intervention. To give a better picture, both efficiency and the remaining SoC_{uav} are plotted in Figure 18. In both cases, it is assumed to bring an energy quantity of 1000 J to the IoT device location. The remaining assumptions are summarized in Table 6, with the UAV battery that is still 9.62 W h and the hovering time t_{hover} equal to t_{ch} or $t_{replacement}$ for charging or swapping processes, respectively. Furthermore, the parameters $P_{ch,max}$, C_{rate}, and η_{wpt} remain crucial for estimating the energy consumption of the charging approach. Moreover, the weight energy density of the carried battery is essential to estimate the intervention energy for UAVs with swapping capabilities.

In addition to the fact that the swapping approach appears more efficient, this graph also represents the remaining energy of the UAV battery. In some cases, especially for closer located devices, the UAV could serve multiple IoT devices without recharging itself. This can be beneficial for the overall FoM value, since more energy is supplied to the drained IoT devices during one intervention. Additional analysis on serving multiple devices with energy is not elaborated on in this manuscript.

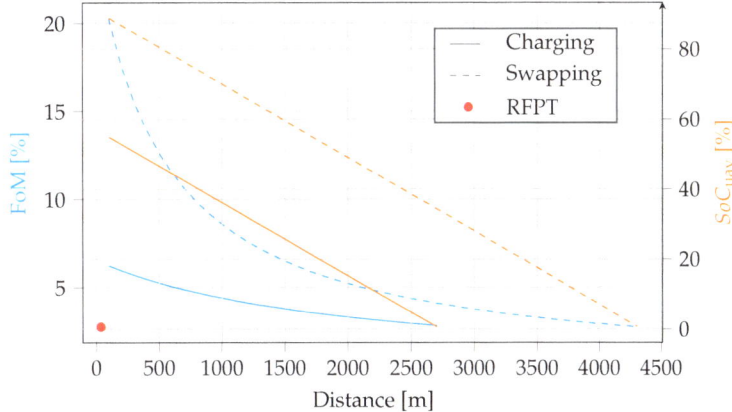

Figure 18. FoM comparison according to the parameters from Table 6.

Table 6. Proposed parameters to compare charging process against swapping process.

Parameters		Charging	Swapping
Size of depleted IoT battery	[J]	1000	1000
No. UAV batteries	(−)	1	1
$P_{ch,max}$	[W]	10	N/A
C_{rate}	(−)	10 (LTO)	N/A
η_{wpt}	(%)	50	N/A
Weight energy density IoT battery	(Wh/kg)	N/A	237 (NCA)
$m_{payload}$	(g)	0	1.17
t_{hover}	(s)	324	60

Previous assessments neglected the weight of the charging mechanism and the swapping mechanism. Results can differ when assuming a different hovering time for the battery replacement. The duration of 1 min to swap a battery is currently a rough estimation. Even when considering a battery-swapping time of 2 min, the swapping process is still advantageous, e.g., in (Cf. 1), the FoM becomes 66.7%, and in (Cf. 4), the FoM becomes 7.9%. In future work, the swapping process can be further studied, and the impact of mechanical options can be identified based on actual implementations.

4.3. UAV Optimizations

The UAV considered in this study is well-suited to give a proper estimate of the possibilities and challenges of UAV-based servicing of IoT nodes. Since this a very small UAV due to legislation restrictions, power consumption is quite low. There are more efficient UAVs models available, which could result in an even lower E_{pm}. Larger blade spans combined with slower turning motors can yield a more efficient flying setup, but more optimal solutions exist based on propeller and motor selection [41]. The frame was chosen for its versatility and adaptability. A 3D-printed version of the frame was realized, which resulted in less than ideal flight characteristics. There were minor vibrations throughout the entire structure, since the frame was not as rigid as the carbon fiber version. Moving to a stiffer frame could greatly reduce vibrations. In a small UAV, the power consumption of the flight electronics has a significant impact on the total flight time, i.e., for this model 20% of the total power consumption. Lithium polymer (LiPo) batteries are typically used in UAVs due to their high discharge rate [42]. Lithium-ion (Li-ion) batteries can prolong flight times by offering higher energy-to-weight densities [33]. The relatively low discharge rate of Li-ions makes them only suitable in UAVs with a low power consumption to weight ratio.

5. Other Ecological Considerations and Future Work

Related studies, such as [43], only focus on energy consumption as an indicator of environmental impact. While this is an important metric, it misrepresents the true ecological impact of a particular device if no consideration is given to the environmental impact of the battery production and manufacturing of the UAV components. In this section, other ecological impacts to consider are introduced, and topics that require or deserve further research are summarized in the future work subsection.

The direct environmental implications of physical devices across their life cycle are frequently disregarded. LCA can here be a great tool for transitioning from a traditional cradle-to-gate to a cradle-to-grave analysis, accounting for all necessary steps while quantifying the true impact of IoT systems. A device's environmental impact typically consists of two main contributions. First, the actual environmental impact of a particular product, for which the Global Warming Potential (GWP) metric is commonly used [44]. Second, the finite availability of resources, in particular resource depletion, which we further elaborate on in Table 7.

Only energy for operation has been taken into account so far in this work. While assessing the environmental footprint related to energy usage, the location of manufacturing and, consequently, the location of operation are also important to consider. The environmental impact will greatly differ due to the manner in which energy is generated. Energy from China (544 gCO2eq/kWh) or India (637 gCO2eq/kWh) has a significantly higher GWP than energy generated in Europe (mean 290 gCO2eq/kWh), e.g., Sweden (44 gCO2eq/kWh) [45].

An initial reflection is provided to better understand the ecological impact of a UAV-based recharging system. Since a complete LCA analysis is not the main purpose of this paper, this work is only a lead to create awareness for the reader that there is a plethora of aspects to consider for a sustainable IoT story, and many R&D questions are open in this domain. We are happy to refer the reader to upcoming papers concerning this topic in future work. Our UAV-based energy provisioning system contributes to a more ecological IoT in two ways. First, our approach reduces the ecological impact compared with previous fire-and-forget approaches. Second, we have the possibility of actively extending the battery life almost limitlessly.

5.1. Impact Comparison

A preliminary LCA analysis is conducted using publicly available carbon footprint data. Since GWP data for lithium-based batteries differ quite a lot in the literature, data from four sources [46–49] were collected and averaged, yielding a quite good estimate for lithium battery manufacturing GWP. For non-rechargeable (alkaline) batteries, data from [50] and [51] are used. The UAV model is based on a small sub-250 g model from Pirson and Bol [44], assuming a typical lifetime of 400 flight hours [52], a power consumption of 50 W (resulting from Figure 4b), assuming European grid energy [53], and a flight time of 13 min (as depicted in Figure 13). The GWP of an extra four batteries, assuming a life cycle of 300 cycles, are included due to battery degradation during the lifetime of the UAV. An IoT energy consumption of 20 J a day for 10 years and a lithium based rechargeable battery with a capacity of 2 kJ are assumed.

Figure 19 depicts the total impact of the UAV battery replacement compared with using a non-rechargeable battery. Only a single battery is assumed to be used during a 10 year lifetime of a single IoT node. Since a much smaller rechargeable than non-rechargeable battery can be used, the GWP thereof can be reduced significantly. Assuming a battery capacity of 2 kJ, a total of 36.5 UAV flights are needed over the course of 10 years to top up the battery. The lifetime of the UAV is assumed to be 400 flight hours (13 min/flight), incorporating the GWP of two battery replacements during its lifetime. The IoT node power consumption is assumed to be 20 J/day. The total GWP of the UAV-based servicing approach (in green) can be environmentally beneficial for powering the IoT node compared with using non-rechargeable batteries (in red). The total environmental impact of using non-rechargeable batteries comes down to 940 gCO2eq, while the UAV-based servicing approach yields a GWP of only 555 gCO2eq. An even lower GWP solution can be designed when optimizing the battery capacity at the IoT node and UAV, taking into account an optimal number of flights.

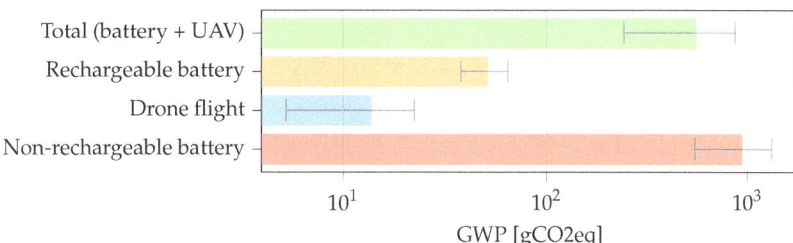

Figure 19. GWP of non-rechargeable fire-and-forget vs UAV-based servicing approach. Assumptions: 13 min flight at 50 Wh with lithium-based IoT node battery capacity of 2 kJ and an energy consumption of 20 J/day for 10 years. Note the logarithmic scale.

5.2. Scarcity of Elements

Only GWP data are available on the most commonly used battery types, e.g., lithium–polymer, lithium-ion, and alkaline batteries. On the GWP of certain battery types, such as LTO or LIC, as considered in this work, there is currently little to no information available. Therefore, we consider the scarcity of elements [54] needed for the manufacturing of the batteries as a way of qualifying the most environmentally friendly battery type. Table 7 gives an overview of the scarce elements required in the considered battery types.

Nickel manganese cobalt (NMC), NCA, LTO, and the older nickel metal hydride (NiMH) and nickel cadmium (NiCd) batteries are considered a bad environmental choice, since they use the most amount of scarce elements, i.e., one or two element(s) in the rising threat category and three or four elements in the future risk category. Lithium iron phosphate (LFP) is considered a better choice, since it uses no elements from the rising threat category and only two elements from the future risk category.

Table 7. Scarcity of elements used in battery production: elements with rising threat from increased use (orange), limited availability future risk to supply (yellow), and plentiful supply (green) [54–59].

		Cobalt	Cadmium	Chromium	Lithium	Nickel	Manganese	Phosphor	Vanadium	Zirconium	Fluor	Iron	Kalium	Aluminum	Titanium
Nickel manganese cobalt	NMC	x			x	x	x	x			x				
Lithium cobalt oxide	LCO	x			x			x			x				
Nickel cobalt aluminum	NCA	x			x	x		x			x			x	
Lithium iron phosphate	LFP				x			x			x	x			
Lithium titanate	LTO/NMC	x			x	x	x	x			x				x
Nickel cadmium	NiCd	x	x			x							x		
Nickel metal hydride	NiMH	x		x		x			x	x		x	x		x

5.3. Manual Labor Comparison

IoT system maintenance is currently performed manually, which requires a lot of human effort and will not be practical in the future due to the fast-growing number of edge devices [1]. Table 8 compares traditional manual maintenance scenarios with the automated approach presented in this work. The required energy per unit kilometer for each intervention type is estimated. The gasoline car results are based on available energy per gallon, assuming a typical efficiency of 20–25% [60]. For electric cars, the energy consumption can be directly used, assuming charging is 100% efficient. The amount of 'food energy' (kcal) is taken into account for walking and cycling. For the big UAV energy consumption, a 3 kg system is assumed [61], while for the small UAV, values from Figure 5 are used, assuming an E_{pm} of 5 J/m. A gasoline-based car uses the most energy per kilometer, i.e., 562 Wh/km. An important improvement can be made using small UAVs in autonomous servicing systems, consuming only 1.39 Wh/km. Additionally, fully automated servicing systems require no user intervention during the lifetime of the product, saving time and human effort, while improving energy efficiency.

Table 8. Comparison of UAV-based operation vs manual intervention [61,62].

	Car (Gasoline)	Car (Electric)	Walking	Cycling	Big UAV	Small UAV (Current Work)
Wh/km	562	187.5	65	17.5	22	1.39

5.4. Future Work

The analyses on the total UAV consumption in this paper are based on reasonable estimates. In future work, a validation, by means of actual measurements, would significantly improve current estimates. Local circumstances, such as weather, can lower WPT efficiency levels or lead to an increase in UAV power consumption. Additional elements contributing to the avionic power, e.g., a swapping mechanism, can also lead to reduced flight times. An extension of the concept from Section 3.3 does not need to be limited to just replacing batteries. Other IoT node parts, such as sensors, central motherboards, or even full devices can be replaced in case of malfunctioning or failure. Here, a new research topic on designing the next generation of serviceable IoT devices arises. This can be supplemented by research into biodegradable materials to implement certain IoT functions, only engaging the UAV for the retrieval of remaining harmful parts.

This work shows a UAV-based servicing approach has many benefits compared with manual replacement. It eliminates manual labor, reduces energy consumption, and lowers the GWP compared with using non-rechargeable batteries. Still, the full impact of the proposed concepts on the environment needs to be further evaluated, since here only an overview of the rare elements in batteries was elaborated. By reducing battery capacity, we can also lower the ecological impact of a typical system. However, this will have ecological repercussions due to the more frequently required UAV flights to deliver the same amount of energy. Hence, optimizations can be performed. It can also be examined whether the environmental impact of an IoT application is more likely to increase or decrease by implementing an additional harvester circuit. Furthermore, the environmental impact of the energy required to fly to the IoT node can be reassessed, taking into account that the share of renewable energy still increases annually.

6. Conclusions

In this work, we presented options, assessments, and potential optimizations of a UAV-based servicing approach. We compared the RFPT system with a UAV approach and clarified the convenient yet very low efficiency of RFPT in short-range energy transfers, while UAVs can achieve a much higher efficiency, and moreover, can be utilized for long-range energy provisioning. A UAV energy consumption model was derived to provide accurate estimates of flight energy consumption. The comparison of battery types at the IoT node in terms of energy density and ecological impact shows that LTO is a promising rechargeable battery technology for the on-site recharging approach, while NCA has the highest energy density, which benefits the efficiency of the discussed swapping process. When examining UAV approaches, our results show that it is more energy efficient to swap instead of recharge the IoT battery. For the case of wireless in situ charging, an actual system implementation is feasible based on known designs. Conversely, autonomous replacement will require a more complex mechanical design.

This work shows a UAV-based servicing approach has many benefits compared with manual replacement. It eliminates manual labor, reduces energy consumption, and lowers the GWP compared with using non-rechargeable batteries.

We can conclude from the work presented in this paper that UAV-based servicing of IoT nodes clearly is of interest from an operational energy efficiency perspective. Moreover, other ecological benefits could be achieved, requiring dedicated study and opening many new research questions.

Author Contributions: Conceptualization, J.V.M., J.C., L.D.S., and L.V.d.P.; SOTA, J.V.M., J.C., and S.G.; validation, J.V.M. and J.C.; writing—original draft preparation, J.V.M., J.C., and L.V.d.P.; writing—review and editing, L.D.S. and L.V.d.P.; visualization, J.C. and J.V.M.; supervision, L.V.d.P. and L.D.S.; funding acquisition, L.V.d.P. and L.D.S. All authors have read and agreed to the published version of the manuscript.

Funding: The REINDEER project has received funding from the European Union's Horizon 2020 research and innovation programme under grant agreement No. 101013425.

Institutional Review Board Statement: Not applicable.

Informed Consent Statement: Not applicable.

Data Availability Statement: The data presented in this study are available on request from the corresponding author.

Conflicts of Interest: The authors declare no conflict of interest.

Abbreviations

The following abbreviations are used in this manuscript:

AUV	Autonomous Underwater Vehicle
BOM	Bill of Materials
BLDC	Brushless DC
CC	Constant Current
CPT	Capacitive Power Transfer
CV	Constant Voltage
DC	Direct Current
DE	Drain Efficiency
EASA	European Union Aviation Safety Agency
EDLC	Electrostatic Double-Layer Capacitors
EIRP	Effective Isotropic Radiated Power
EoL	End of Life
ERP	Effective Radiated Power
ESC	Electronic Speed Control
ETSI	European Telecommunications Standards Institute
FoM	Figure of Merit
GWP	Global Warming Potential
IoT	Internet of Things
IPT	inductive power transfer
ISM	Industrial, Scientific, and Medical
LCA	life cycle assessment
LCO	lithium cobalt oxide
LIC	lithium-ion capacitor
LFP	lithium iron phosphate
Li-ion	lithium-ion
LiPo	lithium polymer
LoRaWAN	long-range wide-area network
LoS	line-of-sight
LPT	laser power transfer
LS	least squares

LTC	lithium thionyl chloride
LTO	lithium titanate
MCU	Microcontroller Unit
NCA	nickel cobalt aluminum
NiCd	nikkel cadmium
NiMH	nikkel metal hydride
NMC	nickel manganese cobalt
PA	power amplifier
PAE	power-added efficiency
RF	radio frequency
RFID	radio frequency identification
RFPT	radio frequency power transfer
SISO	single-input single-output
SoC	state of charge
UAV	unmanned aerial vehicle
UGV	unmanned ground vehicle
UV	unmanned vehicle
WPT	wireless power transfer

References

1. Blog, A.C. White Paper: Economics of a Trillion Connected Devices. 2022. Available online: https://community.arm.com/arm-community-blogs/b/internet-of-things-blog/posts/white-paper-the-route-to-a-trillion-devices (accessed on 26 November 2022).
2. Callebaut, G.; Leenders, G.; Van Mulders, J.; Ottoy, G.; De Strycker, L.; Van der Perre, L. The Art of Designing Remote IoT Devices—Technologies and Strategies for a Long Battery Life. *Sensors* **2021**, *21*, 913. [CrossRef]
3. Moore, S. Lessons From IoT Early Adopters. 2018. Available online: https://www.gartner.com/smarterwithgartner/lessons-from-iot-early-adopters (accessed on 1 December 2022).
4. Van Mulders, J.; Delabie, D.; Lecluyse, C.; Buyle, C.; Callebaut, G.; Van der Perre, L.; De Strycker, L. Wireless Power Transfer: Systems, Circuits, Standards, and Use Cases. *Sensors* **2022**, *22*, 5573. [CrossRef]
5. *Frequency Identification Equipment Operating in the Band 865 MHz to 868 MHz with Power Levels Up to 2 W and in the Band 915 MHz to 921 MHz with Power Levels Up to 4 W*; Harmonised Standard for Access to Radio Spectrum; European Telecommunications Standards Institute: Sophia-Antipolis, France, 2020.
6. Clerckx, B.; Bayguzina, E. Waveform design for wireless power transfer. *IEEE Trans. Signal Process.* **2016**, *64*, 6313–6328. [CrossRef]
7. Li, Z.; Zeng, M.; Tan, H.Z. A multi-band rectifier with modified hybrid junction for RF energy harvesting. *Microw. Opt. Technol. Lett.* **2018**, *60*, 817–821. [CrossRef]
8. Consortium, R. REsilient INteractive Applications through Hyper Diversity in Energy Efficient RadioWeaves Technology. Available online: https://reindeer-project.eu/ (accessed on 7 December 2022).
9. Van der Perre, L.; Larsson, E.G.; Tufvesson, F.; De Strycker, L.; Björnson, E.; Edfors, O. RadioWeaves for efficient connectivity: Analysis and impact of constraints in actual deployments. In Proceedings of the 2019 53rd Asilomar Conference on Signals, Systems, and Computers, Pacific Grove, CA, USA, 3–6 November 2019; pp. 15–22.
10. Van Mulders, J.; Leenders, G.; Callebaut, G.; De Strycker, L.; Van der Perre, L. Aerial Energy Provisioning for Massive Energy-Constrained IoT by UAVs. In Proceedings of the 2022 IEEE International Conference on Communications (ICC): IoT and Sensor Networks Symposium (IEEE ICC'22—IoTSN Symposium), Seoul, Republic of Korea, 16–20 May 2022; pp. 3574–3579. [CrossRef]
11. EASA. Open Category—Civil Drones. Available online: https://www.easa.europa.eu/en/domains/civil-drones/drones-regulatory-framework-background/open-category-civil-drones. (accessed on 3 November 2022).
12. Tiurlikova, A.; Stepanov, N.; Mikhaylov, K. Wireless power transfer from unmanned aerial vehicle to low-power wide area network nodes: Performance and business prospects for LoRaWAN. *Int. J. Distrib. Sens. Netw.* **2019**, *15*, 1550147719888165. [CrossRef]
13. Li, Y.; Zhou, C.; Shi, S.; Xu, Z. Time Allocation in Multi-UAV Energy Harvesting Network. In Proceedings of the International Conference on Artificial Intelligence for Communications and Networks, Xining, China, 23–24 October 2021; pp. 8–19.
14. Cheng, C.; Adulyasak, Y.; Rousseau, L.M. Drone routing with energy function: Formulation and exact algorithm. *Transp. Res. Part B Methodol.* **2020**, *139*, 364–387. [CrossRef]
15. Haider, S.K.; Nauman, A.; Jamshed, M.A.; Jiang, A.; Batool, S.; Kim, S.W. Internet of Drones: Routing Algorithms, Techniques and Challenges. *Mathematics* **2022**, *10*, 1488. [CrossRef]
16. Pugliese, L.D.P.; Guerriero, F.; Macrina, G. Using drones for parcels delivery process. *Procedia Manuf.* **2020**, *42*, 488–497. [CrossRef]
17. Zhang, J.; Campbell, J.F.; Sweeney, D.C., II; Hupman, A.C. Energy consumption models for delivery drones: A comparison and assessment. *Transp. Res. Part D Transp. Environ.* **2021**, *90*, 102668. [CrossRef]
18. Dorling, K.; Heinrichs, J.; Messier, G.G.; Magierowski, S. Vehicle Routing Problems for Drone Delivery. *IEEE Trans. Syst. Man Cybern. Syst.* **2017**, *47*, 70–85. [CrossRef]

19. Ferrandez, S.M.; Harbison, T.; Weber, T.; Sturges, R.; Rich, R. Optimization of a truck-drone in tandem delivery network using k-means and genetic algorithm. *J. Ind. Eng. Manag.* **2016**, *9*, 374–388. [CrossRef]
20. Kirschstein, T. Comparison of energy demands of drone-based and ground-based parcel delivery services. *Transp. Res. Part D Transp. Environ.* **2020**, *78*, 102209. [CrossRef]
21. Liu, Z.; Sengupta, R.; Kurzhanskiy, A. A power consumption model for multi-rotor small unmanned aircraft systems. In Proceedings of the 2017 International Conference on Unmanned Aircraft Systems (ICUAS), Miami, FL, USA, 13–16 June 2017; pp. 310–315. [CrossRef]
22. e-Peas Semiconductors. E-Peas RF Energy Harvesting Datasheet AEM40940 Rev. 1.1. 2018. Available online: https://e-peas.com/wp-content/uploads/2020/04/E-peas_RF_Energy_Harvesting_Datasheet_AEM40940.pdf (accessed on 7 November 2022)
23. Kazimierczuk, M.K. *RF Power Amplifiers*; John Wiley & Sons: Hoboken, NJ, USA, 2014.
24. Ronsmans, J.; Lalande, B. Combining energy with power: Lithium-ion capacitors. In Proceedings of the 2015 International Conference on Electrical Systems for Aircraft, Railway, Ship Propulsion and Road Vehicles (ESARS), Aachen, Germany, 3–5 March 2015; pp. 1–4. [CrossRef]
25. Mauger, A.; Julien, C. Critical review on lithium-ion batteries: Are they safe? Sustainable? *Ionics* **2017**, *23*, 1933–1947. [CrossRef]
26. Tecate GROUP. TYPE TPLC HYBRID CAPACITOR (LIC). 2022. Available online: https://www.tecategroup.com/products/data_sheet.php?i=TPLC-3R8/50MR10X20 (accessed on 7 November 2022).
27. Wireless Power Consortium. Qi Specification Power Delivery. 2021. Available online: https://www.wirelesspowerconsortium.com/knowledge-base/specifications/download-the-qi-specifications.html (accessed on 7 November 2022).
28. Alliance, A. AirFuel Is Cutting the Cord. Available online: https://airfuel.org/wireless-power/ (accessed on 10 November 2022).
29. Rindorf, L.; Lading, L.; Breinbjerg, O. Resonantly coupled antennas for passive sensors. In Proceedings of the SENSORS, 2008 IEEE, Lecce, Italy, 26–29 October 2008; pp. 1611–1614.
30. Mei, J.; Cheng, E.K.; Fong, Y. Lithium-titanate battery (LTO): A better choice for high current equipment. In Proceedings of the 2016 International Symposium on Electrical Engineering (ISEE), Santa Clara, CA, USA, 8–10 June 2016; pp. 1–4.
31. GWL. LTO1865-13 Rechargeable Cell. Available online: https://files.gwl.eu/inc/_doc/attach/StoItem/7015/GWL_LTO1865_Rechargeable.pdf (accessed on 7 November 2022).
32. CUSTOMCELLS. LTO / NMC 622—High Power Cells. 2017. Available online: https://www.customcells.de/fileadmin/customcells/Dokumente/Newsletter/CUSTOMCELLS-Newsletter-2017-09-01-LTO-cells.pdf (accessed on 7 November 2022).
33. SAMSUNG SDI. INR18650-29E. 2012. Available online: https://eu.nkon.nl/sk/k/29E.pdf (accessed on 7 November 2022).
34. NX. 18650 LIFEPO4 BATTERY 1500mAh 3.2V. 2014. Available online: https://docs.rs-online.com/1683/0900766b812fdd11.pdf (accessed on 7 November 2022).
35. Panasonic. Lithium Ion NCR18650B. 2012. Available online: https://www.batteryspace.com/prod-specs/NCR18650B.pdf (accessed on 7 November 2022).
36. Sanyo Cadnica. Cell Type N-700AAC. Available online: https://www.omnitron.cz/_dokumenty/2782019153700772/n-700aac-77.pdf (accessed on 7 November 2022).
37. RS PRO. NI-MH LOW-SELF DISCHARGE. Available online: https://docs.rs-online.com/686b/0900766b81585e4a.pdf (accessed on 7 November 2022).
38. Lee, D.; Zhou, J.; Lin, W.T. Autonomous battery swapping system for quadcopter. In Proceedings of the 2015 international conference on unmanned aircraft systems (ICUAS), Denver, CO, USA, 9–12 June 2015; pp. 118–124. [CrossRef]
39. Suzuki, K.A.; Kemper Filho, P.; Morrison, J.R. Automatic battery replacement system for UAVs: Analysis and design. *J. Intell. Robot. Syst.* **2012**, *65*, 563–586. [CrossRef]
40. Johansson, T.; Fritzin, J. A review of watt-level CMOS RF power amplifiers. *IEEE Trans. Microw. Theory Tech.* **2013**, *62*, 111–124. [CrossRef]
41. Blouin, C. How to Increase Drone Flight Time & Lift Capacity. 2021. Available online: https://www.tytorobotics.com/blogs/articles/how-to-increase-drone-flight-time-and-lift-capacity (accessed on 7 November 2022).
42. Ahmad, K. Lithium-Ion Batteries vs. Lithium-Polymer: Which One's Better? 2022. Available online: https://www.makeuseof.com/lithium-ion-vs-lithium-polymer-which-is-better/ (accessed on 7 November 2022).
43. Long, T.; Ozger, M.; Cetinkaya, O.; Akan, O.B. Energy Neutral Internet of Drones. *IEEE Commun. Mag.* **2018**, *56*, 22–28. [CrossRef]
44. Pirson, T.; Bol, D. Assessing the embodied carbon footprint of IoT edge devices with a bottom-up life-cycle approach. *J. Clean. Prod.* **2021**, *322*, 128966. [CrossRef]
45. Ember. Data Explorer. Available online: https://ember-climate.org/data/data-explorer/ (accessed on 7 November 2022).
46. Sadhukhan, J.; Christensen, M. An In-Depth Life Cycle Assessment (LCA) of Lithium-Ion Battery for Climate Impact Mitigation Strategies. *Energies* **2021**, *14*, 5555. [CrossRef]
47. Melin, H.E. *Analysis of the Climate Impact of Lithium-Ion Batteries and How to Measure It*; Circular Energy Storage-Research and Consulting: London, UK, 2019; pp. 1–17.
48. Philippot, M.; Alvarez, G.; Ayerbe, E.; Van Mierlo, J.; Messagie, M. Eco-efficiency of a lithium-ion battery for electric vehicles: Influence of manufacturing country and commodity prices on ghg emissions and costs. *Batteries* **2019**, *5*, 23. [CrossRef]
49. Sánchez, D.; Proske, M.S.J.B. Life Cycle Assessment of the Fairphone 4. 2022. Available online: https://www.fairphone.com/wp-content/uploads/2022/07/Fairphone-4-Life-Cycle-Assessment-22.pdf (accessed on 7 November 2022).

50. Masanet, E.; Horvath, A. *Single-Use Alkaline Battery Case Study The Potential Impacts of Extended Producer Responsibility (EPR) in California on Global Greenhouse Gas (GHG) Emissions*; California Department of Resources Recycling and Recovery: Sacramento, CA, USA, 2012.
51. Olivetti, E.; Gregory, J.; Associates, C. *Life Cycle Assessment of Alkaline Battery Recycling a Report for the Corporation for Battery Responsibility Executive Summary*; Massachusetts Institute of Technology: Cambridge, MA, USA, 2018.
52. Ciobanu, E. Drone Life Expectancy (How Long Do Drones Last?). Available online: https://www.droneblog.com/drone-life-expectancy/ (accessed on 7 November 2022).
53. Agency, E.E. Greenhouse Gas Emission Intensity of Electricity Generation. Available online: https://www.eea.europa.eu/data-and-maps/daviz/co2-emission-intensity-12/#tab-googlechartid_chart_11 (accessed on 7 November 2022).
54. Science, L. Europe's 'New' Periodic Table Predicts Which Elements Will Disappear in the Next 100 Years. Available online: https://www.livescience.com/64596-new-periodic-table-shows-helium-scarcity.html (accessed on 7 November 2022).
55. Panasonic. *Lithium-ion Batteries Product Information Sheet*; Panasonic: Osaka, Japan, 2017.
56. Kurzweil, P. Chapter 19 - Electrochemical Double-layer Capacitors. In *Electrochemical Energy Storage for Renewable Sources and Grid Balancing*; Moseley, P.T., Garche, J., Eds.; Elsevier: Amsterdam, The Netherlands, 2015; pp. 377–379. [CrossRef]
57. University, B. BU-205: Types of Lithium-Ion. 2021. Available online: https://batteryuniversity.com/article/bu-205-types-of-lithium-ion (accessed on 7 November 2022).
58. Ebin, B.; Petranikova, M.; Ekberg, C. Physical separation, mechanical enrichment and recycling-oriented characterization of spent NiMH batteries. *J. Mater. Cycles Waste Manag.* **2018**, *20*, 2018–2027. [CrossRef]
59. Rydh, C.J.; Karlström, M. Life cycle inventory of recycling portable nickel–cadmium batteries. *Resour. Conserv. Recycl.* **2002**, *34*, 289–309. [CrossRef]
60. Murphy, T. 100 MPG on Gasoline: Could We Really? Available online: https://dothemath.ucsd.edu/2011/07/100-mpg-on-gasoline/ (accessed on 7 November 2022).
61. Rodrigues, T.A.; Patrikar, J.; Oliveira, N.L.; Matthews, H.S.; Scherer, S.; Samaras, C. Drone flight data reveal energy and greenhouse gas emissions savings for very small package delivery. *Patterns* **2022**, *3*, 100569. [CrossRef] [PubMed]
62. Murphy, T. MPG of a Human. Available online: https://dothemath.ucsd.edu/2011/11/mpg-of-a-human/ (accessed on 7 November 2022).

Disclaimer/Publisher's Note: The statements, opinions and data contained in all publications are solely those of the individual author(s) and contributor(s) and not of MDPI and/or the editor(s). MDPI and/or the editor(s) disclaim responsibility for any injury to people or property resulting from any ideas, methods, instructions or products referred to in the content.

Article

Human and Small Animal Detection Using Multiple Millimeter-Wave Radars and Data Fusion: Enabling Safe Applications

Ana Beatriz Rodrigues Costa De Mattos [1], Glauber Brante [2], Guilherme L. Moritz [2] and Richard Demo Souza [1,*]

1. Department of Electrical and Electronics Engineering, Federal University of Santa Catarina (UFSC), Florianopolis 88040-900, Brazil; ana.beatriz.mattos@posgrad.ufsc.br
2. Academic Department of Electrotechnics, Federal University of Technology—Paraná (UTFPR), Curitiba 80230-901, Brazil; gbrante@utfpr.edu.br (G.B.); moritz@utfpr.edu.br (G.L.M.)
* Correspondence: richard.demo@ufsc.br

Abstract: Millimeter-wave (mmWave) radars attain high resolution without compromising privacy while being unaffected by environmental factors such as rain, dust, and fog. This study explores the challenges of using mmWave radars for the simultaneous detection of people and small animals, a critical concern in applications like indoor wireless energy transfer systems. This work proposes innovative methodologies for enhancing detection accuracy and overcoming the inherent difficulties posed by differences in target size and volume. In particular, we explore two distinct positioning scenarios that involve up to four mmWave radars in an indoor environment to detect and track both humans and small animals. We compare the outcomes achieved through the implementation of three distinct data-fusion methods. It was shown that using a single radar without the application of a tracking algorithm resulted in a sensitivity of 46.1%. However, this sensitivity significantly increased to 97.10% upon utilizing four radars using with the optimal fusion method and tracking. This improvement highlights the effectiveness of employing multiple radars together with data fusion techniques, significantly enhancing sensitivity and reliability in target detection.

Keywords: millimeter-wave; data fusion; multiple radars; small animals; human tracking; wireless power transfer

1. Introduction

The Internet revolution has significantly altered the way people access, search, and disseminate information through the interconnection of devices around the world [1]. The emergence of the Internet of Things (IoT) is creating a bridge between the virtual and the real world, necessitating scalable mobile networks to accommodate the demands of an estimated dozen billion connected devices [2]. Moreover, the processing capabilities must evolve to handle the vast amount of information generated by these digital entities [2]. Projections anticipate a staggering 75 billion connected devices by the end of 2025 [3]. Using advanced sensors, as mmWave radars, embedded within everyday objects, the IoT empowers intelligent data-driven decision-making in various industries [1]. MmWave technology is instrumental in advancing IoT applications across various domains, including smart homes, wearables, and smart cities by enhancing, e.g., intelligent surveillance, automated transportation, and security measures with unparalleled accuracy and efficiency [4]. This evolution underscores the growing importance of integrating sensing technologies, such as mmWave radars, into the IoT infrastructure to realize its full potential in enabling smart environments and applications [5].

With the advances in IoT technologies, the demand for high-precision, secure, and private location monitoring has increased significantly. Location monitoring and movement tracking are of critical importance in various scenarios, such as smart homes, indoor navigation, security surveillance, disaster management, and smart healthcare [6]. Among the array of sensors used to detect people, gestures, and objects, cameras and radars are known for their cost effectiveness while maintaining commendable precision levels compared to other sensor technologies [7]. Current research on detection and tracking employs various sensing approaches and algorithms, such as passive infrared sensors (PIR) [8,9], light detection and ranging (LIDAR) [10,11], and digital cameras [12–14]. However, each of these technologies faces challenges related to accuracy, privacy, and environmental robustness [15,16].

Millimeter-wave (mmWave) radars employ short-length electromagnetic waves, resulting in high precision. Unlike technologies such as cameras and LIDAR, radar measurements are less affected by environmental factors such as rain, fog, and dust [15], while also preserving privacy. Additionally, radar can achieve high-range and high-speed object detection [15]. A prominent example of commercial radar systems is the IWR6843 mmWave sensor from Texas Instruments (TI) [17]. These sensors produce point clouds: three-dimensional datasets that convey object positions in three axes, Doppler data, angles for each point, and other relevant information, providing comprehensive environmental data [18]. A common use case of mmWave radar is in the detection and tracking of humans.

However, the literature is scarce on methods capable of detecting and tracking humans and animals in the same environment. By identifying the presence of animals, sensors facilitate early alerts to drivers, machine operators, security personnel, and activation of safety measures, thus reducing the risks of potential incidents [19]. Furthermore, the detection of animals through sensors promotes safe cohabitation in shared environments [19]. Taking into account that there are around a billion pets worldwide (https://www.healthforanimals.org/, accessed on 25 February 2024), the ability to detect and track humans and small animals may result in many novel applications.

1.1. Related Work

Several works have explored detecting or tracking people using mmWave radar [7,16,18,20,21]. The work in [7] presents an identification system named mID, utilizing mmWave radar technology. Meanwhile, the authors in [21] introduce an extended object-tracking Kalman filter capable of estimating the position, shape, and extension of the subjects. It integrates a novel deep-learning classifier designed specifically for efficient feature extraction and rapid inference from radar point clouds. Additionally, the work in [22] implements an mmWave radar-based multi-person tracking system utilizing a single radar.

Moving forward, combining sensors through data fusion has emerged as a promising approach to gaining additional information in various applications [23]. The data fusion process involves multiple stages, including detection, association, correlation, estimation, and combination [23]. It encompasses the fusion of data from similar or dissimilar sensors. For instance, in a multi-sensor system comprising identical sensors, a target detected by several sensors provides estimation states to the fusion center for target tracking [23]. Additionally, the work in [24] showcases the effective fusion of information from multiple radars, resulting in improved area coverage, probability of detection, localization, and tracking performance.

In this line, multi-radar tracking can be seen as a way of obtaining a view of an object from two or more angles simultaneously [25]. According to [25], the use of multiple radars has some advantages and disadvantages. Some of the advantages are the better resolution in the presence of noise, detection uncertainties, and more reliable tracking data [25]. The disadvantages would be the constant communication between the radar platforms and the increased amount of data processing [25]. Various radar fusion techniques are presented in [23], employing a coefficient calculation method based on the trace of the error covariance matrix. The use of the strong tracking filter (STF) is introduced in the estimation of the target state, demonstrating superior performance compared to conventional or extended Kalman filters. This integration improves the overall target tracking performance.

In [26], a simulation utilizing the fusion of multiple radars suggests that employing two radars results in a higher detection probability and higher precision compared to a single radar. Furthermore, [27] introduces a multi-radar calibration method by tracking pedestrian trajectories. The fusion of multiple radars has shown utility in estimating human posture. In [28], two mmWave radars were strategically placed: one detecting (x, y) data and the other capturing (x, z) data to collect reflection points. A neural network was used for data fusion. In [29], an algorithm called Pontilism was introduced for a system of multiple radars. This algorithm addresses specular reflections, sparsity, and noise in radar point clouds, enhancing radar perception with 3D bounding boxes. The study demonstrated that the use of multiple radars resulted in a reduced error compared to using a single radar.

There are some works in the literature that use point clouds generated from multiple radars to specifically detect and track people, as in [16,18,30,31], where different radar positioning scenarios were proposed. For example, the work in [30] introduced a human tracking system based on mmWave radar, employing two radars placed along the walls of a room. This setup enabled the detection of moving humans by sparse point clouds. Similarly, the authors of [16] investigated the use of two mmWave radar sensors for accurate people detection and tracking. However, their radar positioning differed, with the radars located at the corners of the room. Furthermore, a real-time system framework is proposed in [31] to merge radar signals to track human position and body status. Unlike previous studies, the authors utilized a configuration involving three radars, one placed on the ceiling and the other two on the walls, ensuring precise tracking accuracy.

In the pioneering study using point clouds from multiple mmWave radars presented in [18], a software framework capable of communicating with multiple radars and applying a customized data-processing chain is introduced. These radars were placed on the walls of a room. The conclusion shows that the proposed system achieves over 90% sensitivity in indoor human detection. In particular, using a two-radar configuration significantly improves precision from 46.9% to an impressive 98.6%. However, in this case, the sensitivity decreased from 96.4% to 90.4%. Depending on the application, as those concerned with security or safety aspects, a reduction in sensitivity may be unacceptable. Moreover, the authors discuss the potential interference among multiple radars, showing that the probability of interference when using four radars is less than 1%. However, such a probability increases considerably with more than ten radars, which would then require explicit synchronization between radars or an interference-detection algorithm.

Unlike human detection, the detection of animals presents a distinct challenge, given the variations in size and shape. The work presented in [32] explores in-phase and quadrature (IQ) radar data on humans and animals, focusing on the extraction of radar-data-distinguishing features to classify animals versus humans based on micro-Doppler signatures. Additionally, in [33], the use of radar micro-Doppler signatures for the automatic contactless identification of lameness is presented, showing preliminary results for dairy cows, sheep, and horses. Furthermore, the classification system in [34] utilizes an mmWave dual receiver to distinguish between humans and animals. This system uses feedback signal responses from targets with a dual-receiver mmWave radar, utilizing a neural network based on synthetic 2D tensor data to categorize human and animal features [34]. However, none of these studies have utilized point clouds from mmWave radars for the simultaneous detection and tracking of people and animals. Although mmWave radars commonly collect data in IQ format, the point cloud format is advantageous in terms of external radar processing. Processing IQ data demands a large communication bandwidth and high computing power [35]. Moreover, receiving data directly in the point cloud format allows for the application of advanced data processing techniques like clustering and filtering with enhanced efficiency and speed.

While some existing literature explores the use of multiple radars to detect people or objects, such as [16,18,30,31], and there are also studies that focus on animal classification, such as [33,34], none of the previous works address the simultaneous detection and tracking of people and small animals, such as dogs and cats, using multiple radar systems. Note that a system optimized for detecting people may be very inefficient in detecting small animals. A relevant application of simultaneous detection of humans and small animals is in autonomous vehicles. Moreover, another essential application of human and small animal detection is in the domain of wireless power transfer [36], ensuring safety in settings that involve wireless charging for electronic devices located in areas with the frequent presence of animals, the latter being the case with a modern living room, as illustrated in Figure 1. The appeal of wireless power transfer lies in its various benefits [37]. Notably, the convenience of avoiding connectors during device charging contributes to its attractiveness [37]. Additionally, a contactless solution proves more reliable, sidestepping issues like corrosion, dust intrusion, and moisture exposure [37]. To address potential health risks associated with electromagnetic fields, the wireless charging system can be intelligently deactivated upon detecting the presence of humans or animals in the environment [36].

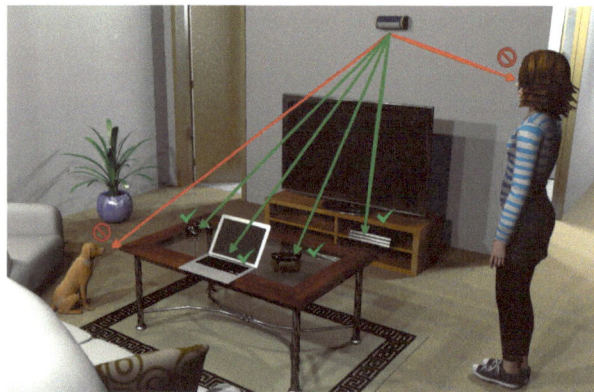

Figure 1. An illustrative scenario of the application of mmWave radars for safety-aware wireless energy transfer. A power beacon charges several devices. If the presence of humans or animals is detected by means of mmWave radars, potentially unsafe electromagnetic exposure can be avoided by turning off the power beacon or even by informing the power beacon to redesign the beams accordingly.

1.2. Novelty and Contribution

This paper introduces a strategy for the simultaneous detection of humans and small animals employing multiple mmWave radars. The decision to limit our study to these two targets is with the intention of exploring one of the challenging scenarios for radar detection, particularly when potential targets significantly differ in terms of energy signatures, which depend on their size. It is considerably challenging to optimize a radar system to effectively detect targets with large deviations in energy signatures. If the radar setting is optimized to minimize false negatives for a target with a small energy signature, such as a small animal, this optimization could compromise the accuracy of detecting larger targets, like humans, thereby increasing the risk of false positives for the latter. Detecting multiple humans is in principle less challenging than detecting a small animal and a human in the same scene. Thus, our selection of a small animal and a human as targets stems from viewing them as an ideal benchmark for testing the limits of radar sensitivity and detection capabilities. Then, the primary objective is to demonstrate the enhanced detection efficacy achievable by multiple radars. It is shown that algorithms relying solely on a single radar may not capture sufficient reflection points from small animals, potentially leading to their misclassification as noise or remaining undetected. To the best of the authors' knowledge, this is the first work to use point clouds from up to four mmWave radars to detect and track people and small animals in the same environment. We explore two different radar positioning scenarios and present a comparative analysis of their respective results. Furthermore, this study includes an examination of three data-fusion methodologies. Importantly, our focus is on target detection, not on the classification of the target. The goal is to highlight the increased efficacy in target detection using multiple radars, showcasing how this approach overcomes limitations associated with the use of a single radar for targets with diverse shapes and sizes.

The proposed system achieves 97.1% sensitivity and up to 91.4% precision in the detection of humans and small animals in an indoor environment, considering the best fusion strategy. The contribution of this article can be summarized as follows.

- We investigate the use of multiple mmWave radars to detect people and small animals, analyzing the impact of different data fusion and radar position strategies.
- We show that data fusion from multiple radars can significantly improve sensitivity and precision, enabling the simultaneous detection of small animals and humans.

The rest of this paper is structured as follows. The principles of mmWave radar are reviewed in Section 2. Section 3 describes the proposed approach, while Section 4 introduces the implementation details and the test setup. Section 5 evaluates the system, while Section 6 concludes the paper.

2. mmWave Radar Preliminaries

Radar systems emit electromagnetic waves that interact with objects in their path. By capturing reflected signals, these systems extract valuable information about the range, Doppler velocity, and angular positioning of the objects. Radars can be categorized into two types based on the signal they employ: frequency-modulated continuous wave (FMCW) radar and pulsed radar [15]. The radar used in this study, the IWR6843 industrial starter kit (ISK) 2.0 from TI, is an FMCW radar operating in the mmWave 60 GHz to 64 GHz band, equipped with four reception channels and three transmission channels [17].

In the case of the FMCW radar, the transmitted signal is called a chirp, which is a sinusoidal signal characterized by a linear increase in frequency over time [15]. A chirp is characterized by initial frequency f_c, bandwidth B, and duration T_c. The slope S of the chirp defines the rate at which the frequency increases with time. A sequence of chirps forms a frame [15]. The illustration in Figure 2 presents the block diagram that describes the operational principle of an FMCW radar: a chirp is generated by a synthesizer, sent through the transmit (TX) antenna, and partially reflected by a target, and it finally reaches a set of receive (RX) antennas [15]. After mixing and low-pass filtering, an object in front of the radar generates an IF signal with a constant frequency [15]. Then, such an IF signal is sampled by an analog-to-digital converter (ADC), so that the ADC data are processed [15]. In the processor, the standard mmWave radar processing chain initially accepts ADC data as input. It then executes range and Doppler fast Fourier transform (FFT) operations, subsequently engaging in non-coherent detection through the implementation of the constant false alarm rate (CFAR) algorithm [38]. The final step involves estimating the angle using a 3D-FFT technique, which results in the generation of detected points termed point cloud data [38].

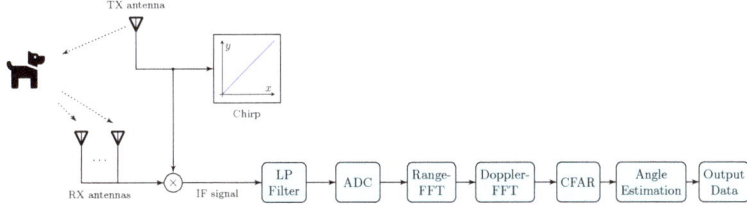

Figure 2. Diagram of the detection process of a target using an FMCW radar.

Moreover, the constant false alarm rate (CFAR) algorithm is one of the key technologies of radar signal processing [39]. It estimates the average power of the background according to the reference cells around the cell under test (CUT) as the threshold to detect targets, which maximizes the target's detection probability while maintaining a constant probability of false alarm.

2.1. Output Data

The main information in the payload output by the radar is the point clouds, which contain reflections from the radar, positioning on the (x, y, z) axes, velocity, and signal strength. The term "radar point cloud" universally defines a compilation of detected objects reflected by the radar processing chain [40]. Originally, the concept of a point cloud emerged to characterize multi-dimensional data points derived from sensors like LIDAR and range cameras [40]. In some studies, point cloud data are described as a flexible information model commonly used to condense object signatures [40]. Essentially, point cloud data comprises numerous sets of individual points positioned uniquely in Euclidean space [40].

A primary advantage of this representation lies in its ability to convey crucial object information while demanding minimal computational and memory resources [40]. This quality renders it suitable for devices with limited resources, such as the TI mmWave radar [40]. Additionally, point clouds represent target signatures in point form, enabling the representation of complex targets using only a few data points. In contrast, a typical LIDAR point cloud data frame, sampled from scene surfaces, may contain thousands or millions of data points. This quantity surpasses the data points collected from a scene via mmWave radar, where streaming raw IQ data without additional hardware is impossible due to memory and hardware constraints in the single-chip radar [40].

2.2. Radar Configuration

The mmWave radar used, IWR6843ISK [17], provides a high degree of flexibility in the configuration of chirp parameters and also allows multiple chirp configurations within a single frame [38]. Among the many configurable parameters are the maximum and minimum detection distances of a radar sensor, range resolution (the ability to distinguish nearby objects), and parameters for maximum velocity, velocity resolution, and angular resolution. The threshold of the CFAR algorithm is also configurable, making it possible to filter out detected points outside the specified limits in the range domain or the Doppler domain. Initially, the configuration file is transmitted to the radar via a serial port, which requires a connection to a central processor and consumes a short period of time. However, once established, the configuration can be hard-coded, allowing the device to autonomously boot, configure, and emit chirps and transmit output data through a serial port without additional user intervention.

3. Proposed Approach

This work considers the use of point clouds generated by M different mmWave radars to detect both people and small animals. The proposed approach consists of three sequential modules: data acquisition, data fusion, and tracking.

(1) **Data Acquisition.**
Each FMCW radar transmits mmWave chirps and records reflections from the scene. Subsequently, it processes the dynamic point clouds, identifying and eliminating points corresponding to static objects.

(2) **Data Fusion.**
The data obtained by each of the radars are transformed into a common coordinate system so that a method for data fusion and clustering can be implemented.

(3) **Tracking.**
The system associates the same human/small animal in consecutive frames and uses a multiple-object tracking algorithm to maintain their trajectories.

3.1. Data Acquisition

As previously stated, the FMCW radar operates by transmitting mmWave signals and capturing their reflections within a scene at a moment in time. The returned signal undergoes preliminary processing on the sensor, which then computes the point clouds. Reflections from static elements such as the ground, door frame, ceiling, walls, and furniture introduce a notable challenge [41]. To enhance the distinction between objects of interest and the background scene, a calibration step is incorporated into the system. In the installation phase, the device captures radar returns from the background, establishing a reference dataset. This recorded background information is then subtracted from the current frame during operation, facilitating the identification of newly introduced objects in the scene [41].

The resulting data are transmitted into a central processor, where rotation and translation matrices are computed individually for each radar, incorporating their specific orientations and positions within the system. This process is facilitated by the known spatial coordinates and orientations of each radar unit. Subsequently, the data acquired from each radar undergo a transformation to align with a unified coordinate system, ensuring a consistent and coherent spatial reference across all radar sources.

3.2. Data Fusion

The generated points of each radar are placed into one coordinate system, and the data go through a clustering process. Three data-fusion methodologies are evaluated.

3.2.1. Method 1 [18]—Intersection of Detected Data

The first approach is based on the method introduced in [18] and is illustrated in Figure 3, considering $M = 4$ radars. In Figure 3a, we present the raw data from each radar, as points in different colors. The data gathered by each radar, stored in (x, y, z) coordinate formats, is processed via the density-based spatial clustering of applications with noise algorithm (DBSCAN). In the realm of density-based clustering algorithms, DBSCAN stands out as a widely embraced algorithm within this classification [21], having demonstrated successful application in clustering radar point clouds, as indicated in [7,16,18,21]. A major feature is that it does not require the number of clusters to be specified a priori [7]. Furthermore, DBSCAN detects clusters of arbitrary shapes, while it can automatically mark outliers to cope with noise, enhancing its effectiveness in handling noisy data [7,42].

The assignment of a point to a cluster in the DBSCAN algorithm depends on the neighborhood of the point around a radius ϵ [42]. Then, this algorithm classifies points into three distinct categories: core, a point within a cluster that boasts a minimum of *minpts* neighbors within its ϵ-neighborhood; border, a point within a cluster that possesses fewer than *minpts* neighbors in its ϵ-neighborhood; and noise, an outlier that does not align with any cluster [42].

The clusters detected by each radar are illustrated by ellipses in Figure 3a. After evaluating the clusters' dimensions and positions, the system proceeds to compute the eigenvectors specific to each cluster [18]. Subsequently, the algorithm estimates distances and identifies overlapping regions between clusters from different radars, preserving the groups where the centroids align closely and most of their areas intersect [18]. Unlike the methodology in [18], which uses only two radars, here, we extend the method for up to four radars. Consequently, in this case, a positive decision necessitates detection from all radars; otherwise, the input is classified as noise. Figure 3b illustrates the final result of this method by another ellipse. Note that all radars must detect the target; otherwise, it is not detected in the final step. This can be a problem for detecting small animals, as they generate fewer points than humans and may not be detected by all radars simultaneously, thus potentially missing detection. Thus, one should expect a decrease in sensitivity with the increase of M. This issue can be alleviated by considering a relatively small value of *minpts* in DBSCAN, but at the potential cost of increasing the occurrence of false positives.

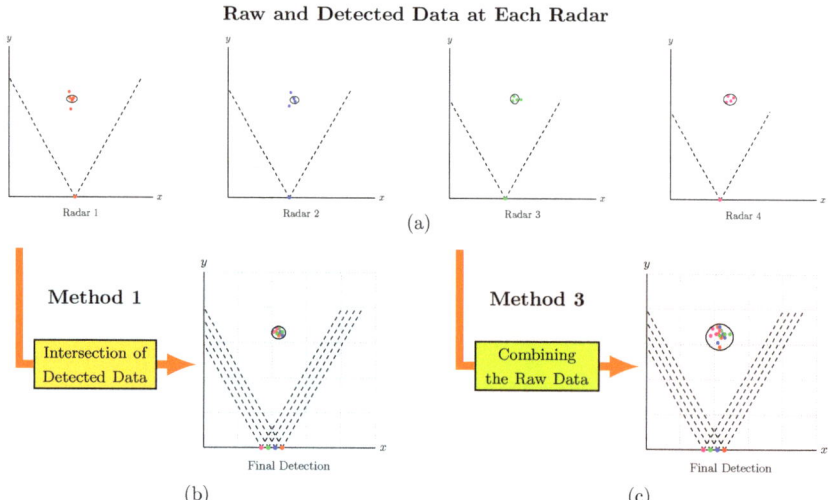

Figure 3. The raw and detected data at each radar are shown in (**a**). Method 1, based on the intersection of individually detected data, is illustrated in (**b**). Method 3, based on the combination of raw data, is shown in (**c**).

3.2.2. Method 2—R out of M

The second fusion method is a modified version of Method 1 [18]. Unlike the original methodology, where detection relied on the intersection of individual detections of all M radars, this adapted method introduces flexibility by varying R, the minimum number of detecting radars to confirm a detection event. The approach of Method 1 is applied in each possible combination of R out of M radars, leading to $M!/(R!(M-R)!)$ possible intersections. For instance, in the case of $M = 4$ radars and $R = 2$, the method proposed in [18] is applied separately in each possible pair among the four radars. Taking Figure 3a as an example, Method 2 would consider the following set of intersections: {(Radar 1, Radar 2), (Radar 1, Radar 3), (Radar 1, Radar 4), (Radar 2, Radar 3), (Radar 2, Radar 4), (Radar 3, Radar 4)}. In this example, it is sufficient for a target to be successfully detected if any $R = 2$ of the $M = 4$ radars detect it. Clearly, when $R < M$, the sensitivity should be increased with respect to Method 1, but at the cost of precision.

3.2.3. Method 3—Combining the Raw Data

In the third and final method, clustering is not applied individually in the raw data of each radar, as in Methods 1 and 2 above. Rather than using individual radar data independently, the collected data from all radars undergo processing in a unified coordinate system through the DBSCAN algorithm. Consequently, the point clouds from each radar are collectively considered for clustering. The procedure is illustrated again with the aid of Figure 3, where the final result of Method 3 is shown in Figure 3c. Therefore, unlike Method 1, when the number of radars M increases, the sensitivity also tends to increase due to the availability of more points, making undetected targets much less frequent.

3.3. Tracking

To enhance the detection rate, a tracking algorithm is implemented, similar to the one proposed in [7]. The tracking module takes as input a vector of cluster measurements, including positioning on the (x, y, z) axes and velocity information from the radars. Tracking both a human and a small animal through the continuous capture of individual point clouds requires the efficient temporal association of detection, alongside noise correction and prediction in sensor data. The flow of the multi-target tracker system is illustrated in Figure 4.

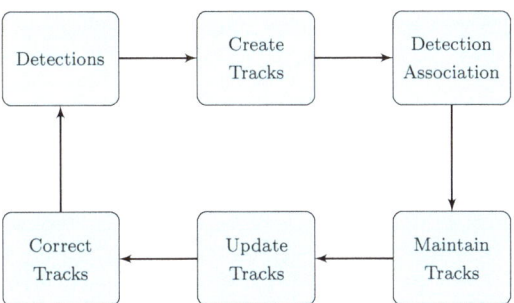

Figure 4. Block diagram of the proposed tracking process.

In this work, tracks are established to detect multiple individuals, whether people or small animals, in each frame. While [7] utilizes the Hungarian algorithm for target association across frames, this work opts for James Munkres's variant of the Hungarian assignment algorithm [43]. The Hungarian algorithm represents a combinatorial optimization method [7]. It operates through a distance matrix, which holds the Euclidean distances between every pair of tracks along the matrix rows and detections in the columns [44]. These distances are computed from the centroids of predicted and detected objects, where smaller distances correspond to a greater likelihood of correctly associating detections with predictions [44].

The main difference between the Hungarian algorithm and the Munkres variant lies in how it iterates through the cost matrix to find the optimal solution for assignment problems. The Munkres algorithm employs alternating path and labeling techniques to identify and update assignments more efficiently, reducing computational complexity compared to the original version of the Hungarian algorithm [43]. Similar to [7], a new track is initiated for each detection, originating either from the first incoming frame or those not associated with an existing track. Tracks that remain undetected for a continuous duration of U frames are flagged as inactive and excluded from subsequent associations. Furthermore, a Kalman filter is employed for trajectory forecasting and adjustments. Further elaboration on these processes is provided below.

3.3.1. Tracks Creation and Association

At the beginning of the tracking process, an empty track is created, with each track being a structured representation of a target detected by the radars. This structured format aims to maintain the state of a tracked target. After data fusion, centroids and bounding boxes are returned if any target is detected. To maintain continuous tracking of individual point clouds for people or animals, an effective temporal association of detection is crucial.

The association method assigning detections tracks is facilitated through the application of James Munkres's variant of the Hungarian algorithm, which manages the assignment problem between existing tracks and new detections [45]. The association process employs a cost matrix $\mathbf{C}$, with rows representing tracks and columns representing detections [43]. The element C_{ij} in the matrix delineates the cost of assigning detection j to track i [43], and it is calculated using the Euclidean distance between the predicted location of the track and the detected object's centroid:

$$C_{ij} = \sqrt{(x_{\text{track},i} - x_{\text{detect},j})^2 + (y_{\text{track},i} - y_{\text{detect},j})^2}, \tag{1}$$

where $x_{\text{track},i}$ and $y_{\text{track},i}$ are the coordinates of the i-th track's predicted position, and $x_{\text{detect},j}$ and $y_{\text{detect},j}$ are the coordinates of the j-th detection. The algorithm then processes this cost matrix to determine the optimal assignment of detections to tracks, minimizing the total cost [45]. This yields the indices for both assigned and unassigned tracks and detections, allowing for the updating of existing tracks and the creation of new ones for unassigned detections. Through this method, the tracking system ensures the continuous monitoring of targets by dynamically managing the creation of new tracks and the association of detections to existing tracks, optimizing the tracking process over time [45].

3.3.2. Track Prediction, Update and Correction

To effectively track an object's movement across frames, predicting its future location is important. The previous motion patterns feed the predictions [45], which are performed using a Kalman filter. The Kalman filter is essential for predicting an object's future location, accounting for process noise ($\mathbf{Q}$) and measurement noise ($\mathbf{R}$). It maintains a state ($\mathbf{x}$) for each track, comprising location and velocity along the (x, y, z) axes. The state for each track at time k is updated based on the previous state at time $k-1$ and the current measurements [46,47]. The state prediction equation is given by

$$\hat{\mathbf{x}}_{k|k-1} = \mathbf{F}_k \mathbf{x}_{k-1} + \mathbf{B}_k \mathbf{u}_k, \tag{2}$$

where $\hat{\mathbf{x}}_{k|k-1}$ is the predicted state estimate at time k, given all available information up to time $k-1$, $\mathbf{F}_k$ is the state transition model applied to the previous state $\mathbf{x}_{k-1}$, $\mathbf{B}_k$ is the control input model applied to the control vector $\mathbf{u}_k$, which represents any known external influences on the state, while $\mathbf{x}_{k-1}$ is the state estimate at time $k-1$ [46].

The covariance prediction equation is

$$\mathbf{P}_{k|k-1} = \mathbf{F}_k \mathbf{P}_{k-1} \mathbf{F}_k^\top + \mathbf{Q}_k, \tag{3}$$

where $\mathbf{P}_{k|k-1}$ is the predicted estimate covariance [46].

In the tracking algorithm, the function responsible for updating each assigned track seamlessly incorporates the corresponding detection information. It accurately calls the method to correct the location estimate, storing the new bounding box in the process. This update is performed in a frame-by-frame manner during the post-processing stage. When a new measurement $\mathbf{z}_k$ is received, the update and correction steps are performed as follows [46]:

$$\mathbf{K}_k = \mathbf{P}_{k|k-1} \mathbf{H}_k^\top (\mathbf{H}_k \mathbf{P}_{k|k-1} \mathbf{H}_k^\top + \mathbf{R}_k)^{-1} \tag{4}$$

$$\mathbf{x}_k = \hat{\mathbf{x}}_{k|k-1} + \mathbf{K}_k (\mathbf{z}_k - \mathbf{H}_k \hat{\mathbf{x}}_{k|k-1}) \tag{5}$$

$$\mathbf{P}_k = (I - \mathbf{K}_k \mathbf{H}_k) \mathbf{P}_{k|k-1} \tag{6}$$

where $\mathbf{K}_k$ is the Kalman gain, $\mathbf{z}_k$ is the measurement vector, and $\mathbf{H}_k$ is the measurement model. In addition, $\mathbf{P}_k$ is the updated estimate covariance, and I is the identity matrix.

These steps ensure that the tracker accurately predicts the object's movement across frames, incorporating both model predictions and real measurements to refine the position and velocity estimates [45].

3.3.3. Track Maintenance

Within each frame, detections are either linked to existing tracks or remain unlinked, leading to what we term "invisible" tracks for those without corresponding detections. New tracks are initiated from unassigned detections. Importantly, we manage each track's visibility by incrementally tracking the number of consecutive frames it remains unlinked. This count is crucial for determining when a track should be considered inactive and subsequently removed, indicating that the object has likely exited the observable area.

For a given track T_i, let us denote its visibility count as V_i, a method to avoid confusion with the cost matrix C. This visibility count is updated as follows:

$$V_i = \begin{cases} 0 & \text{if } T_i \text{ is linked to a detection,} \\ V_i + 1 & \text{if } T_i \text{ is not linked to a detection.} \end{cases} \tag{7}$$

A track is considered for removal if its visibility count V_i exceeds a predefined threshold θ, indicating prolonged absence from the field of view:

$$\text{if } V_i > \theta, \text{ then remove } T_i. \tag{8}$$

This mechanism underlines the dynamic nature of tracking, where the sensitivity and accuracy are notably enhanced by the process [45].

Figure 5 displays the trajectory of an identified target. In this scenario, a small animal enters the scene and moves toward the positive y-axis. Four radars are employed for detection. The blue dots in Figure 5 represent the detections made by the radars that were confirmed by the tracking process. The red dots are the detections that were missed by the radars but that were included after the tracking process. Note the relevance of tracking in this application, as it clearly increases the sensitivity.

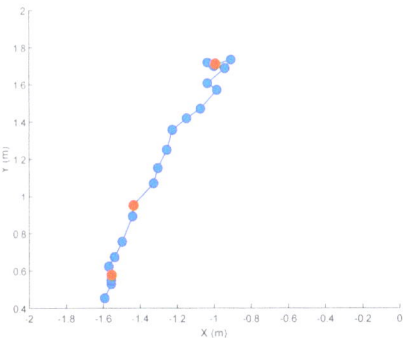

Figure 5. Two-dimensional superior view plot example of an animal being tracked using the tracking algorithm. One small animal was present in the scene, and four radars were utilized. The blue dots are the radars' detections that were confirmed by the tracking, while the red dots are those detections missed by the radars but that were included by the tracking process.

4. Implementation

4.1. Radar Configuration

As the purpose of this work is to detect not only humans but also small animals, the typical configurations provided by the manufacturer need to be adjusted to fit the project goals. An adequate threshold was set for the CFAR algorithm with the aim of obtaining sufficient data for post-processing. Given the objective to detect both small animals and humans, capturing a greater number of point clouds than those solely for humans is crucial, as animals may generate fewer point clouds due to their smaller size and distinct shapes. The radars are configured for an indoor environment, with a maximum distance of 5 m, a range resolution of 7 cm, a maximum radial velocity of 2.4 m/s, a velocity resolution of 0.15 m/s, and a frame duration of 100 ms. The mmWave radars were configured for an azimuth opening angle of 120° and an elevation angle of 30°.

When employing multiple radars, understanding signal interference becomes crucial. As stated in [18], the probability of interference remains below 1% when utilizing four radars, but this probability escalates with the use of more than ten radars. In such cases, explicit synchronization among the radars or the implementation of an interference detection algorithm becomes necessary. Consequently, it can be inferred that the likelihood of interference is minimal when concurrently operating up to four radars, aligning with the approach proposed in this study.

4.2. Setup

The experiments were conducted in a 4 m × 4 m room. The animal detected in the test was a small dog, weighing approximately 3 kg and 40 cm tall. During tests, a camera was used to monitor the environment, and the radars operated in an unsynchronized manner. The camera acted as an auxiliary means of observation to validate the presence or absence of targets within the environment. Therefore, its role was important in the sense of corroborating the radar's detection capabilities. By comparing the visual evidence captured by the camera with the radar's detection outcomes, we could assess the accuracy and reliability of our radar system more effectively. As detecting a small animal is more challenging than detecting a human, we set it that during 66.67% of the time, only the animal was in the area, and 33.33% of the time, there was a human and an animal. The system was operated for 3000 frames.

Once configured, the radars were placed in two different scenarios for the tests. In the first scenario, the radars were horizontally aligned, as shown in Figure 6a, separated by the same distance between them. In the second scenario, the radars were each positioned close to one of the four walls, at the center of each wall, all pointing towards the center of the room, as depicted in Figure 6b. These two scenarios were proposed to analyze the detection performance in different radar positions and to obtain data at different angles. In the images, the dashed line illustrates the opening angle of the radars. Given the radar's lower elevation angle compared to the azimuth angle as a result of antenna disposition, the radars were placed at the typical animal's height. Following the manufacturer's suggestions, none of the radars were placed on the ceiling [48]. These placement variations were intended to explore different perspectives and angles for data collection, providing a comprehensive analysis of the system performance in various configurations. In our setup, each radar was positioned to ensure there were no obstructions directly in front of it, facilitating unimpeded detection capabilities. Moreover, careful consideration was given to positioning the radars in locations where targets are most likely to be detected within their field of view while minimizing the presence of obstructive elements. Despite the potential for challenges posed by physical obstructions, the advanced nature of current radar systems enables the detection of targets even through certain barriers, such as glass surfaces. This capability significantly increases the flexibility and effectiveness of radar systems in complex environments. However, optimizing radar placement transcends simply overcoming physical obstructions; it also entails strategic positioning to maximize the field of view of the radar array, ensuring comprehensive coverage and enhanced detection accuracy. This strategic approach to radar placement, combined with a calibration technique, provides a robust solution to the challenges of data fusion in practical applications.

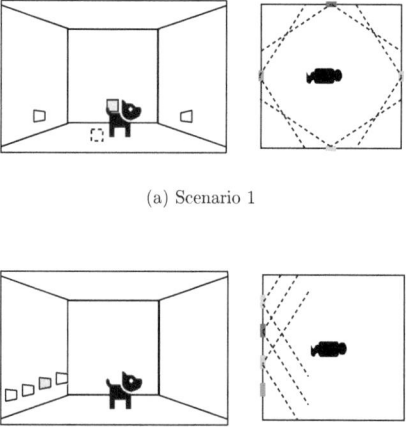

(a) Scenario 1

(b) Scenario 2

Figure 6. The two scenarios with their radar placement and field of view for human and animal detection tests.

4.3. DBSCAN Algorithm

For each one of the data fusion methods and for each radar placement, the DBSCAN algorithm was calibrated to detect both people and small animals. The parameters of DBSCAN mentioned in Section 3.2.1 were adjusted to improve target detection and minimize the creation of false targets. Small animals typically generate fewer reflection points, even with a lower threshold in the CFAR algorithm, while humans tend to produce more reflection points, posing a challenge for simultaneous detection.

The adjustments in DBSCAN parameters were made to address this challenge and optimize the algorithm's performance. The goal was to enhance target detection while avoiding the generation of excessive false positives. It is worth noting that the use of multiple radars contributes to generating a sufficient number of points, improving the overall effectiveness of the detection system, especially for small animals.

4.4. Tracking Algorithm

The tracking algorithm was implemented to improve the detection rate and prevent the generation of ghost targets. The algorithm thresholds are configured with the aim of improving the detection rate, especially when using fewer radars. It is important to note that depending on the distance and movement of the target, there might be frames where no point cloud data are transmitted, emphasizing the importance of tracking for successful detection.

4.5. Performance Metrics

The evaluation of the detection system involved the use of the following key metrics (The F1-score of each test was also analyzed; however, it did not yield different conclusions from those shown by precision and sensitivity. Therefore, for the sake of brevity, it is omitted here):

- **Positives**

  (P): Human and/or animal present in the area.
- **True Positives**

  (TP): Human and/or animal in the detection area that is successfully detected by the radar.
- **False Positives**

  (FP): Noise or other objects in the detection area that are falsely detected as humans or animals.
- **Sensitivity**

  (TP/P): The ability to detect humans and/or animals when they are in the detection area.
- **Precision**

  (TP/(TP + FP)): The ability to distinguish human and/or animal from false detection.

An ideal system should exhibit high sensitivity and high precision [18], but that is a very challenging task. Moreover, in safety-related applications, such as those related to wireless energy transfer [36], sensitivity is more relevant than precision.

5. Results

Tests were carried out using the two radar-placement options mentioned in Section 4, and the three data fusion methods in Section 3.2 were applied.

5.1. Single Radar

First, a test was performed using a single radar in order to highlight the motivation to use multiple radars. For the sake of brevity, the results are presented only for the first scenario, where the radars are positioned side by side and utilize the tracking algorithm. The conclusions are very similar for the second scenario. Then, the sensitivity and precision achieved are presented in Table 1. The parameters used for human detection were those proposed in [7], while for the detection of small animals, the number of required point clouds was reduced to around 1/8 of the total points. In the optimized scenario for detecting both humans and small animals, the parameters were fine-tuned to achieve optimal performance, aiming at high sensitivity with balanced precision, avoiding significant discrepancies between the two parameters.

Table 1. System performance with DBSCAN thresholds optimized for animal detection only, for human detection only, or for both, with a single radar in the first scenario.

DBSCAN Optimized for	Precision	Sensitivity
Small animals	52.8%	81.4%
Humans	100%	32.6%
Both	67.1%	75.2%

Analysis of the data in Table 1 reveals notable differences: when using DBSCAN parameters specifically optimized for the detection of small animals, higher sensitivity is achieved, as expected, but a larger incidence of ghost detections is also observed, reducing precision. This situation is illustrated in Figure 7, which shows the results of the DBSCAN algorithm in a situation where both an animal and a human were present in the scene. Note that an additional ghost target was detected. In contrast, when optimizing the parameters for human detection, there is a decrease in true positives, often resulting in the failure to detect the animal and a decrease in sensitivity, leading to the results shown in Table 1, where the sensitivity is severely compromised, but the precision becomes very high. Finally, when the system is optimized to detect both humans and animals, a more balanced performance is achieved, but it is probably still insufficient for many applications, such as those related to safety. A possible solution to increase both the sensitivity and the precision is to use multiple radars.

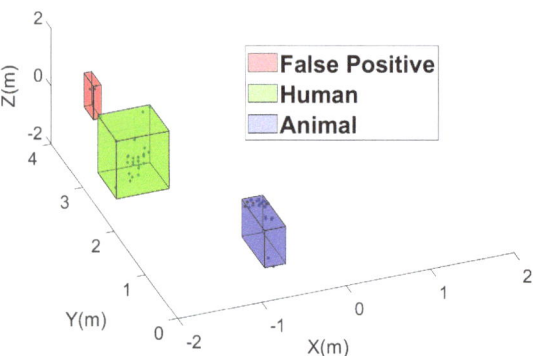

Figure 7. Detection of three targets in a scene that contained only two (a small animal and a human) using optimized DBSCAN parameters for the detection of small animals only.

Another test was conducted switching the tracking algorithm on and off. The test considered the optimized DBSCAN parameters for detecting both humans and animals, and the results are shown in Figure 8. Examining the data makes it evident that the integration of the tracking algorithm significantly increases the sensitivity, which is fundamental for high-performance applications. In the next subsection, the performance of the tracking algorithm with multiple radars is presented.

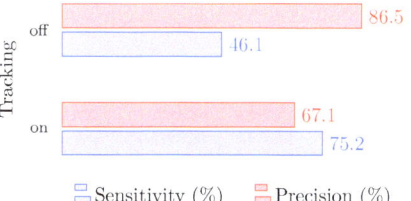

Figure 8. System performance with and without the tracking algorithm for Scenario 1 and a single radar.

5.2. Multiple Radars

Next, we discuss the results of applying the methodology proposed in Section 3, considering the three data-fusion strategies and the two radar-placement scenarios. First, we demonstrate the algorithm performance in tracking a person and an animal, aiming to discern the algorithm behavior with the employment of multiple radar systems. Specifically, in this case, we utilized the third fusion method. The results obtained are displayed in Figure 9. It becomes clear that tracking is enhanced with the use of multiple radars; the analysis reveals that while a single radar setup provides a baseline capability for object tracking, the integration of two, three, or four radars significantly amplifies the sensitivity and accuracy. Notably, it is observed that when employing one and two radars, the system occasionally confuses the tracks, mistakenly swapping the person for the animal and vice versa. This issue, however, is effectively mitigated with the deployment of three and four radar configurations, wherein such inaccuracies do not occur. This progressive enhancement in tracking performance underlines the importance of multi-radar configurations for high-fidelity tracking in complex environments.

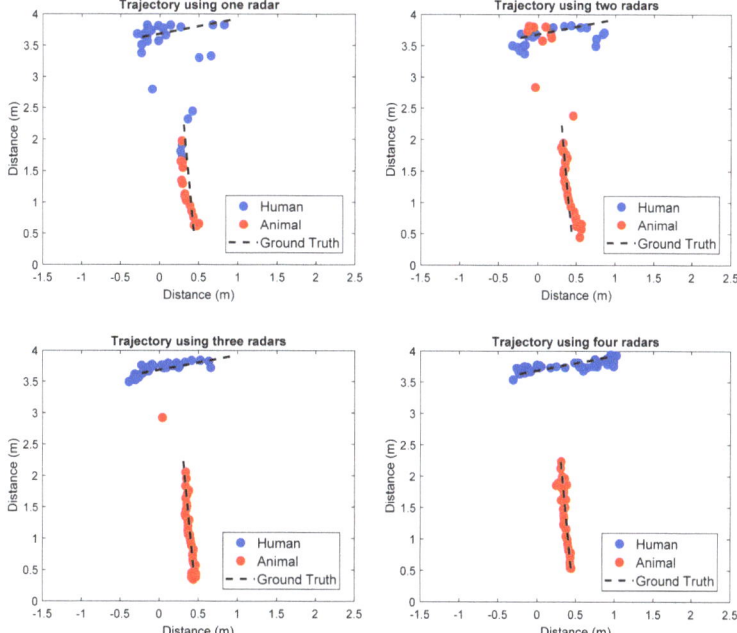

Figure 9. Operation of the tracking algorithm with 1 to 4 radars. Each subfigure illustrates the tracking behavior as the number of radars increases.

The progressive enhancement in tracking performance underlines the importance of multi-radar configurations for high-fidelity tracking in complex environments. It is important to underscore a key advantage of our multi-radar configuration, which is particularly demonstrated in scenarios of visual obstruction, such as when a large human obscures a small animal from the view of one radar. In these cases, the probability remains high that other radars in the system will have an unobstructed view of the animal, ensuring its continuous detection and tracking. This benefit is notably pronounced in our implementation of the second and third fusion methods, where the detection of a target by all radars is not a prerequisite for its positive detection. Such a feature underscores the strategic advantage of employing multiple radars, as it allows for the maintenance of tracking accuracy and system resilience even when individual radars face visual obstructions.

Figure 10 shows the precision and sensitivity results for data fusion Methods 1 and 3, respectively, "Intersection of Detected Data" and "Combining the Raw Data", versus the number of radars M. Clearly, when the number of radars increases, Method 1 performs better in terms of precision but loses considerably in terms of sensitivity. This is because every radar must detect the target to be finally considered detected. In the case of small animal detection, it is plausible that radars might not simultaneously detect the target, reducing the sensitivity. With the same arguments, when all radars detect a target, it is very probable that this is a true positive, increasing the precision.

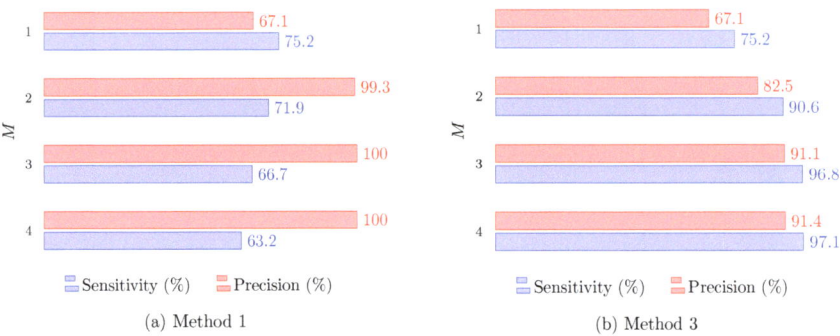

Figure 10. Precision and sensitivity for Methods 1 and 3 in the first scenario versus the number of radars M.

Note that a completely different behavior is observed with Method 3, as both precision and sensitivity increase with M. As this method includes all available raw data in a unique clustering process, having more radars improves performance in both aspects, achieving more than 90% in sensitivity and precision for $M \geq 3$. Such a capability to achieve high sensitivity and high precision at the same time is very interesting from the point of view of safety-related applications. Figure 11 shows similar results, but for the second scenario, where the radars are on each of the walls. The same trends are observable, although it is clear that a better performance was obtained in the first scenario, where the radars are side by side.

Figure 12 illustrates the performance of Method 2 for both scenarios. Recall that Method 2, R out of M, is an alternative to Method 1, where here only R of the M radars have to detect a target to be finally detected. We consider $M = 4$ radars and vary R from 1 to 4. Note that a much more balanced performance than that obtained by Method 1 can be achieved, especially with $R = 2$ and for the first scenario. That is very reasonable since a positive detection can now be achieved even if some of the radars missed the target. However, as illustrated in Table 2, where we consider only the best-performing configurations for each method, the performance of Method 3 is still the best, being able to achieve both higher sensitivity and higher precision than Method 2.

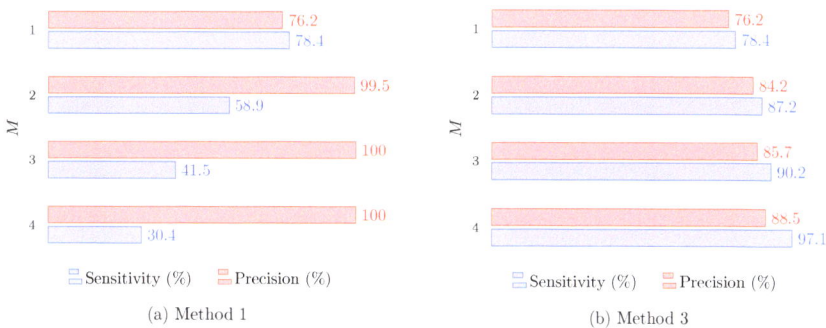

Figure 11. Precision and sensitivity for Methods 1 and 3 in the second scenario versus the number of radars M.

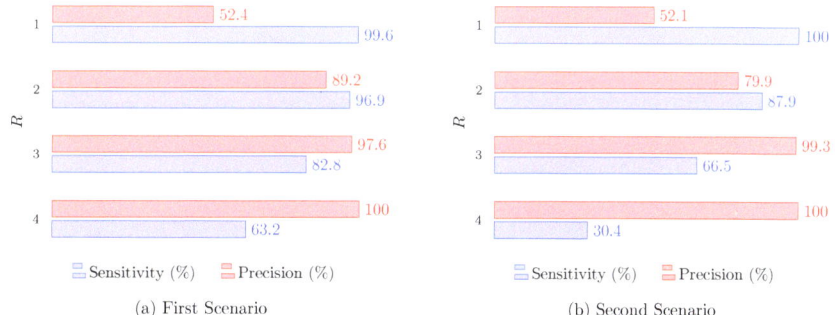

Figure 12. Precision and sensitivity for Method 2 in the first and second scenarios, considering $M = 4$ radars, versus R, the number of radars required for detection.

Table 2. Best performance for Methods 2 and 3 in Scenarios 1 and 2.

Method	Scenario 1		Scenario 2	
	Precision	Sensitivity	Precision	Sensitivity
Method 2	89.2%	96.9%	79.9%	87.9%
Method 3	91.4%	97.1%	88.5%	97.1%

6. Conclusions

In this work, the detection and tracking of humans and small animals was investigated using multiple mmWave radars. First, the sensitivity of using a single radar to detect humans and animals simultaneously was shown to be relatively low, motivating the use of multiple radars. Then, two radar-positioning scenarios and three data-fusion strategies were analyzed. We showed that the data-fusion strategy that combines the raw data before applying a clustering algorithm performs best, achieving high levels of sensitivity and precision. The results demonstrate that the use of multiple radars to detect people and small animals is very promising, even in safety-related applications where sensitivity must be high. A somewhat straightforward extension of this work would be the application of multiple radars to detect both humans and animals in outdoor environments, as well as in settings with animals and people of different sizes. A perhaps more challenging and rewarding future work would be the fusion of radar data with other technologies so that high sensitivity and precision can be achieved with fewer sensors.

Moreover, despite the success in achieving high sensitivity in the detection and tracking of humans and animals, we may encounter limitations when the targets remain in close proximity over extended periods, moving together. Thus, another potential future work is the thorough investigation of the effects, and the corresponding solutions, of grouped targets on tracking accuracy.

Finally, as a practical step forward, we plan the construction of a prototype system for wireless power transfer informed by multiple radars for the detection of both people and animals. This system holds the potential to enhance safety and efficiency in various environments, addressing the unique challenges posed by the coexistence of humans and animals of different sizes.

Author Contributions: Conceptualization, A.B.R.C.D.M. and R.D.S.; methodology, A.B.R.C.D.M. and R.D.S.; software, A.B.R.C.D.M.; validation, A.B.R.C.D.M., G.B., G.L.M. and R.D.S.; formal analysis, A.B.R.C.D.M., G.B., G.L.M. and R.D.S.; investigation, A.B.R.C.D.M.; resources, A.B.R.C.D.M.; data curation, A.B.R.C.D.M.; writing—original draft preparation, A.B.R.C.D.M.; writing—review and editing, G.B., G.L.M. and R.D.S.; visualization, A.B.R.C.D.M., G.B., G.L.M. and R.D.S.; supervision, R.D.S.; project administration, R.D.S.; funding acquisition, A.B.R.C.D.M. and R.D.S. All authors have read and agreed to the published version of the manuscript.

Funding: This work has been partially supported by CNPq (402378/2021-0, 305021/2021-4, 307226/2021-2), RNP/MCTIC 6G Mobile Communications Systems (01245.010604/2020-14), and Agência Nacional de Energia Elétrica and Celesc Distribuição S.A. (PD05697-1323/2023).

Institutional Review Board Statement: Ethical review and approval were not required as this study does not harm, is not physically invasive, and does not pose any potential danger to humans or animals in any way.

Informed Consent Statement: Informed consent was obtained from all subjects involved in the study.

Data Availability Statement: Data are contained within the article.

Conflicts of Interest: The authors declare no conflicts of interest.

Abbreviations

The following abbreviations are used in this manuscript:

ADC	Analog-to-Digital Converter
AoA	Angle of Arrival
CFAR	Constant False Alarm Rate
CUT	Cell Under Test
DBSCAN	Density-Based Spatial Clustering of Applications with Noise
FFT	Fast Fourier Transform
FMCW	Frequency Modulated Continuous Wave
FP	False Positive
IF	Intermediate Frequency
IoT	Internet of Things
IQ	In-Phase and Quadrature
LIDAR	Light Detection and Ranging
mmWave	Millimeter Wave
P	Positive
RX	Receiving Antenna
STF	Strong Tracking Filter
TI	Texas Instruments
TP	True Positive
TX	Transmitting Antenna

References

1. Al-Sarawi, S.; Anbar, M.; Abdullah, R.; Al Hawari, A.B. Internet of things market analysis forecasts, 2020–2030. In Proceedings of the 2020 Fourth World Conference on Smart Trends in Systems, Security and Sustainability (WorldS4), London, UK, 27–28 July 2020; pp. 449–453.
2. Perwej, Y.; Haq, K.; Parwej, F.; Mumdouh, M.; Hassan, M. The internet of things (IoT) and its application domains. *Int. J. Comput. Appl.* **2019**, *975*, 182.

3. Pattnaik, S.K.; Samal, S.R.; Bandopadhaya, S.; Swain, K.; Choudhury, S.; Das, J.K.; Mihovska, A.; Poulkov, V. Future Wireless Communication Technology towards 6G IoT: An Application-Based Analysis of IoT in Real-Time Location Monitoring of Employees Inside Underground Mines by Using BLE. *Sensors* **2022**, *22*, 3438. https://doi.org/10.3390/s22093438.
4. Nath, R.K.; Bajpai, R.; Thapliyal, H. IoT based indoor location detection system for smart home environment. In Proceedings of the 2018 IEEE International Conference on Consumer Electronics (ICCE), Las Vegas, NV, USA, 12–14 January 2018; pp. 1–3. https://doi.org/10.1109/ICCE.2018.8326225.
5. Cui, Y.; Liu, F.; Jing, X.; Mu, J. Integrating Sensing and Communications for Ubiquitous IoT: Applications, Trends, and Challenges. *IEEE Netw.* **2021**, *35*, 158–167. https://doi.org/10.1109/MNET.010.2100152.
6. Lu, C.X.; Rosa, S.; Zhao, P.; Wang, B.; Chen, C.; Stankovic, J.A.; Trigoni, N.; Markham, A. See through smoke: Robust indoor mapping with low-cost mmwave radar. In Proceedings of the 18th International Conference on Mobile Systems, Applications, and Services, Toronto, ON, Canada, 15–19 June 2020; pp. 14–27.
7. Zhao, P.; Lu, C.X.; Wang, J.; Chen, C.; Wang, W.; Trigoni, N.; Markham, A. mID: Tracking and Identifying People with Millimeter Wave Radar. In Proceedings of the 2019 15th International Conference on Distributed Computing in Sensor Systems (DCOSS), Santorini Island, Greece, 29–31 May 2019; pp. 33–40. https://doi.org/10.1109/DCOSS.2019.00028.
8. Hua, Y.; Ono, Y.; Peng, L.; Xu, Y. Unsupervised Learning Discriminative MIG Detectors in Nonhomogeneous Clutter. *IEEE Trans. Commun.* **2022**, *70*, 4107–4120. https://doi.org/10.1109/TCOMM.2022.3170988.
9. Oh, H.; Nam, H. Energy Detection Scheme in the Presence of Burst Signals. *IEEE Signal Process. Lett.* **2019**, *26*, 582–586. https://doi.org/10.1109/LSP.2019.2900165.
10. Garrote, L.; Perdiz, J.; da Silva Cruz, L.A.; Nunes, U.J. Point Cloud Compression: Impact on Object Detection in Outdoor Contexts. *Sensors* **2022**, *22*, 5767. https://doi.org/10.3390/s22155767.
11. Zhang, Y.; Wang, L.; Jiang, X.; Zeng, Y.; Dai, Y. An efficient LiDAR-based localization method for self-driving cars in dynamic environments. *Robotica* **2022**, *40*, 38–55. https://doi.org/10.1017/S0263574721000369.
12. Roy, A.M.; Bose, R.; Bhaduri, J. A fast accurate fine-grain object detection model based on YOLOv4 deep neural network. *Neural Comput. Appl.* **2022**, *34*, 1–27. https://doi.org/10.1007/s00521-021-06651-x.
13. Liu, J.J.; Hou, Q.; Liu, Z.A.; Cheng, M.M. PoolNet+: Exploring the Potential of Pooling for Salient Object Detection. *IEEE Trans. Pattern Anal. Mach. Intell.* **2023**, *45*, 887–904. https://doi.org/10.1109/TPAMI.2021.3140168.
14. Tsai, Y.S.; Modales, A.V.; Lin, H.T. A Convolutional Neural-Network-Based Training Model to Estimate Actual Distance of Persons in Continuous Images. *Sensors* **2022**, *22*, 5743. https://doi.org/10.3390/s22155743.
15. Iovescu, C.; Rao, S. *The Fundamentals of Millimeter Wave Sensors*; Texas Instruments: Dallas, TX, USA, 2017.
16. Huang, X.; Tsoi, J.K.P.; Patel, N. mmWave Radar Sensors Fusion for Indoor Object Detection and Tracking. *Electronics* **2022**, *11*, 2209. https://doi.org/10.3390/electronics11142209.
17. Texas Intruments. *IWR6843, IWR6443 Single-Chip 60-to 64-GHz mmWave Sensor*; SWRS219E, Rev. E; Texas Instruments: Dallas, TX, USA, 2021.
18. Cui, H.; Dahnoun, N. High precision human detection and tracking using millimeter-wave radars. *IEEE Aerosp. Electron. Syst. Mag.* **2021**, *36*, 22–32.
19. Forslund, D.; Bjärkefur, J. Night vision animal detection. In Proceedings of the 2014 IEEE Intelligent Vehicles Symposium Proceedings, Dearborn, MI, USA, 8–11 June 2014; pp. 737–742. https://doi.org/10.1109/IVS.2014.6856446.
20. Lin, J.; Hu, J.; Xie, Z.; Zhang, Y.; Huang, G.; Chen, Z. A Multitask Network for People Counting, Motion Recognition, and Localization Using Through-Wall Radar. *Sensors* **2023**, *23*, 8147. https://doi.org/10.3390/s23198147.
21. Pegoraro, J.; Rossi, M. Real-Time People Tracking and Identification from Sparse mm-Wave Radar Point-Clouds. *IEEE Access* **2021**, *9*, 78504–78520. https://doi.org/10.1109/ACCESS.2021.3083980.
22. Chen, W.; Yang, H.; Bi, X.; Zheng, R.; Zhang, F.; Bao, P.; Chang, Z.; Ma, X.; Zhang, D. Environment-Aware Multi-Person Tracking in Indoor Environments with MmWave Radars. *Proc. ACM Interact. Mob. Wearable Ubiquitous Technol.* **2023**, *7*, 89. https://doi.org/10.1145/3610902.
23. Xu, Y.; Jin, Y.; Zhou, Y. Several methods of radar data fusion. In Proceedings of the 2002 3rd International Symposium on Electromagnetic Compatibility, Beijing, China, 21–24 May 2002; pp. 664–667. https://doi.org/10.1109/ELMAGC.2002.1177518.
24. Yan, J.; Liu, H.; Pu, W.; Jiu, B.; Liu, Z.; Bao, Z. Benefit Analysis of Data Fusion for Target Tracking in Multiple Radar System. *IEEE Sens. J.* **2016**, *16*, 6359–6366. https://doi.org/10.1109/JSEN.2016.2581824.
25. Cowley, D.C.; Shafai, B. Registration in multi-sensor data fusion and tracking. In Proceedings of the 1993 American Control Conference, San Francisco, CA, USA, 2–4 June 1993; pp. 875–879.
26. Yang, X.; Tang, J.; Liu, Y. A novel multi-radar plot fusion scheme based on parallel and serial plot fusion algorithm. In Proceedings of the 2017 2nd International Conference on Frontiers of Sensors Technologies (ICFST), Shenzhen, China, 14–16 April 2017; pp. 213–217.
27. Li, S.; Guo, J.; Xi, R.; Duan, C.; Zhai, Z.; He, Y. Pedestrian trajectory based calibration for multi-radar network. In Proceedings of the IEEE INFOCOM 2021-IEEE Conference on Computer Communications Workshops (INFOCOM WKSHPS), Vancouver, BC, Canada, 10–13 May 2021; pp. 1–2.
28. Sengupta, A.; Jin, F.; Zhang, R.; Cao, S. mm-Pose: Real-time human skeletal posture estimation using mmWave radars and CNNs. *IEEE Sens. J.* **2020**, *20*, 10032–10044.
29. Bansal, K.; Rungta, K.; Zhu, S.; Bharadia, D. Pointillism: Accurate 3d bounding box estimation with multi-radars. In Proceedings of the 18th Conference on Embedded Networked Sensor Systems, Virtual, 16–19 November 2020; pp. 340–353.

30. Li, W.; Wu, Y.; Chen, R.; Zhou, H.; Yu, Y. Indoor Multi-Human Device-Free Tracking System Using Multi-Radar Cooperative Sensing. *IEEE Sens. J.* **2023**, *23*, 27862–27871.
31. Shen, Z.; Nunez-Yanez, J.; Dahnoun, N. Multiple Human Tracking and Fall Detection Real-Time System Using Millimeter-Wave Radar and Data Fusion. In Proceedings of the 2023 12th Mediterranean Conference on Embedded Computing (MECO), Budva, Montenegro, 6–10 June 2023; pp. 1–6.
32. Tahmoush, D.; Silvious, J. Remote detection of humans and animals. In Proceedings of the 2009 IEEE Applied Imagery Pattern Recognition Workshop (AIPR 2009), Washington, DC, USA, 14–16 October 2009; pp. 1–8.
33. Shrestha, A.; Loukas, C.; Le Kernec, J.; Fioranelli, F.; Busin, V.; Jonsson, N.; King, G.; Tomlinson, M.; Viora, L.; Voute, L. Animal lameness detection with radar sensing. *IEEE Geosci. Remote Sens. Lett.* **2018**, *15*, 1189–1193.
34. Darlis, A.R.; Ibrahim, N.; Subiantoro, A.; Yusivar, F.; Albaqami, N.N.; Prabuwono, A.S.; Kusumoputro, B. Autonomous Human and Animal Classification Using Synthetic 2D Tensor Data Based on Dual-Receiver mmWave Radar System. *IEEE Access* **2023**, *11*, 80284–80296. https://doi.org/10.1109/ACCESS.2023.3299325.
35. Pearce, A.; Zhang, J.A.; Xu, R. A Combined mmWave Tracking and Classification Framework Using a Camera for Labeling and Supervised Learning. *Sensors* **2022**, *22*, 8859. https://doi.org/10.3390/s22228859.
36. López, O.L.; Rosabal, O.M.; Azarbahram, A.; Khattak, A.B.; Monemi, M.; Souza, R.D.; Popovski, P.; Latva-aho, M. High-power and safe RF wireless charging: Cautious deployment and operation. *arXiv* **2023**, arXiv:2311.12809.
37. Van Mulders, J.; Delabie, D.; Lecluyse, C.; Buyle, C.; Callebaut, G.; Van der Perre, L.; De Strycker, L. Wireless Power Transfer: Systems, Circuits, Standards, and Use Cases. *Sensors* **2022**, *22*, 5573. https://doi.org/10.3390/s22155573.
38. Texas Instruments, Inc. *MMWAVE SDK User Guide*; Document Version 1.0; Texas Instruments, Inc.: Dallas, TX, USA, 2019.
39. Xu, C.; Wang, F.; Zhang, Y.; Xu, L.; Ai, M.; Yan, G. Two-level CFAR Algorithm for Target Detection in MmWave Radar. In Proceedings of the 2021 International Conference on Computer Engineering and Application (ICCEA), Kunming, China, 25–27 June 2021; pp. 240–243. https://doi.org/10.1109/ICCEA53728.2021.00055.
40. Mafukidze, H.D.; Mishra, A.K.; Pidanic, J.; Francois, S.W.P. Scattering Centers to Point Clouds: A Review of mmWave Radars for Non-Radar-Engineers. *IEEE Access* **2022**, *10*, 110992–111021. https://doi.org/10.1109/ACCESS.2022.3211673.
41. Texas Instruments, Inc. *Static Detection CLI Commands*; Application Note; Texas Instruments, Inc.: Dallas, TX, USA, 2023.
42. The MathWorks, Inc. *Statistics and Machine Learning Toolbox™ User's Guide R2023b*; The MathWorks, Inc.: Natick, MA, USA, 2023.
43. Munkres, J. Algorithms for the assignment and transportation problems. *J. Soc. Ind. Appl. Math.* **1957**, *5*, 32–38.
44. Hamuda, E.; Mc Ginley, B.; Glavin, M.; Jones, E. Improved image processing-based crop detection using Kalman filtering and the Hungarian algorithm. *Comput. Electron. Agric.* **2018**, *148*, 37–44. https://doi.org/10.1016/j.compag.2018.02.027.
45. The MathWorks, Inc. *Get Started with Computer Vision Toolbox*; Online; The MathWorks, Inc.: Natick, MA, USA, 2023.
46. Kalman, R.E. A new approach to linear filtering and prediction problems. *J. Basic Eng. Mar.* **1960**, *82*, 35–45.
47. Hamilton, J. The Kalman Filter. *Time Ser. Anal.* **1994**, *13*, 1–799.
48. Texas Instruments, Inc. *Best Practices for Placement and Angle of mmWave Radar Devices*; Application Brief; Texas Instruments, Inc.: Dallas, TX, USA, 2023.

Disclaimer/Publisher's Note: The statements, opinions and data contained in all publications are solely those of the individual author(s) and contributor(s) and not of MDPI and/or the editor(s). MDPI and/or the editor(s) disclaim responsibility for any injury to people or property resulting from any ideas, methods, instructions or products referred to in the content.

MDPI AG
Grosspeteranlage 5
4052 Basel
Switzerland
Tel.: +41 61 683 77 34

Sensors Editorial Office
E-mail: sensors@mdpi.com
www.mdpi.com/journal/sensors

Disclaimer/Publisher's Note: The title and front matter of this reprint are at the discretion of the Guest Editors. The publisher is not responsible for their content or any associated concerns. The statements, opinions and data contained in all individual articles are solely those of the individual Editors and contributors and not of MDPI. MDPI disclaims responsibility for any injury to people or property resulting from any ideas, methods, instructions or products referred to in the content.

www.ingramcontent.com/pod-product-compliance
Lightning Source LLC
LaVergne TN
LVHW072323090526
838202LV00019B/2343